Praise for *Power Hungry*

"Bryce douses the green energy movement with a cold shower of facts and figures, ones that collectively remind us that a transition to wind and solar power would take decades, that it would be astronomically expensive, that it would make the U.S. reliant on China for turbines, and that it would lead to 'energy sprawl.' For all the intuitive appeal of renewable energy, *Power Hungry* makes a convincing case that decarbonizing the world's primary energy use will mean letting the sun shine and the wind blow while embracing natural gas as a bridge to nuclear energy."

—James McWilliams, Freakonomics Blog

"Bryce has compiled a catalogue of hard facts and statistics that puncture just about every myth you will read in breathless accounts of the coming 'Green Economy'."

—William Tucker, *The American Spectator*

"Bryce deftly sets out to debunk the myths of the ever-popular going green campaign and answers more specific technological difficulties and cost containment issues. . . . His views will undoubtedly be rejected or disbelieved, but he backs up those views with hard evidence provoking readers to do the math for themselves, verify statistics, and basically, check up on him with more than ninety pages of references, statistical appendixes, and energy data notes. This is the must-read book for the twenty-first century."

—M. Chris Johnson, *San Francisco Book Review*

"[S]hould be mandatory reading for U.S. policymakers. . . . The promise of renewables has consistently been oversold by the political class. Solar and wind energy both suffer from major structural deficiencies. . . . Our current national energy debate is heavy on passion and hyperbole; it could use a sizable dose of historical perspective and empirical reality."

—Duncan Currie, *The National Review*

"I have long known that there is nothing remotely 'green' about putting wind farms all over the countryside, with their eagle-slicing, bat-popping, subsidy-eating, rare-earth-demanding, steel-rich, intermittent-output characteristics. But until I read Robert Bryce's superb and sober new book *Power Hungry,* I had not realised just how dreadfully bad for the environment nearly all renewable energy is. . . .

Bryce's book is more than a demolition of renewable energy. It contains a fascinating and detailed account of the shale gas revolution and of the latest developments in modular nuclear technology. It makes a persuasive case that this century will be dominated by 'N2N' energy— natural gas to nuclear—and that the consequence of the rise of both will be continuing steady decarbonisation of the economy. This is the best book on energy I have read. It confirms my optimism—and my rejection of the renewable myth."

—**Matt Ridley,** *The Rational Optimist*

"Bryce is especially good at explaining why fossil fuels have become entrenched as our main energy sources."

—*Philadelphia Inquirer*

"A brutal, brilliant exploration. . . . If *Power Hungry* sounds like a supercharged polemic, its shocks are delivered with forensic skill and narrative aplomb. . . . It is unsentimental, unsparing, and impassioned; and, if you'll excuse the pun, it is precisely the kind of journalism we need to hold truth to power." —*Wall Street Journal*

"His magnificently unfashionable, superlatively researched new book dares to fly in the face of all current conventional wisdom and cant. . . . I have never yet found any book or author who does a more thorough, unanswerable job of demolishing universally held environmental myths than Mr. Bryce does. . . . Mr. Obama is reputed to be an omnivorous reader of serious intellectual volumes. He should drop everything else and put Robert Bryce's invaluable book at the top of his list. So should every senator and Congress member and every self-important, scientifically illiterate pundit in America, right and left alike. They will all learn a lot." —*Washington Times*

"Bryce uses copious facts and research to make a compelling case that renewable sources have their place in our energy future but they aren't the viable panacea we're led to believe." —*Library Journal*

"[A] terrific buy for anyone with a strong interest in the nation's energy supply. . . . A full 54 pages devoted to references illustrate the comprehensive research Bryce has done, as well as the quality of his sources. He is at his best destroying many of the myths regarding renewable energy, providing powerful mathematical proofs that anyone can understand. . . . The primary theme of this book is the importance of power density. As Bryce thoroughly documents, coal, oil, natural gas, and nuclear power provide such power density while wind, solar, and biofuels do not. You will not find a book on energy that makes this important point more strongly than this one."

—Jay Lehr, *Heartland Institute*

"*Power Hungry* provides a grand tour of our energy landscape in the best journalistic tradition of serving the public good, exposing the cant of received wisdom and using the authority and weight of good numbers to put ideas into proper perspective. Bryce's numbers provide giant shoulders upon which to stand, allowing us to see farther and better, increasing our knowledge and improving the odds for institutional wisdom. There are few things more important to the world's life, liberty, and happiness than an enhanced ability to convert abundant energy into high power at affordable cost. Robert Bryce, with buoyant bonhomie, marks the way."

—**Jon Boone**, MasterResource.org

"Robert Bryce is an energy realist. So reading him is refreshing. First, because most people when discussing matters of energy are either ill- or misinformed, naïve, liars, or have a personal stake in the policy outcomes. Second, because every time I read something by Bryce, I learn something new. . . . *Power Hungry* [is a] laser-like dismantling of the myth that so-called green energy can displace fossil fuels anytime in the near future."

—**Sterling Burnett**, *National Review Online*

POWER HUNGRY

POWER
HUNGRY

The Myths of "Green" Energy
AND THE
Real Fuels of the Future

ROBERT BRYCE

PUBLICAFFAIRS
New York

Designed by Brent Wilcox

The Library of Congress has catalogued the hardcover edition as follows:
Bryce, Robert.
    Power hungry : the myths of "green" energy and the real fuels of the future / Robert
Bryce.
        p.   cm.
    Includes bibliographical references and index.
    ISBN 978-1-58648-789-8 (alk. paper)
    1. Clean energy industries.   2. Power resources—Forecasting.   I. Title.
    HD9502.5.C542B79 2010
    333.79—dc22
                                                                    2010001655
Paperback ISBN 978-1-58648-953-3
E-book ISBN 978-1-58648-043-9

10 9 8 7 6 5 4 3 2

"He who refuses to do arithmetic is doomed to talk nonsense."

JOHN McCARTHY
computer pioneer, Stanford University

CONTENTS

LIST OF FIGURES, TABLES, AND PHOTOS

Figures

Tables

Photos

AUTHOR'S NOTE

My daughter, Mary, has many favorite authors. One of them is Shannon Hale, a successful writer of novels for young adults. A while ago, Mary quoted Hale's writing advice: "Write what fascinates you."[1]

Hale's advice reminds me of just how lucky I am. There is no more complex or more fascinating topic than energy. We use that word—based on the Latin *energia*—to describe a myriad of different forces, substances, and ideas. These many meanings—whether it's the chemical energy in a chunk of coal in a Chinese power plant or the kinetic energy in a baseball that's been tagged for a quick ride into the cheap seats by Bo Jackson's bat—are too numerous to be encompassed by a single word.

In addition, the scale of energy use and the complexity and importance of the energy business are unmatched by any other industry. The study of energy includes physics, geology, chemistry, engineering, metallurgy, telemetry, seismology, finance, politics, religion, biology, genetics, botany—the list goes on and on. The energy sector has captivated me since I was a child growing up in Tulsa, and no matter how much I study it, I still feel like a rank amateur. And yet, if we are to make wise choices about energy policy, it is essential for all of us—as voters, as owners and managers of businesses, and as policymakers—to understand what energy is, what power is, how they are measured, and which forms of energy and power production make the most sense environmentally and economically.

I have written this book to help people gain that understanding. I have attempted to explain the fundamentals of energy and power production in a way that will enable readers to understand the energy policy debate and make informed decisions. I believe in the relentless application of logic to our discourse on energy, power, and the future. And so I also wrote this book in the hope that it would help to inform a more careful and reasoned approach to energy use and policy. The need for that approach became evident during the promotion of my last book, *Gusher of Lies*, which explained why the United States cannot—and will not—be "energy independent." In the months after the book was published, I heard one question repeated more than any other: Why don't we use more renewable energy?

That question is of great interest to me because I have invested directly in renewables. I have 3,200 watts of solar photovoltaic panels on the roof of my house in Austin. Although those panels provide about one-third of the electricity that my family and I consume, and provide for a slight reduction in our monthly electricity bill, the capital cost of the panels was quite high, the panels require regular cleaning, and they have not been without problems. I've only owned the panels for about five years, and the inverter, which turns the 12-volt power from the panels into 110-volt power that we can use in the house, has already failed once. Luckily, it was still under warranty. But when I'm up on the roof with a long-handled mop every month or so, swabbing those panels, I wonder if they were really worth it.

My personal experiences, as well as the many studies that have been done on both wind and solar, have led me to conclude that those energy sources will remain niche players for the foreseeable future. And yet, many Americans simply don't want to hear that. The romance of renewable energy is such that we are ignoring logic and common sense as well as hard facts and figures. We must bring more depth to the discussion, more reasoned analysis, more evidence-based decisionmaking, and less emotion and biased thinking.

In this book, I have attempted to make the mathematics as accessible as possible by including plenty of graphics and by showing my calculations in the endnotes. But let me be clear: Deconstructing the vagaries of the world's biggest industry requires digesting a lot of data. I have chewed on lots of numbers over the past few months, and you will need to gnaw on a few digits, too, if you are to truly understand the issues. If I have made errors in my calculations, or in the text, graphics, or endnotes, please forgive them. These mistakes are mine and mine alone. If you do find an error, please let me know so that I can correct it in the next edition.

Readers will likely notice that this book contains a number of references to books and articles by Vaclav Smil. I make no apology for that. Smil, who has published about thirty books, has spent most of his long career focusing on energy, and in a sector where cant and hyperbole often dominate, his work stands out for its erudition and clarity. I also make frequent references to the work of Jesse Ausubel. Again, I make no apology. Ausubel is among the foremost energy thinkers in the United States, and his work has helped to shape my approach to energy issues. When I met Ausubel for the first time in Manhattan in September 2008, I asked him to name his favorite authors on energy-related topics. He named Smil and then added, "I am not very interested in what other people are writing. I am interested in data." Ausubel's point resonated and I began mining energy data so that I could make my own calculations, draw my own conclusions, and create my own graphics. Too much of our energy discussion is dominated by glib pundits who do not do their own research. In addition to his savvy analytical skills, Ausubel can turn a phrase. A few months after our first meeting, he told me that "other people's data, like other people's money, can be perilous."

Among the main data sources that I mined for this book was the BP Statistical Review of World Energy. BP publishes its data in the form of an Excel spreadsheet, which facilitates the kind of number crunching that is essential in discerning trends. Although every data source has its limitations, the BP Statistical Review has become a standard reference for the energy industry and is trusted by researchers and forecasters around the world.

I would like to thank my friend and colleague Seth Myers for helping to create the figures. Seth is a journalist with a masters degree from the University of Missouri who knows how to make graphics that tell a story.

I would also like to thank my new friend Stan Jakuba, who volunteered to educate me in energy conversions, SI, and the differences between power and energy. He was also a marvelously scrupulous reader who never tired of reading yet another draft of the manuscript—or of advising me to cut yet more words. My longtime friend Robert Elder Jr.

offered encouragement, read many drafts, and continually demanded that I make my arguments more lucid. I appreciate his assistance.

I have been extremely fortunate to have the help of my father-in-law and favorite chemist, Paul G. Rasmussen, a professor emeritus at the University of Michigan, who provided constructive comments on multiple drafts of the manuscript. He patiently tutored me in thermodynamics and taught me about batteries, the periodic table, and the peculiarities of the lanthanides.

I must acknowledge Michael J. Economides, Christine Economides, and Alex Economides, who have been supportive of my work at *Energy Tribune* and elsewhere, and my friend Mimi Bardagjy, who graciously and punctiliously helped me with fact-checking. My agent, Dan Green, continues to be a wonderful sounding board and friend.

In addition I would like to thank the people at PublicAffairs, including the publisher, Susan Weinberg, and editor-at-large Peter Osnos. Susan and Peter, along with Tessa Shanks, Whitney Peeling, and Clive Priddle, are real pros. My favorite person at PublicAffairs, Lisa Kaufman, has edited all four of my books with patience and keen insight. She worked me like a sled dog, but she understood how I needed to structure this book to make it more readable. I am extremely lucky to have such a skilled editor and such a dear friend.

I would also like to thank the following: Chris Cauthon, Becca Followill, John Harpole, Art Smith, John Olson, Randy Hulme, Mark Papa, Buddy Kleemeier, George Kaiser, Tad Patzek, Mark Mills, J. Paul Oxer, Bryan Shahan, Violet and Ronald Cauthon, Hans Mark, Vic Reis, Pierre-Rene Bauquis, Bertrand Barré, Jarret Adams, Patricia Marie, Joe Bryant, Porter Bennett, Swadesh Mahajan, Joe Craft, Eric Anderson, Terry Thorn, A. F. Alhajji, Fred LeGrand, Donald Sadoway, Harold Weitzner, Martin Snyder, and Bill Reinert.

Since this book is about energy, here are a few factoids that might be of interest: It was written with a MacBook Pro (equipped with a 2.5 GHz Intel Core 2 Duo processor) attached to a 30-inch Apple monitor. Together, the computer and monitor draw about 180 watts (0.24 horsepower). I use a Brother laser printer that draws about 12 watts in standby mode. The primary software programs were Microsoft Word for Mac 2004, Excel, and NoteTaker. I've bragged about NoteTaker before. It's indispensable. During the course of writing this book, I conducted more than 200 interviews, created about 150 Excel spreadsheets, read and clipped about 500 news articles, created about 200 graphic files, and purchased and read (or skimmed) about four dozen books.

Last but not least, I must acknowledge my trophy wife, Lorin, and our trophy children, Mary, Michael, and Jacob, who were frequently ignored during the writing of this book. I have to say it in every book, so here goes: Lorin, children, I love you more than chocolate.

Austin, TX
31 January 2010

The Cardinal Mine
A Point of Beginning

WHEN PETE HAGAN hits the right seam, he can mine a dozen tons of coal in 45 seconds. Working an array of toggle switches mounted on a radio-controlled panel hanging from a dusty strap around his neck, he stands a few feet behind a snarling orange mining machine as it assaults an 8-foot-high wall of bituminous coal.

Hagan deftly toggles a switch and the massive, low-slung machine made by Joy Mining Machinery lurches a few feet forward.[1] Sparks fly off the wall of coal as the huge rotating drum of carbide-steel claws rips in. The dark workspace boils over with dust and noise. The narrow beams from the electric lamps on our hard hats bounce around the cavern, barely piercing the surging cloud of coal dust. Within seconds the dust subsides as water hisses from jets on the mining machine, dousing the coal shooting through that voracious maw. The conveyor belt on the machine's tail slams hundreds of pounds of coal rearward through its gullet onto a "shuttle car"—a long, big-wheeled, electric-powered vehicle that ferries the coal from the mining machine to a string of conveyor belts that whisk the fuel to the surface.

The shuttle car overflows with black rocks. The vehicle's driver, sitting in a windowless cab slung on the side, snaps a silver lever, and the machine lurches into reverse and quickly vanishes around a corner in the barely lit underground labyrinth. For 20 or 30 seconds, while waiting for another shuttle car to appear, Hagan has a chance to talk.

1

Visitors are rare here, 600 feet below the rolling woodlands and farmland of western Kentucky. A quick interview yields the relevant facts: Hagan has been mining coal underground for thirty-six years—and he likes it. In a soft, slow drawl, he explains the various buttons and switches on the control panel for the mining machine. The brick-sized battery clamped to his belt powers the control panel for the mining machine as well as for the lamp clipped to the front of his hard hat. "This one controls the height of the rotor," he explains, flipping a switch that sends the massive, steel-toothed rotor roaring to life. As he wipes the dust off the switches to display the labels on the panel, an empty shuttle car whooshes into view. Without a word, Hagan returns to work, turning the fury of the mining machine back on the coal seam. Within a minute, the new car is filled to overflowing, and, like the one before it, disappears to disgorge its load.

It's a loud, dusty, claustrophobic ballet of horsepower, hydraulics, and brute force. And it is producing what may be the U.S. economy's single most essential commodity: inexpensive energy.

Given the way the energy business is portrayed by politicians, environmental advocates, and various promoters of "green" energy, Hagan and his fellow miners, the mine, the machinery—the entire operation—should be an anachronism. We've repeatedly been told that the modern world of Google, GPS, and HD video will be powered by statuesque wind turbines and shimmering solar panels. The Cardinal Mine is a relic of the nineteenth century, not the vanguard of the twenty-first—or at least that's what the politicos, environmental activists, and promoters have been claiming.

But far from being outdated, the mine, owned by Tulsa-based Alliance Resource Partners, is among the most productive underground coal mines in the United States. The mine, the thirty-fifth largest in America, produces about 6 tons of coal per miner work-hour.[2] That's about two times the national average for underground coal mines.[3]

Shortly before Eric Anderson, the tall, boyish-looking manager of the Cardinal Mine, took me underground, we sat in his office running through the mine's numbers. "We typically mine coal for sixteen hours every day, Monday through Friday," said Anderson, a burly, friendly guy who got his degree in mine engineering from West Virginia University.

About one hundred employees are working underground at any given time. In 2008, the mine produced about 15,350 tons of high-sulfur bituminous coal per day, most of it burned by electric utilities within the state of Kentucky.[4]

As Anderson cleared maps and other papers from the table adjacent to his desk, I asked him for the heat content of the coal. His reply: about 12,500 Btu (British thermal units) per pound. I pulled out my laptop and converted the mine's output into its equivalent in barrels of oil. The numbers were surprising, even to Anderson. The mine produces the raw energy equivalent of 66,000 barrels of oil per day.[5] And that number—66,000 barrels of oil equivalent—provides a useful metric for understanding what too few of the people who are preaching the glories of the green future seem to grasp: the enormous scale of our energy consumption.

On an average day, the energy output of the Cardinal Mine is nearly equal, in raw energy terms, to the daily output of *all* the solar panels and wind turbines in the United States. It's hard to imagine—and it's probably a bit painful to accept, particularly given the coal industry's lousy public image and the ongoing campaign by environmental groups to reduce coal use and carbon dioxide emissions. But it's true.

Before I demonstrate why, readers should be forewarned that this book contains a lot of numbers. Rest assured, the calculations involved are straightforward and are based on easily verifiable data. Fancy math skills are not required; you need have only a willingness to engage in basic arithmetic. But if we are going to understand our energy challenges, then we must be willing to delve into the data and fearlessly confront the numbers.

So here is the first set of numbers: In 2008, the United States produced 52,026,000 megawatt-hours of electricity from wind and 843,000 megawatt-hours from solar, for a total of 52,869,000 megawatt-hours.[6] That's equivalent to about 88,300 barrels of oil per day.[7] Thus, on an average day, by itself, the Cardinal Mine, which has about 400 people on its payroll, produces about 75 percent as much raw energy as all of the wind turbines and solar panels in the United States.[8]

Now let's be clear, the energy coming out of the mine—that 66,000 barrels of oil equivalent in the form of black rocks—is not the same as

the highly ordered electrical energy that comes out of those wind turbines and solar panels. About two-thirds of the heat energy in coal gets
lost when it's burned to produce electricity. Nor does the 66,000 barrel
figure reflect the mine's energy inputs. The Cardinal Mine has a big appetite for electricity, diesel fuel, water, steel rods, and cinder blocks.
Therefore, the net energy produced by the mine is substantially less than
66,000 barrels of oil equivalent per day. Furthermore, that figure doesn't
account for the devastation that the global coal industry inflicts on the
surface of the Earth through strip mines, mountaintop removal, or the massive ash ponds at power plants. Nor do the figures account for the miners who die each year in the world's coal mines, or the pollutants—sulfur
dioxide, soot, and mercury, to name just a few—that are emitted when
coal is burned.

I'm not providing the numbers from the Cardinal Mine as a defense—
or criticism—of coal. Instead, the 66,000 barrels of oil equivalent figure
provides us with a metric—a place that land surveyors call a Point of
Beginning—that allows us to begin separating the energy rhetoric from
the energy reality.[9] And it is important to make clear just how different
rhetoric and reality can be when it comes to energy production and use,
because Americans are woefully uninformed about the subject, despite
the intense interest that energy and the environment have been getting
over the past few years.

We use hydrocarbons—coal, oil, and natural gas—not because we
like them, but because they produce lots of heat energy, from small
spaces, at prices we can afford, and in the quantities that we demand.
And that's the absolutely critical point. The energy business is ruthlessly
policed by the Four Imperatives: power density, energy density, cost, and
scale. The purpose of this book is to bring those factors alive; in doing so,
to explain why the transition away from hydrocarbons will be a costly
and protracted affair; and to point the way toward viable energy policies
and priorities for the next few decades.

Over the past century or so, the United States has built a $14-trillion-
per-year economy that's based almost entirely on cheap hydrocarbons.[10] No matter how much the United States and the rest of the
world may desire a move away from those fossil fuels, the transition to
renewable sources of energy—and to no-carbon sources such as nuclear

FIGURE 1 Annual U.S. Energy Production: Comparing Wind and Solar with Other Energy Sources

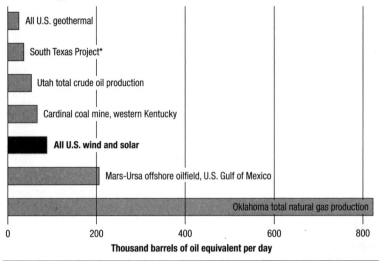

*2,700 megawatt nuclear plant

Sources: Energy Information Administration, South Texas Nuclear Operating Company, and Alliance Resource Partners.

power—will take most of the twenty-first century and require trillions of dollars in new investment. So, given the Four Imperatives and the stark realities posed by the long energy transition that lies ahead, what are we to do?

That question brings us to the other purpose of this book: to debunk some of the energy myths that have come to dominate our political discussions and to lay out the best "no-regrets" energy policy for the United States and the rest of the world. Analysts have coined the term "no regrets" to describe policies that benefit the economy while also reducing emissions of carbon dioxide and other greenhouse gases. After explaining why so many of the "green" technologies now being promoted simply won't work, I will look at the sources that can provide large amounts of energy while also benefiting the economy and the environment.

Of course, there's tremendous political appeal in "green jobs," a "green collar economy," and in what U.S. President Barack Obama calls a "new energy future."[11] We've repeatedly been told that if we embrace those ideas, provide more subsidies to politically favored businesses, and launch more government-funded energy research programs, then we would resolve a host of problems, including carbon dioxide emissions, global climate change, dependence on oil imports, terrorism, peak oil, wars in the Persian Gulf, and air pollution. Furthermore, we're told that by embracing "green" energy we would also revive our struggling economy, because doing so would produce more of those vaunted "green jobs."

These claims ignore the hard realities posed by the Four Imperatives. It may be fashionable to promote wind, solar, and biofuels, but those sources fail when it comes to power density. We want energy sources that produce lots of power (which is measured in horsepower or watts) from small amounts of real estate. And that's the key problem with wind, solar, and biofuels: They require huge amounts of land to generate meaningful amounts of power. And although the farm lobby loves biofuels such as corn ethanol, that fuel fails on two counts: power density and energy density. Corn ethanol production requires vast swaths of land, and the fuel that it produces is inferior to gasoline because it is corrosive, it is hydrophilic (meaning it loves water, and adding water to your motor fuel is not a good idea), and it contains just two-thirds of gasoline's heat content.

I'll discuss the Four Imperatives throughout this book, but for now, suffice it to say that power density and energy density are directly related to the other two imperatives: cost and scale. If a source has low power density, it invariably has higher costs, which makes it difficult for that source to scale up and provide large amounts of energy at reasonable prices.

Despite these realities, the deluge of feel-good chatter about "green" energy has bamboozled the American public and U.S. politicians into believing that we can easily quit using hydrocarbons and move on to something else that's cleaner, greener, and, in theory, cheaper. The hard truth is that we must make decisions about how to proceed on energy very carefully, because America simply cannot afford to waste any more money on programs that fail to meet the Four Imperatives. And that's particularly true now. The economy is weak, millions of Americans are unemployed, and record numbers of homes are in foreclosure. We've already wasted plenty of cash—and time—on the corn ethanol scam and other boondoggles. Congressional mandates forcing motorists to buy ethanol-blended gasoline were supposed to reduce America's dependence on foreign oil and bring energy independence to the United States. Instead, these measures only worsened air quality, increased food costs, damaged untold numbers of engines, and slashed the amount of grain available in the global marketplace.[12]

People in the United States and around the world are hungry for power. They want it for their cars, motorcycles, and lawnmowers, and they want it for their flat-screen TVs, mobile phones, computers, and Cuisinarts. They want power because power drives those devices and in doing so creates wealth and increases personal happiness. And although this book will expose many of the myths about "green" energy, it will deliver some good news about America's situation. It will demonstrate that the smartest, most forward-looking U.S. energy policy can be summed up in one acronym: "N2N"—natural gas to nuclear.

Natural gas and nuclear power are the fuels of the future because they have high power density, are relatively low cost, and can provide the enormous quantities of energy we need. In addition, they produce lower carbon-dioxide emissions than oil and coal and almost zero air pollution. N2N means using natural gas in the near term as we transition to nuclear power over the long term.

N2N will work because the United States sits atop gargantuan natural gas resources. Over the past few years, the U.S. natural gas industry has developed technologies that wring natural gas from shale beds. Thanks to those technological breakthroughs in the production of what the industry calls "unconventional gas," estimates of U.S. gas resources are larger than they've ever been before. How large? U.S. gas resources are thought to contain the energy equivalent of more than 350 billion barrels of crude oil, or roughly as much as the known oil reserves of Saudi Arabia and Venezuela combined.

Of course, much of that gas won't be produced, because of its cost or its distance from pipelines. Nevertheless, the revolution in shale gas production, along with continuing discoveries of offshore gas deposits and drilling success for onshore conventional gas, has led gas analysts to increase their estimates not only of U.S. gas resources but of global gas resources as well. In November 2009, the International Energy Agency

(IEA) estimated that recoverable global gas resources now total about 30,000 trillion cubic feet.[13] At current global rates of consumption, that's enough to last for 280 years.[14]

Though natural gas offers enormous near-term potential, our long-term energy plans must include nuclear—and before going further, let me be clear about where I stand on nuclear power: If you are anti–carbon dioxide and anti-nuclear, you are pro-blackout.

There is no other low- or no-carbon form of electricity generation that can provide relatively large amounts of new power generation at a relatively agreeable cost and do so relatively soon. And the key word in that sentence is "relatively." There is no question that nuclear power will be expensive. It will also require substantial governmental involvement if it is to be safe and affordable. And though nuclear power has big up-front costs, particularly in comparison with several other forms of electricity generation, it also provides the essential always-on power that our society demands. It is the only existing source that offers a long-term, large-scale, zero-emissions alternative to coal- and natural-gas-fired power generation. Obama understands this—or at least he appears to. In an April 2009 speech in Prague, he said, "We must harness the power of nuclear energy on behalf of our efforts to combat climate change."[15]

Despite Obama's statement, and despite the ongoing political push for more low- or no-carbon forms of energy, environmental activists remain adamantly opposed to nuclear power generation. In 2005, some three hundred environmental groups—including Greenpeace, the Sierra Club, and Public Citizen—signed a manifesto that said, "We flatly reject the argument that increased investment in nuclear capacity is an acceptable or necessary solution. . . . Nuclear power should not be a part of any solution to address global warming."[16]

That position ignores nuclear power's many benefits. Nuclear power not only has zero carbon-dioxide emissions, it also produces very small amounts of waste, requires little real estate, and provides large quantities of always-on power. It does, however, come with two significant drawbacks: Nuclear power plants are expensive, and they take years to build and put into operation. Thus, achieving a substantial increase in nuclear power production in the United States will take decades. In the meantime, natural gas provides the most attractive option.

Together, natural gas and nuclear are essential to the ongoing decarbonization of the world's primary energy use, a trend that has been ongoing for about two hundred years. Decarbonization, the trend favoring fuels with lower carbon content, is occurring because energy consumers are always seeking cleaner, denser forms of energy that allow them to do work cleaner, faster, and more precisely. Embracing N2N offers a no-regrets energy policy that will lead to further decarbonization while providing multiple benefits to the United States and the rest of the world.

The structure of this book follows the basic outline contained in the title. In Part 1, I discuss our hunger for power, how much we use, where it comes from, and why our desire for power of all kinds continues to increase. I show that the deluge of criticism about how Americans "use too much energy" is off base, and I describe the sheer scale of our power consumption, explain why we use hydrocarbons such as coal, oil, and natural gas, and show why we'll keep using them for a long time to come.

In Part 2, I debunk many of the myths that people believe about "green" energy by showing that renewables are not the solution to our environmental problems. I demonstrate that wind power is the electricity sector's equivalent of the corn ethanol scam. Like ethanol, wind power is a subsidy-dependant juggernaut that is the antithesis of "green." I show that wind power has not and likely will not make substantial cuts in carbon dioxide emissions, and I take a hard look at a country that—in theory, at least—is supposed to provide a model for the United States and other nations embarking on wind power: Denmark. I have some fun at the expense of T. Boone Pickens by exposing the false claims contained in "Pickens Plan," and I show why the hype over ideas such as carbon capture and sequestration, cellulosic ethanol, and electric cars is just that—hype.

In Part 3, I demonstrate why N2N makes so much sense. I look at the megatrends that favor natural gas and nuclear, provide a brief history of the U.S. gas business, and explain how the U.S. natural gas industry has unlocked galaxies of methane from rocks that were once thought to be impossible to tap. I also explore the technologies that could enable us to address the problem of nuclear waste and those that could help us to revive the U.S. nuclear sector.

That's the outline. But before going further, let me add one other important point about my perspective in this book: There's no political agenda at work here. I am neither Republican nor Democrat. I'm a charter member of the Disgusted Party, a raging centrist, and a recovering liberal. My energy policy is simple: I'm in favor of air conditioning and cold beer. My motivation for writing this book comes from a desire to break through the energy happy-talk so that the United States can have a serious discussion about its future. Energy realities are not dictated by the ideologies of the Left or the Right. They are determined by the laws of physics and the brutal realities of big numbers.

In one of the best-selling business books of all time, *Good to Great: Why Some Companies Make the Leap . . . and Others Don't,* author Jim Collins explained the factors that separate great companies from the also-rans. Among the most essential was that companies that do well confront the brutal facts.[17] In a section called "Facts Are Better Than Dreams," Collins said that successful organizations had a rigorous process of disciplined thought. While making strategic decisions, smart companies "infused the entire process with the brutal facts of reality."[18]

We must confront the brutal facts about energy. But as we do so, there's plenty of reason for optimism. If we are smart, we face a bright future; that future need not be one of energy shortages and energy crises, but instead can be one of energy abundance. The key words in that last sentence are "if we are smart." And that's the challenge. Over the past few years, American voters have been bombarded with nonsense about energy, and much of that nonsense has been embraced. One reason the American public has believed the nonsense is that many of us are woefully ignorant about science and math. Add in a dollop of guilt and a few drams of fear, and it becomes apparent why the United States is in such a quandary.

But first things first. Before we can debunk the nonsense, we must understand the difference between energy and power.

PART I
OUR QUEST FOR POWER

CHAPTER 1

Power Tripping 101

WE DON'T GIVE A DAMN about energy. What we want is power.

Differentiating between the two is essential. Here's the simplest way to do so: Energy is the ability to do work; power is the rate at which work gets done.[1] The more power we have, the quicker the work gets done. And in our speed-obsessed world, we are constantly finding ways to use more power—from a handful of electrons racing through the thinner-than-a-human-hair circuits of a microprocessor to the vast quantities of thrust developed by the jet turbines on a Paris-bound 767.

We don't care what energy *is*. We want what energy *does*. We would gladly fill our fuel tanks with jelly beans, marbles, or Hostess Twinkies if they could deliver the power needed to propel our Camrys and Suburbans to places like Wasilla or Waxahachie. We aren't after energy, we are after what energy provides. And what energy provides is power. We use energy to make power. We convert energy—measured in barrels of oil, tons of coal, and cubic feet of gas—into power, which we tabulate in watts or horsepower.

Much, if not all, of human history can be seen as the pursuit of increasing amounts of power. Whether it's the Egyptians and the Mesoamericans coordinating and harnessing hordes of human muscle to build the pyramids, the use of horses in warfare, Hannibal's use of elephants to cross the Alps in 218 B.C., the first use of steam to drive a piston, the gallop of General George Patton's tanks across France in the summer of 1944, or the Saturn V rockets (which produced 160 million horsepower, or 120 gigawatts) that catapulted American astronauts to the Moon, there has been a consistent, millennia-long effort to find and utilize more power.[2]

Power allows us to do things that make us happy, wealthy, and comfortable. Power gets us up in the air and down the road. Power fetches the e-mail, makes the coffee, and bakes the cake. Power allows us to cut the grass, roast the turkey, cool the beer, fly to Rome, and, of course, keep the lights on. While those facts may seem self-evident, here's the essential truth about our power-driven world: The overwhelming majority of the power we use comes from hydrocarbons because they can provide us with the reliable and abundant power that we desire.

Today, 90 percent of the horsepower we use (or, if you prefer, 9 out of every 10 watts) comes from the burning of oil, natural gas, and coal.[3] And the key attribute of hydrocarbons is their reliability. Renewable energy is dandy, but it simply cannot provide the gargantuan quantities of always-available power that we demand at prices we can afford. The production of electricity from the wind and the sun will continue growing rapidly in the years ahead. But those sources are incurably intermittent. As Stewart Brand, the environmental activist and creator of the *Whole Earth Catalog*, put it during a lecture in mid-2009, "wind and solar can't help because we don't have a way to store that energy."[4] Given our inability to store the energy that comes from wind and solar, those sources will remain bit players in our overall energy mix for the foreseeable future.

After two decades of studying the energy business, I believe those points about energy and power are self-evident. They are not based on ideology; instead, they are grounded in basic physics and basic math. I'm not opposed to environmental protection. Far from it: I'm a birdwatcher and a beekeeper. But now that I've reached middle age, I've finally learned how to use a calculator. Using that device—as well as a bunch of Excel spreadsheets and basic textbooks on physics—has forced me to become a realist on energy issues. And therein lies my frustration: As I've become more pragmatic, our public discourse about energy and energy policy has gone the other way. Discussions about energy matters—which are usually accompanied by arguments about climate change—have devolved into a vitriolic, divisive mess where facts and reasoned argument are largely ignored. In their place we hear vituperative attacks on the "deniers," adolescent arguments about whose fault it is or isn't that the climate is changing, and outlandish claims about the speed with which the United States can (or should) transform its multi-trillion-dollar energy and power delivery systems into "greener" ones. Scientists, journalists, and analysts who dare question the apocalyptic predictions of the global warming alarmists are likely to feel the electronic wrath of bloggers such as Joe Romm, the self-appointed Savonarola of the Al Gore acolytes.[5]

Since September 11, 2001, the United States and the rest of the Western world has been inundated with claims that we should radically change our energy (and power) diet, and do so immediately. We're told that we should abandon our existing systems for something new, something that's low-carbon, solar-powered, wind-powered, or, better yet, powered by the energy sector's single most desired element: unobtanium.

It doesn't seem to matter where the new power will come from as much as it does that we all agree that moving to something else—anything else—is a really good idea. We must, we're told, make a hurried energy transition, because:

- The United States should be "energy independent." Doing so will free us from the vagaries of the world energy market and increase employment here in America.
- We can no longer rely on oil from the Middle East. The suppliers in the region—and Saudi Arabia, in particular—are not our friends. And the Second Iraq War provides further evidence of our unhealthy obsession with the region.
- Cutting oil use will reduce terrorism. This is a favorite claim of neoconservative politicos such as former CIA director James Woolsey and his fellow traveler Frank Gaffney, the head of the Center for Security Policy, a right-wing think tank based in Washington, D.C. In January 2007, Gaffney declared that "some of the hundreds of billions of dollars we transfer each year to various petroleum-exporting nations wind up in the hands of terrorists. This is not simply an addiction. It is a death wish."[6]
- Hydrocarbons are bad. Using them, says one Sierra Club official, is "fossil fuelish."[7] If we don't kick the hydrocarbon habit, we're told,

disastrous climate change will result. Burning coal, oil, and natural gas releases carbon dioxide, which causes global warming. Therefore, hydrocarbons must be replaced with something else, or we will all burn in hell, or something just like it.

- We must quit using oil because we are running out of it. The world will soon reach—or has already passed—its ability to produce increasing amounts of petroleum. This peak in oil production presages a global economic meltdown because we have no substitutes for oil.

These claims seem plausible, and in some cases, they are being put forward by credible people. But they are largely based on faulty assumptions. The promoters of these arguments have gained traction in recent years because the overwhelming majority of the American public simply isn't equipped with the facts. The arguments are often designed as flag-waving, emotional appeals that are accompanied by a big dose of fear, and as a consequence the U.S. public has been primed to believe that an overhaul of our energy system is not only essential, it's patriotic and spiritually righteous, it's good business, and it will once and for all cure the problems of halitosis and premature baldness.

Here's the reality: Whether the issue is oil imports, carbon dioxide emissions, or a peak in global petroleum production, we live in an increasingly interdependent world.[8] With regard to oil and imports, the promoters of energy autarky ignore a myriad of inconvenient truths. Among them: During the first six months of 2009, the United States exported—yes, exported—an average of 1.9 million barrels of oil per day.[9] At that level, U.S. oil exports are on par with countries such as Angola and Venezuela.[10] Of course, the vast majority of those exports are refined products, not crude. Why has the United States become a major player in the international oil market for refined products? Because U.S. refineries are among the best in the world, and they can produce the types of fuels the global market demands. Thus, the United States, the world's biggest importer of oil, is also one of its biggest exporters, and it has been a major exporter for years. So here's a tip: The next time you hear somebody promoting "energy independence," grab your wallet. Whatever they are proposing to achieve that delusional goal will surely cost you money.

As for terrorism, the very nature of the global oil market—the biggest, most integrated, most transparent market ever created—undermines the claim that using less oil will somehow result in a reduction in the tactics of terror. Although it's true that some petrostates have ties to terrorism—Iran being an obvious example—it's just as true that Iran and other oil exporters cannot be isolated from the global oil market. Terrorism isn't an ideology, it's a tactic, a cheap tactic, and it doesn't depend on petrodollars. In May 2009, the Rand Corporation, one of the oldest defense-focused think tanks in Washington, released a report concluding that America's "reliance on imported oil is not by itself a major national security threat." The report went on to debunk the claim that oil and terrorism are related, saying, "Terrorist attacks cost so little to perpetrate that attempting to curtail terrorist financing through measures affecting the oil market will not be effective."[11]

Many people may be worried about peak oil, but those concerns frequently ignore the fundamentals of the marketplace. Prices and technology are always combining to unlock hydrocarbons that were once thought unreachable. Let's look at just one month: September 2009. During that month alone, several companies announced major oil and gas finds. For instance, BP announced that its Tiber prospect in the Gulf of Mexico may hold more than 3 billion barrels of oil.[12] That well was drilled in 4,100 feet of water to a depth of 35,000 feet.[13] On September 11, the Spanish energy firm Repsol announced the biggest natural gas discovery in Venezuela's history. The discovery, located in the Gulf of Venezuela in a water depth of about 200 feet, may contain 8 trillion cubic feet of gas, the energy equivalent of about 1.4 billion barrels of oil.[14] Five days later, Anadarko Petroleum announced the Venus find offshore Sierra Leone, and the company said that the geology of offshore West Africa appears favorable for hydrocarbons along a line some 700 miles long that goes through Liberia, Cote d'Ivoire, and Ghana.[15] In addition, Petrobras, the Brazilian national oil company, announced yet another major offshore discovery. This one involved a big pool of hydrocarbons underneath more than 7,000 feet of water in the Santos Basin, an area south of Rio de Janeiro.[16]

Just for grins, let's suppose that Petrobras and the other big companies decided to suddenly stop looking for more oil. Even if that unlikely event occurred, the world still has about 1.25 trillion barrels of proved reserves waiting to be tapped. That's a lot of petroleum—about forty-two years' worth at current rates of production.[17] Sure, the world will one day hit its peak in oil production—or perhaps that peak has already passed. Whatever the case, we will keep using oil for decades to come, and our consumption will rise or fall depending on the price. Commodities have always been rationed by price. Oil is a commodity, and as the price of that commodity increases, the rationing of oil will become more pronounced and we will be forced to use petroleum more efficiently.

Global climate change and carbon dioxide emissions are the causes *du jour*. There is a widespread belief that if the people of the world do not unite to drastically reduce their carbon dioxide output, then catastrophic climate change will occur. In fact, some environmental activists have decided that the optimum level of carbon dioxide in the global atmosphere should be 350 parts per million. (By late 2009, the concentration was about 390 parts per million.) On October 24, 2009, the supporters of the 350 parts per million target conducted more than 4,000 synchronized demonstrations around the world. Their aim: to build a "global community" to support the 350 ppm goal.[18] The chairman of the Intergovernmental Panel on Climate Change, Rajendra Pachauri, has said he is "fully supportive" of the 350 ppm goal.[19] In November 2009, former vice president Al Gore, appearing on the *Late Show with David Letterman*, declared that unless the people of the world took drastic action to curb carbon dioxide emissions, it could be "the end of civilization as we know it."[20]

Gore may be right. It's also possible that he's wrong. In many ways, Gore's opinion doesn't matter, because no matter how much the United States may want to lead the effort to reduce carbon emissions, it cannot, and will not, be able to substantially slow the increasing global use of coal, oil, and natural gas. Why? There are simply too many people living in dire energy poverty for them to forgo the relatively low-cost power that can be derived from hydrocarbons. (I will discuss carbon dioxide emissions at length in Part 2.) For proof of that, consider the per-capita carbon dioxide emissions in the world's most populous countries. From 1990 to 2007, the per-capita emissions of carbon dioxide in the United States fell by 1.8 percent. But during that same time period, per-capita emissions soared in Brazil, China, India, Indonesia, and Pak-

FIGURE 2 Percentage Change in CO$_2$ Emissions Per Capita in the Six Most Populous Countries, 1990 to 2007

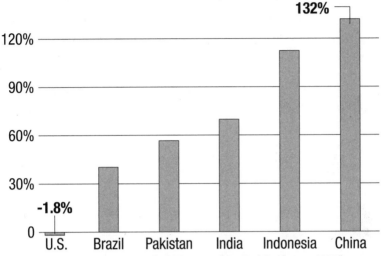

Source: International Energy Agency, "CO$_2$ Emissions from Fuel Combustion 2009," http://www.iea.org/co2highlights/co2highlights.pdf, 90–91.

istan. Those five countries contain more than 3 billion people, and their energy-consumption patterns are being replicated in nearly every major developing country on the planet.

That reality was reflected in Copenhagen in December 2009 when leaders from 192 countries met for what the Associated Press called "the largest and most important UN climate change conference in history."[21] After two weeks of wrangling and lofty rhetoric, the meeting ended with an eminently predictable result: no legally binding agreement on any reductions in carbon emissions, only a promise to reduce emissions "individually or jointly," and an agreement to meet again a year later in Mexico City to discuss all of the same contentious issues one more time.[22]

In short, all of these concerns, from worries that we have reached (or will soon reach) a peak in oil production and are (or will soon be) entering a period of inevitable decline, to the alarmist cries over impending global warming—and the supposed solutions to them—hinge on the belief that the transition away from hydrocarbons to renewable resources can be done quickly, cheaply, and easily.

That. Is. Not. True.

Tomorrow's energy sources will look a lot like today's, because energy transitions are always difficult and lengthy. "There is one thing all energy transitions have in common: they are prolonged affairs that take decades to accomplish," wrote Vaclav Smil in November 2008. "And the greater the scale of prevailing uses and conversions, the longer the substitutions will take."[23] Smil, the polymath, prolific author on energy issues, and distinguished professor at the University of Manitoba, wrote that while a "world without fossil fuel combustion is highly desirable . . . getting there will demand not only high cost but also considerable patience: coming energy transitions will unfold across decades, not years."[24]

Indeed, energy transitions unfold slowly and are always under way whether we recognize them or not. Between 1973 and 2008, the amount of electricity generated in the United States with nuclear reactors increased by more than 800 percent. Nuclear power now accounts for about 20 percent of the electricity generated in America. But for the average homeowner whose immediate interest is in sweeping the carpet or baking a pie, that transition has been invisible. For the consumer, the electricity that comes out of the wall socket is a commodity. How it is generated is of little interest. The key concern for the consumer was, and continues to be, that electricity remain cheap and always available.

The $5-trillion-per-year global energy business dwarfs all other sectors of the economy.[25] Given its size, and given that any major energy transition will take decades, we must carefully analyze the various energy sources to determine which ones can satisfy the Four Imperatives: power density, energy density, cost, and scale. Using those metrics will help us to confront the brutal facts, winnow out the pretenders, and increase the consumption of the winners.

But before we begin that winnowing process, we must take a look back in order to understand how we got to this place in U.S. energy history. That requires calling out some of the energy posers who claim to have the answers while also taking a hard look at the underlying causes of America's energy unease. And much of that unease comes from three factors: guilt, fear, and ignorance—the deadly trio that has been incarcerating the human mind for millennia.

CHAPTER 2

Happy Talk

THE TELEVISION NEWS industry has a great term: "happy talk." Producers work to make sure that every second of TV airtime is filled with scripted content. But try as they might, they often find themselves with several seconds of unfilled air time that must be made to look purposeful. On some occasions, in an effort to make their newsreaders appear more likeable to viewers, TV producers may ask the talking-heads to engage in some friendly banter to fill the airtime between news segments and commercial breaks. These chummy bits of patter are called—you guessed it—happy talk.

Over the past few years, Americans have been inundated with energy happy talk. And it has come from personalities ranging from Dallas billionaire T. Boone Pickens and former vice president Al Gore to *New York Times* columnist Thomas Friedman and media darling Amory Lovins, the chairman and chief scientist at the Rocky Mountain Institute, a Colorado-based think tank.

For Pickens, the bogeyman to be slain is foreign oil. For Gore, the villain is carbon dioxide. And while the sin to be cured varies with the preacher, the message of deliverance is largely the same: Repent. Give up those evil hydrocarbons and embrace the virtues of renewable energy before you face the eternal damnations of foreign oil, global warming, and a carbon footprint that's bigger than Boone Pickens' ego.

Lovins is among the most quoted purveyors of energy happy talk. In 2007, he wrote a short piece called "Saving the Climate for Fun and

Profit" in which he said that curbing carbon dioxide emissions "will not cost you extra; it will save you money, because saving fuel costs less than buying fuel."[1] In 2008, he claimed that the issues of "climate change, oil dependence, and the spread of nuclear weapons—go away if we just use energy in a way that saves money, and since that transition is not costly but profitable, it can actually be led by business."[2] Venture capitalist Vinod Khosla is another veteran happy talker. In May 2006, Khosla claimed that making motor fuel out of cellulose was "brain dead simple to do" and that commercial production of cellulosic ethanol was "just around the corner."[3] Ten months later, Khosla was once again hyping cellulosic ethanol, saying that biofuels could completely replace oil for transportation and that cellulosic ethanol would be cost-competitive with corn ethanol production by 2009.[4] Alas, Khosla's crystal ball turned out to be somewhat cloudy. By late 2009, despite hundreds of millions of dollars in venture-capital investment in cellulosic ethanol companies, not one of those efforts had been successful in producing significant quantities of the fuel for commercial use.

In July 2008, Al Gore, winner of the Nobel Peace Prize, declared that the United States should "commit to producing 100% of our electricity from renewable energy and truly clean carbon-free sources within 10 years."[5] Four months later, in an op-ed in the *New York Times*, Gore said the nation must replace "dangerous and expensive carbon-based fuels with 21st-century technologies that use fuel that is free forever: the sun, the wind and the natural heat of the earth."[6]

About that same time, Gore, along with a coalition of environmental groups called the Alliance for Climate Protection, launched a $300 million media campaign designed to stop global climate change.[7] That campaign is backed by a number of websites, including Wecansolveit.org, Climateprotect.org, and RepowerAmerica.org. By early 2009, more than 2 million people had joined and had agreed to the statement, "I want to Repower America with 100% clean electricity within 10 years." The "grassroots partners" behind the effort include the National Audubon Society, the Evangelical Environmental Network, and other groups.[8]

And then there's Pickens. The Dallas-based energy mogul is one of a long line of super-wealthy Texans endowed with a Messiah complex who have luxuriated in the national limelight by promising to deliver—pick one or more of the following—better football (Jerry Jones, Dallas Cowboys); better basketball (Mark Cuban, Dallas Mavericks); a better president (H. Ross Perot, 1992 and 1996); better football (Boone Pickens, Oklahoma State University); and better energy policy (Pickens, again). On July 4, 2008, Pickens launched a $58 million media campaign aimed at promoting the "Pickens Plan."[9] The campaign launch included a barrage of TV ads starring the Texas energy baron, who begins the pitch with a syrupy drawl: "I've been an oilman all my life. . . . " The centerpiece of the Pickens Plan: "By generating electricity from wind and solar and conserving the electricity we have, we will be free to shift our use to natural gas to where it can lower our need for foreign oil."[10]

The public loves the idea of renewable energy. On November 19, 2008, WorldPublicOpinion.org released a poll of nearly 21,000 people from twenty-one nations. The findings: Seventy-seven percent of respondents believed that their country should "emphasize more" use of solar and wind energy.[11]

In December 2008, the League of Conservation Voters (LCV) sent out an e-mail asking voters to sign a petition that was to be sent to President-Elect Barack Obama. The LCV's board is a Who's Who of American environmental groups, including representatives from the Natural Resources Defense Council, Friends of the Earth, the Wilderness Society, and the Environmental Defense Fund.[12] The petition told Obama that it is:

> Time to Repower, Refuel, and Rebuild America. We need to get our economy moving by building a clean energy future. We applaud your efforts to make energy a top priority, and look forward to working with you to achieve these goals:
> - Move to 100% electricity from clean sources such as wind and solar;
> - Cut our dependence on oil in half;
> - Create 5 million new clean energy jobs; and
> - Reduce global warming pollution by at least 80%.[13]

In April 2009, during the telecast of the Miss USA pageant, the show's emcee, Billy Bush, and cohost Nadine Velasquez declared that the silicone-and-swimsuits soiree was, in fact, environmentally friendly and was therefore part of NBC's initiative, "Green Is Universal." The slogan is a play on the name of the TV outfit's parent company, NBC Universal, a subsidiary of industrial giant General Electric.[14] Following the corporate plug, Velasquez said that "Miss USA will be awarded a brand new, more eco-friendly green crown, because green reigned here."[15]

Nobel Prize–winning physicist Steven Chu, who now serves as the U.S. secretary of energy, has made his own glib pronouncements.[16] In mid-2009, Chu appeared on *The Daily Show with Jon Stewart* and said, "We want energy but we want it carbon-free."[17]

"Carbon-free" energy appears to be such a selling concept that even pimps have begun hawking it. In 2005, Heidi Fleiss, the "Hollywood Madam" who gained notoriety in the mid-1990s after she was arrested and convicted on attempted pandering charges, announced that she was planning to open a "stud farm" in Nevada that would cater to female customers.[18] But in 2009, Fleiss announced that she had dropped plans for the bordello and was instead focusing her talents on alternative energy. "That's where the money is," she said. "That's the wave of the future."[19]

From Gore to Chu and Miss USA to the Hollywood Madam, Americans are being carpet-bombed with energy happy-talk. And that happy talk has contributed to a widespread sense of guilt.

□□□

Here's an exercise: The next time you hear someone say "We are addicted to oil" or "We are addicted to coal," try this: Substitute the word "prosperity" for "oil" or "coal."

I don't offer that idea to be flippant, but rather to point out just how disconnected America's rhetoric about energy is from the perspective of the 2 to 3 billion people on the planet who live in dire energy poverty. At the same time that many of those people are still relying on biomass (such as wood, straw, or dung) for their cooking needs, and spending large chunks of their time and labor procuring those fuels every day, most Americans live in a world of energy abundance with access to cheap fuels that their counterparts in places like South Africa, Sudan, Laos, Afghanistan, Vietnam, and Pakistan can only dream of.

While most of us certainly appreciate the many blessings of prosperity, there's a growing sense that U.S. citizens should sign up for Jenny Craig or an Atkins Diet for gasoline and electricity. Conflating energy use with addiction—sex addiction, gambling addiction, alcohol addiction, Internet addiction—has facilitated a growing sense of anxiety in Americans.[20] Add in fears about global warming—which many scientists believe is being caused by, or at least exacerbated by, the burning of hydrocarbons—along with claims about energy shortages and terrorism, and that guilt becomes ever more easily exploitable by politicians, pundits, and erstwhile capitalists looking to suckle at the federal teat. On top of all this, Americans feel guilty about their prosperity, particularly when compared with the grinding poverty that is common throughout the world.

In mid-2009, a Canadian energy analyst, Peter Tertzakian, published a book called *The End of Energy Obesity* that tapped into these themes of guilt and addiction. In the first chapter, he declared that "we have become increasingly addicted to energy because we thoroughly enjoy the standard of living that energy-consuming devices and services make possible."[21] Tertzakian's claim echoes the worldview espoused by Barack Obama in early 2007 when he officially announced that he was running for president. The United States, Obama proclaimed, must break free of the "tyranny of oil."[22]

Huh? Billions of people would dearly love to be tyrannized by oil in exactly the same way most Americans are. Consider India, a country of 1.1 billion people, where the average resident consumes about 0.11 gallons of oil per day.[23] The average American consumes about twenty-four times as much. And yet, over the past few years, many Americans have become increasingly ambivalent about their energy use. In some circles, people who drive SUVs are subjected to ridicule; conversely, fuel-efficient cars such as the Toyota Prius confer on their drivers a certain amount of environmental cachet, or "eco-bling."

The growing Western obsession with carbon dioxide has even led some consumers to buy "carbon credits"—a type of get-out-of-jail-free card, an environmental indulgence—that theoretically allows them to offset a certain amount of the carbon dioxide they are responsible for emitting. The promoters of these indulgences promise buyers that their money will go to "green" projects, such as a system that captures methane gas from a Chinese landfill, or perhaps the construction of a dam in India.[24] But by 2008, the market for carbon indulgences had grown to some $54 million per year, and the Federal Trade Commission was advising consumers to be wary of the potential for fraud when buying them.[25] And in late 2009, a British travel company, Responsibletravel.com, announced that it had quit offering carbon offsets because, in the words of the company's founder, they had become the equivalent of a "medieval pardon" that allowed buyers "to continue polluting."[26]

Along with carbon credits, Americans have been barraged with claims about the desirability of being "carbon neutral." In 2007, Al Gore's followers held Live Earth, a global series of concerts that claimed it was "carbon neutral" because, among other things, it had purchased carbon credits to offset the air travel done by concert organizers and performers.[27] At the 2008 Democratic National Convention, the Democratic National Committee created a "green delegate challenge" that asked each of the 5,000 delegates who were going to the convention in Denver to pay $7.50 for a "carbon offset." The money was to be funneled to NativeEnergy, a Vermont outfit that promised to invest the money in various renewable energy projects.[28]

It's not just the Greens and the Democrats. One of history's most prolific purveyors of indulgences, the Roman Catholic Church, has begun equating carbon dioxide emissions with sin. In September 2007, the *New York Times* reported that the Vatican was aiming to become "the world's first carbon-neutral state." In pursuit of that concept, the Vatican paid to plant a convent-load of trees on a 37-acre tract of land in Hungary. The plot was to be renamed the Vatican Climate Forest, and once the trees were in place, the Vatican would, in theory, have an atmospheric dispensation for all the carbon dioxide emissions that came from its cars, its offices, and, presumably, the hallowed lungs of the Holy Father himself, Pope Benedict XVI.

The *Times* quoted a Vatican official, Monsignor Melchor Sánchez de Toca Alameda, who averred that buying carbon credits was akin to penance. The monsignor did not advocate sackcloth and ashes, but he implied that believers may avoid eternal carbon damnation by "not using heating and not driving a car, or one can do penance by intervening to offset emissions, in this case by planting trees."[29] Of course, there's no sin in planting trees. But can the trees in Hungary really offset all of the Vatican's carbon output? And for how long?

It's interesting to ponder what Abraham Maslow, the American psychologist, might have thought of all this. Maslow originated the idea of the "hierarchy of needs," the concept that humans, as they increasingly satisfy their physiological needs—food, water, sleep, clothing, shelter, sex, and so on—begin seeking to fulfill more complex needs, such as love, esteem, and "self-actualization." Using Maslow's template, it appears that many Americans have become so wealthy, so sated with living well—in a way that is made possible by using large quantities of cheap hydrocarbon-driven power—that their successful self-actualization depends in part on how much guilt they feel about consuming the very commodities that allow them to prosper.

An example: A few months ago, a friend of mine, a well-compensated M.D., bought a large Toyota SUV. The nearly 3-ton behemoth was equipped with a 381-horsepower engine, a DVD player, leather seats, and a rear cargo area big enough for a quick game of badminton. While escorting me on a trek around the perimeter of the vehicle, the physician admitted some feelings of guilt about his new wheels—and yet the mere act of sharing those pangs of iniquity appeared to assuage some of his feeling of environmental sinfulness.[30]

The disconnect between America's energy use and its guilt over that energy use is even being featured in an ad campaign sponsored by—get this—one of the world's biggest energy companies. That's right: California-based Chevron (2008 revenues: $273 billion) has been running a campaign called "Will You Join Us?"[31] The advertising barrage includes billboards and print ads with photos of handsome people with text lines imposed over, or near, their faces.[32] One says, "I will leave the car at home more." Another reads, "I will finally get a programmable thermostat." But my all-time favorite is the ad that proclaims, "I will use less energy."[33]

Pardon my insolence, but how many people in Uganda, Cambodia, or Peru wake up in the morning and declare, "By golly, I'm going to use less energy today"? Not many, I'd wager. And yet, this notion of guilt, combined with rhetoric about addiction and the idea of "using less," has become a powerful theme in American politics.

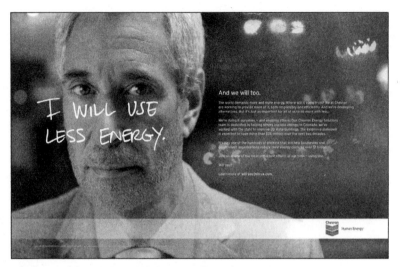

PHOTO 1 Ad from Chevron's 2009 "Will You Join Us?" campaign

Source: Chevron, http://www.willyoujoinus.com/assets/downloads/media/Chevron_Iwill _use%20less%20energy.pdf.

Furthermore, consider Chevron's corporate strategy: It is advising its customers to use less of the stuff it sells. Imagine what might happen if other companies followed suit: Microsoft would encourage people to use its software less or forgo the updates; Ford Motor could run TV ads advising drivers to continue piloting their old clunkers; Whole Foods could advise grocery buyers that what they really need isn't fresh produce and a warm baguette, it's . . . nothing at all. Given Chevron's lead, corporate America may now forsake selling anything ever again. Imagine the environmental benefits! Just think of the reduction in carbon dioxide emissions!

Now, some of Chevron's motivation for its ad campaign may be that it wants to soften the company's image. And when it comes to image, energy companies are often seen as only slightly more likeable than Lucifer himself. A 2006 Gallup poll found that just 15 percent of Americans had a positive view of the oil and gas industry, whereas 77 percent had a negative image. Out of twenty-five sectors that Gallup asked about, the oil and gas industry ranked dead last. Even the federal government ranked ahead (but just barely) of the oil and gas industry in the collective opinion of the general public.[34] Add in the huge profits that the industry has made in recent years, including Exxon Mobil's record $45.2 billion profit in 2008, at a time when Americans were paying record-high prices for gasoline, and the industry's need for an image makeover becomes even more apparent.[35]

Thus, along with their feelings of guilt, Americans are angry at the companies that provide them with the energy they require. And to top it off, many Americans are fearful. Their fears are evident in the findings of an early 2009 Zogby International survey that was conducted for the Manhattan Institute, a conservative think tank. Zogby interviewed 1,000 randomly selected adults from across the United States about issues regarding energy and the environment. Of the questions that focused solely on energy, the most lopsided response came from a question dealing with potential shortages of hydrocarbons, where 70.6 percent of the respondents agreed that the United States "must move to renewable energy because we are rapidly running out of oil, natural gas, and other fossil fuels."[36]

If that Zogby poll were an election, it would have been a landslide. Perhaps this fear is understandable. Over the past few years, books, magazines, and newspapers have continually hyped the dangers of peak oil. Catastrophists such as author James Kunstler, who wrote the 2005 book *The Long Emergency*, have predicted that once we reach that peak, rapid declines in production will follow, and then, warns Kunstler, "epidemic disease and faltering agriculture will synergize with energy scarcities to send nations reeling."[37]

Americans are fearful about energy because of the lingering images of the 1973 Arab oil embargo and the 1979 oil price shock. More recently, they have endured the supply disruptions in the wake of Hurricane Katrina and the price shocks of mid-2008 that sent gasoline prices to more than $4 per gallon.

Politicians frequently use those events to stoke the fear that the United States could somehow be "cut off" from the global oil market. For instance, in 2006, Bill Clinton gave a speech in California during which he said, "Think of the instability and the impotence you feel knowing that every day we have to have a lifeline from places half a world away that could cut us off in a minute."[38] Of course, it's worth noting any time that Bill Clinton mentions "impotence." But he's hardly the only one stoking the fears of a possible embargo or shortage. In mid-2009,

Michael Moore, the liberal documentary filmmaker, published an essay in *The Daily Beast* in which he forecast a real-life edition of the *Mad Max* movies complete with oil-crazed survivors of a scorched planet battling each other for the last few liters of gasoline. In his article, titled "Goodbye, GM," Moore assailed the "war" that he said was "being waged by the oil companies against you and me." He went on to say that the evil oil barons "are not telling the public what they know to be true—that there are only a few more decades of useable oil on this planet. And as the end days of oil approach us, get ready for some very desperate people willing to kill and be killed just to get their hands on a gallon can of gasoline."[39]

The drumbeat of fear abounds in discussions about global warming. In June 2009, the Obama administration released a report on climate change that called for massive reductions in U.S. emissions of carbon dioxide. Without a major change in energy use, the report said, "Likely future changes for the United States and surrounding coastal waters include more intense hurricanes with related increases in wind, rain, and storm surges."[40] It went on to say that because of global warming, "crop and livestock production will be increasingly challenged," and that "coastal areas are at increasing risk from sea-level rise and storm surge."[41] The report concluded that our only choice is to cut carbon dioxide emissions and that, "unless the rate of emissions is substantially reduced, impacts are expected to become increasingly severe for more people and places."[42]

Although guilt, anger, and fear are key elements of Americans' gullibility when it comes to energy matters, the most important factor is ignorance. Most people simply don't understand how energy and power are produced. And that lack of knowledge, combined with widespread scientific illiteracy and innumeracy, makes for a deadly combination.

□□□

In 2007, I interviewed Vaclav Smil about energy issues.[43] I asked him why Americans are so easily swayed about energy matters. His response: scientific illiteracy and innumeracy. "Without any physical, chemical, and biological fundamentals, and with equally poor understanding of basic economic forces, it is no wonder that people will believe anything," he told me.[44] Verifying Smil's claim is all too easy. A 2007 study by Michigan State University determined that just 28 percent of American adults could be considered scientifically literate.[45] In February 2009, the California Academy of Sciences released the findings of a survey which found that most Americans couldn't pass a basic scientific literacy test. The findings:

- Just 53 percent of adults knew how long it takes for the Earth to revolve around the Sun.
- Just 59 percent knew that the earliest humans did not live at the same time as dinosaurs.
- Only 47 percent of adults could provide a rough estimate of the proportion of the Earth's surface that is covered with water. (The academy decided that the correct answer range for this question was anything between 65 and 75 percent.)
- A mere 21 percent were able to answer those three questions correctly.[46]

In July 2009, the Pew Research Center for the People and the Press released the results of a survey of 2,001 adult Americans regarding science issues. Among the findings: Just 46 percent knew that electrons are smaller than atoms.[47]

Those findings shouldn't be surprising. Ignorance of the sciences and the natural world has plagued the world for centuries. This centuries-long suspicion of science, which continues today with regular attacks on Charles Darwin and his theory of evolution, was recognized by British scientist and novelist C. P. Snow in the 1950s when he delivered a lecture called "The Two Cultures." Snow argued that there was a growing disconnect between the culture of the sciences and the culture of the humanities, and that bridging that gap was critical to understanding and addressing the world's problems. Snow placed "literary intellectuals at one pole—at the other scientists," and noted that in between there was "a gulf of mutual incomprehension."[48] Snow then laid out a critical point about the general public's lack of understanding of energy and thermodynamics. As Snow put it:

A good many times I have been present at gatherings of people who, by the standards of the traditional culture, are thought highly educated and who have with considerable gusto been expressing their incredulity at the illiteracy of scientists. Once or twice I have been provoked and have asked the company how many of them could describe the Second Law of Thermodynamics. The response was cold: it was also negative. Yet I was asking something which is about the scientific equivalent of: Have you read a work of Shakespeare's?[49]

Indeed, although most moderately cultured people will be familiar with *A Comedy of Errors* or *The Merchant of Venice*, the laws of thermodynamics are considered by many of these same people to be the domain of nerds and wonks. Thus, the first law of thermodynamics—energy is neither created nor destroyed—and the second law—energy tends to become more random and less available—are relegated to the realm of too much information.[50] This apathy toward science makes it laughably easy for the public to be deceived, or for people to deceive themselves.

Alas, the apathy toward science in America is matched—or perhaps even exceeded—by the lack of interest in mathematics. Over the past few years, the United States has been inundated with depressing data about the state of the country's mathematical skills. And unfortunately, the data appears to reflect a grim reality.

A 2008 study published by the American Mathematical Society put it bluntly: "It is deemed uncool within the social context of USA middle and high schools to do mathematics for fun."[51] The study went on to explain that "very few USA high schools teach the advanced mathematical skills, such as writing rigorous essay-style proofs, needed to excel."[52] Another report issued in 2008, this one from the U.S. Department of Education's National Mathematics Advisory Panel, declared that math education in the United States "is broken and must be fixed."[53] The report found "that 27% of eighth-graders could not correctly shade 1/3 of a rectangle and 45% could not solve a word problem that required dividing fractions."[54] The report also found poor math skills among adults:

- 78 percent of adults could not explain how to compute the interest paid on a loan.

- 71 percent could not calculate miles per gallon on a trip.
- 58 percent were unable to calculate a 10 percent tip for a lunch bill.[55]

This scientific illiteracy and innumeracy gets exacerbated in energy discussions by an equally thorny problem: the many different ways in which we measure units of energy. We use several sources of energy, and each is measured and sold in a mind-boggling variety of units. Oil is measured and sold in barrels, tons, gallons, and liters. Natural gas is measured and sold in cubic meters, millions of Btu, therms, dekatherms, and cubic feet. Coal comes in long tons and short tons, but its pricing depends on several other factors, including heat content, ash content, sulfur content, and, most important, the distance between the coal mine and the power plant. Electricity is sold in kilowatt-hours, but electricity terminology spans other units, including volts, amperes, and ohms.[56] Add in joules, watts, ergs, and calories, and things get even more complex. Furthermore, different entities use different metrics. For instance, the BP Statistical Review of World Energy publishes much of its data in millions of tons of oil equivalent. The Energy Information Administration prefers quadrillion Btu, or "quads." (One quad is approximately equal to 172 million barrels of oil equivalent, or about 1 exajoule.) Meanwhile, the International Energy Agency, as well as many countries, uses joules.

This googol of energy metrics complicates energy discussions. It also makes it more difficult to move past feel-good ideas that will do little or nothing to actually address our future energy and power needs.

In order to move past the happy talk, as well as the guilt, fear, and ignorance, we have to address the issues of energy and power in a rigorous manner. We must take an approach that includes numbers, units, and precise terminology. Understanding the difference between energy and power, for example, requires a bit of elementary physics, as well as proper definitions of key terms. In the next chapter, I will walk you through the physics and the terminology so that you can see why America's discussions about "energy" are so misguided. Power is what we want. And lots of it.

CHAPTER 3

Watt's the Big Deal?
(Power Tripping 102)

ENERGY GETS THE HEADLINES and the attention. It's the buzzword that pundits and politicos count on to pack a punch. Thus, we've been barraged by the ever-present "energy crisis" as well as other combinations: energy security, energy scarcity, energy management, energy policy, and dozens more. In the consumer world, we have energy bars, energy drinks, and, for consumer electronics, Energy Star.

My antique copy of the *Shorter Oxford English Dictionary* (printed in 1936) contains half a dozen definitions for "energy." Meanwhile, its definitions for "power" cover nearly half a page. The word "power" now gets used in numerous contexts—political power, electrical power, brain power, black power, Chicano power, star power, flower power, power trip, power walking, power lunch, and computing power, to name just a few.

Wrestling the two terms to the ground requires real effort, particularly given that fact that many people make the mistake of using "power" and "energy" interchangeably. But we must persevere. Definitions matter. In order to properly address a problem, we must first define it and agree on a common set of terms. And given that our effort requires basic physics, the first stop on our power quest is the work done by a Scotsman whose last name has become synonymous with power: James Watt.

We use Watt's name on a near-daily basis. But few people know what a "watt" is or why Watt's work was so important. Here are the essential facts: Watt, born in 1736, made critical improvements to the steam engine. Those inventions raised the efficiency of steam engines so much that Watt, having patented the improvements, became a wealthy man.[1] But Watt knew that improvements to the steam engine were not enough. He needed a metric that could help his customers understand the amount of work done by his steam engines in an hour or in a day. Given the centrality of horse-pulled power to eighteenth-century industry, and his ability to measure the work done by horses, it's not surprising that he dubbed his new unit a "horsepower." The result of his various measurements: 1 horsepower = 33,000 foot-pounds per minute.[2]

The idea of foot-pounds per minute is hardly an intuitive metric, but in Watt's day it made sense. Watt did a lot of work with coal mines, where horses were the draft animal of choice. Of course, a horse couldn't lift a bucket of coal weighing 33,000 pounds. But that same horse could likely raise 330 pounds of coal 100 feet in 1 minute. Or it might be able to lift 33 pounds of coal 1,000 feet in that same time frame. The combination of feet and pounds can be whatever numbers you choose. But if the product of the two numbers equals 33,000 foot-pounds per minute, then you are producing 1 horsepower.[3]

Since Watt's day, horsepower has coexisted with other measures of power, including Btu per hour, calories per day, kilocalories per minute, and ergs per second, to name just a few. For decades, this welter of power metrics confused even the most savvy of scientists and laymen. In 1960, the International System of Units, commonly called SI, was established. SI units are the result of a centuries-long effort to create a uniform system of measurement for distance, mass, time, current, power, pressure, and temperature—you name it—as well as symbols for numbers in the thousands, millions, billions, and so on. SI facilitates analysis and discussion, particularly among people from different cultures, languages, and fields of interest. And, know it or not, SI parlance has become part of our everyday speech. Kilo (k) means thousand, as in kilogram or kilowatt. Similarly, mega (M) means million, as in megajoule or megawatt, and giga (G) means billion, as in gigabyte or gigaflop. (See Appendix B for a full listing of SI numerical designations.)

SI simplified discussions of energy and power. The joule (J), named after the British scientist James Prescott Joule, is the only unit of measure

in SI for any kind of energy, regardless of its form.[4] The watt (W)—
named for James Watt some six decades after his death—is the only unit
of measure in SI for any kind of power.

To differentiate between joules and watts, it may help to think of
them thusly: The total amount of energy produced is measured in joules;
power generation is measured in watts. Put another way, the quantity of
energy consumed is measured in joules; how quickly that energy gets
consumed is measured in watts.[5] Thus, operating a 60-watt light bulb
requires power, which, as just discussed, is measured in watts. After an
hour, when you switch the light off, you can then measure the amount of
energy that was consumed by the light, which is measured in joules—or
in kilowatt-hours or in Btu—all of which are measures of energy.

Watts and joules are often used together. Calculating power requires
knowing the amount of energy and the time over which it was used. This
calculation has become a basic formula in physics. The equation is simple:

$$\text{power} = \frac{\text{energy}}{\text{time}}$$

One watt is equal to 1 joule per second. The corollary is just as important: 1 joule = 1 watt-second.[6] With the notable exception of the
United States, the entire world uses SI when discussing energy and
power. Though SI units are valuable and laudable, it doesn't mean that
everyone who uses them comprehends them. Indeed, although the
watt has become a standard unit for measuring power, horsepower continues to be part of our everyday discussions, particularly when we are
talking about cars, chainsaws, and lawnmowers. Why? It's a centuries-
old metric that's easy for most laymen to grasp. Everyone can imagine
a horse pulling a plow or a carriage and the work that that job entails.
So which metric makes more sense? During the course of writing this
book, I asked dozens of people which term they understood better—
watts or horsepower. Some people replied with a blank stare and chose
neither. A handful (including nearly all of the engineers and scientists)
preferred watts. But the majority preferred horsepower. Asked why,
they said that they were more familiar with horsepower ratings on cars
than they were with power ratings listed in watts. Men generally preferred horsepower. Women generally picked watts. Thus, my sample

may have been skewed because the population I surveyed had more males than females.

This book will use both horsepower and watts. Pick whichever unit you prefer—just remember that both are measures of power, not energy, and keep in mind that 1 horsepower is equal to 746 watts.

Now that the mini-lecture on physics and SI is done, we must return to the task at hand: defining *energy* and *power*. One of the best explanations I've heard comes from my good friend Stan Jakuba, a whip-smart engineer who has spent decades advocating the virtues and simplicity of SI. "Energy has many forms," he explains, "such as electricity, heat, work, kinetic energy, potential energy, chemical energy, nuclear energy, etc. Energy can be visualized as an amount of something. Power is the energy *flow*."

Jakuba's vivid explanation underscores an essential concept: Energy is an amount, while power is a measure of energy flow. And that's a critical distinction. Energy is a sum. Power is a rate. And rates are often more telling than sums.

To illustrate that fact, let's express energy and power in oil terms. Energy is measured in barrels. Power is measured in barrels *per day*. Suppose you have discovered an oil field containing 1 billion barrels of oil. That's a lot of energy, sure. But that energy is worthless unless it can be brought out of the ground. And, generally speaking, the faster you can get it out, the better. Thus, an oil field that holds, say, 100 million barrels of oil that can produce 10,000 barrels per day is worth a whole lot more— we can even say it is more *powerful*—than one that produces 10,000 barrels per week.

The key word here is "per." When buying a car, we want to know the rates: How many miles *per* gallon does it get, how fast—in miles *per* hour—can it go?

Yet another good analysis comes from Richard Muller's 2008 book *Physics for Future Presidents*, in which he writes, "For power and energy, the kilowatt is the rate of energy delivery (the power) and the kilowatt-hour is the total amount of energy delivered."[7] Combining Muller's explanation with Jakuba's provides yet one more way to conceive of energy and power: The kilowatt-hour gives us a tally of the energy provided, whereas the kilowatt measures the rate of energy flow. And that rate of energy flow can be measured in watts, kilowatts, megawatts, gigawatts, terawatts, or, of course, in horsepower, thousands of horsepower, millions of horsepower, and so on.

Energy doesn't produce wealth. Energy *use* produces wealth, and the majority of the energy we use is fed into engines, turbines, and motors to produce power. It's converting energy—of whatever type—into motion that gives it value. And that's what engines do: They convert energy into mechanical motion that can be used for doing work. As Jakuba cleverly phrased it, the more we increase *the energy flow* through those engines, the more power we get. And the more power we have, the more work we can do.

That leads to another key point: The very word "power" implies control. When it comes to doing work, we insist on having power that is instantly available. We want the ability to switch things on and off whenever we choose. And that desire largely excludes wind and solar from being major players in our energy mix, because we can't control the wind or the sun. Weather changes quickly. A passing thunderstorm or high-pressure system can take wind- and solar-power systems from full output to zero output in a matter of minutes. The result: We cannot reliably get or deliver the power from those sources at the times when it is needed.

Renewable energy has little value unless it becomes *renewable power*, meaning power that can be dispatched at specific times of our choosing. But achieving the ability to dispatch that power at specific times means solving the problem of energy storage. And despite decades of effort, we still have not found an economical way to store large quantities of the energy we get from the wind and the sun so that we can convert that energy into power when we want it.

Which renewable sources can provide clean renewable power? One of the best is geothermal—which can provide a constant flow of predictable power that can be dispatched when needed. By October 2009, the United States had about 3,100 megawatts of geothermal production capacity. And geothermal promoters were predicting that they could triple that quantity to about 10,000 megawatts of baseload power capacity.[8] That could help, but it would still be a trifle when compared to the total U.S. generating capacity of 1 million megawatts.

The hype over renewables can only be debunked by thoughtfully walking through the numbers and the terms. And the most important of the terms are the first two items of the Four Imperatives: power density and energy density.

Power density refers to the amount of power that can be harnessed in a given unit of volume, area, or mass.[9] Examples of power-density metrics include horsepower per cubic inch, watts per square meter, and watts per kilogram. (In Part 2, I will show why watts per square meter may be the most telling of these. Using watts per square meter allows us to make a direct comparison between renewable energy sources such as wind and solar and traditional sources such as oil, natural gas, and nuclear power.)

Energy density refers to the amount of energy that can be contained in a given unit of volume, area, or mass. Common energy density metrics include Btu per gallon and joules per kilogram.[10]

When it comes to questions about power and energy, the higher the density, the better. For example, a 100-pound battery that can store, say, 10 kilowatt-hours of electricity is better than a battery that weighs just as much but can only hold 5 kilowatt-hours. Put another way, the first battery has twice the energy density of the second one. But both of those batteries are mere pretenders when compared with gasoline, which, by weight, has about eighty times the energy density of the best lithium-ion batteries.

As our society develops and urbanizes, we are seeking to use power in ever-greater quantities in ever-smaller places, and that is particularly true in our cities. Watt's breakthroughs increased the efficiency of the steam engine. Put another way, he increased the power density of the engine by designing it to produce more power from the same amount of space and from the same amount of coal. Ever since Watt's day, the world of engineering has been dominated by the effort to produce ever-better engines that can more quickly and efficiently convert the energy found in coal, oil, and natural gas into power. And that effort to increase the power density of our engines, turbines, and motors has resulted in the production of ever-greater amounts of power from smaller and smaller spaces.

The evolution of power density can be visualized by comparing the engine in the Model T with that of a modern vehicle. In 1908, Henry

Ford introduced the Model T, which had a 2.9-liter engine that produced 22 horsepower, or about 7.6 horsepower per liter of displacement.[11] A century later, Ford Motor Company was selling the 2010 Ford Fusion. It was equipped with a 2.5-liter engine that produced 175 horsepower, which works out to 70 horsepower per liter.[12] Thus, even though the displacement of the Fusion's engine is about 14 percent less than the one in the Model T, it produces more than nine times as much power per liter.[13] In other words, over the past century, Ford's engineers have made a nine-fold improvement in the engine's power density.

Now let's consider energy density. An easy way to understand energy density is to consider the amount of energy contained in a 5-gallon bucket that is filled with gasoline. Now consider that same bucket filled with dried leaves. Obviously, the energy density in the bucket filled with gasoline is far greater than the energy density in the one filled with leaves. Or consider corn ethanol. Although farm-state politicians and agribusiness promoters have been able to foist their fuel on motorists in non-farm states, ethanol contains just two-thirds of the heat energy of gasoline, meaning that motorists who use ethanol-blended gasoline must refill their tanks more often.

Our quest for power density provides another argument against a return to renewable energy sources. The kinetic energy of the wind and the solar radiation from the sun are diffused. Some companies, such as General Electric and Vestas, manufacture huge turbines to turn the diffused kinetic energy of the wind into highly concentrated energy in the form of electricity. Photovoltaic cells capture diffused light energy and concentrate it into electricity, which is then fed into wires. Concentrated thermal solar-energy systems employ huge arrays of mirrors to concentrate sunlight so that it can be used to heat a fluid that can then be used to run a generator. But with both wind and solar, and with corn ethanol and other biofuels, engineers are constantly fighting an uphill battle, one that requires using lots of land, as well as resources such as steel, concrete, and glass, in their effort to overcome the low power density of those sources.

For millennia, humans relied almost completely on renewable energy. Solar energy provided the forage needed for animals, which could then be used to provide food, transportation, and mechanical power. Traveling on lakes, oceans, or canals was made possible by the wind, human muscle, or animal muscle. And though today's wind turbines are viewed as the latest in technological achievement, land-based systems that captured the power of the wind have been recorded through much of human history. About 1,000 years ago, a visitor to Seistan, a region of eastern Iran, wrote that the wind "drives mills and raises water from streams, whereby gardens are irrigated. There is in the world (and God alone knows it) nowhere where more frequent use is made of the winds."[14] The use of hydropower, likewise, goes way back. The ancient Greeks used waterwheels; so did the Romans, who recorded the use of waterwheels in the first century B.C.[15] The use of mechanical power from water continued to the beginning of the Industrial Revolution. And while solar, wind, and water power all provided critical quantities of useful energy, they were no match for coal, oil, and natural gas. Hydrocarbons provided huge increases in power availability, allowing humans to go from diffused and geographically dispersed power sources to ones that were concentrated and free of specific geographic requirements. Hydrocarbons were cheap, could be transported, and most important, had greater energy density and power density.

That increasing availability of power has allowed us to do ever-greater amounts of work in less time. And because we need power for many different applications, we have lots and lots of engines, turbines, and motors. In fact, the engines of our economy are, in fact, just that: engines. And some of those engines are enormous.

At its most basic level, the $5-trillion-per-year global energy sector—the world's biggest single business—exists primarily to feed the engines that permeate our towns and cities. Big Oil, Big Coal, and the Big Utilities are servants of the world's engines. Whether those engines are fueled with oil, coal, natural gas, or enriched uranium doesn't really matter to consumers. What matters to them is that they continue to have a plentiful supply of fuel that can be fed into those engines so that the engines can continue to turn the heat energy in the various fuels into motion.

In the process of turning that heat energy into motion, engines now generally lose about two-thirds of the heat content in the various fuels. But once again, that matters little to consumers, who are primarily interested in power that is cheap, abundant, always available, and as clean

PHOTO 2 This massive diesel engine, designed by Finland-based Wärtsilä, is used on large ships (see www.wartsila.com). Each cylinder has a diameter of about 1 meter and displaces about 1,800 liters. The Wärtsilä engines, which can turn about 50 percent of the thermal energy in diesel fuel into useful power, are among the most efficient engines ever produced.

as possible. For someone living in midtown Manhattan or central Tokyo, the idea of using coal or firewood to cook dinner is absurd. The only fuels that meet the clean air standards of those urban settings are natural gas and electricity. Furthermore, the more cheap, abundant, clean power those consumers get, the more they use. The result: Over the past few decades, energy consumption among city dwellers has increased, a fact that can be proven by peeking inside the average apartment. Three decades ago, that apartment might have had the standard kitchen appliances—toaster, stove, refrigerator, and mixer. Today, that same kitchen will almost certainly have all of those appliances as well as a microwave oven, bread maker, coffeemaker, juicer, convection oven, dishwasher, and food processor. And a few steps away, where there once was only a small black-and-white television, there is now a giant-screen TV, a DVD player, and digital video recorder, as well as a laptop computer and ink-jet printer. In 1980, the average U.S. household had just three consumer electronic products. Today, it has about twenty-five of those devices.[16]

All those electronics have had a clear result: The amount of power that we are able to consume in our homes has dramatically increased. And that spike in power use is not just happening in Manhattan and Tokyo; it's happening all over the world, accelerated by the ongoing worldwide migration toward city living. In 2008, according to the International Energy Agency, about half of the world's population was living in cities. By 2030, that percentage is expected to rise to 60 percent.[17] And that will mean a corresponding rise in demand for power, because city dwellers use more power than their rural counterparts.[18]

The inexorable quest for power—whether in the form of computing power, a bigger engine in a new car or a better vacuum cleaner—will continue apace. Why? Because consumers and entrepreneurs are always seeking better, more efficient technologies that allow them to do more things faster. Computer makers such as Apple or Lenovo wouldn't be in business for very long if they started selling computers that were slower and had less computing power than the ones they had built two years earlier. Or imagine what would happen if a carmaker such as BMW or Mercedes Benz announced that its newest convertible took longer to go from 0 to 60 miles per hour than the model it built the previous year. The company's market share would vanish faster than Dick Cheney's hunting partners.

Power is like sex and Internet bandwidth: The more we get, the more we want. And that's one of the biggest problems when it comes to energy transitions. We have invested trillions of dollars in the pipelines, wires, storage tanks, and electricity-generation plants that are providing us with the watts that we use to keep the economy afloat. The United States and the rest of the world cannot, and will not, simply jettison all of that investment in order to move to some other form of energy that is more politically appealing.

Yes, we will gradually begin moving toward other forms of energy. But that move will be just that: gradual. And for those who doubt just how lengthy energy transitions can be, history offers some illuminating examples.

Power Equivalencies of Various Engines, Motors, and Appliances, in Horsepower (and Watts)

Saturn V rocket: 160,000,000 (120 billion W)[19]

Boeing 757: 86,000 (64.1 million W)[20]

Top fuel dragster: 7,500 (5.6 million W)[21]

M1A1 tank: 1,500 (1.1 million W)[22]

Formula 1 race car: 750 (560,000 W)[23]

2009 Ferrari F430: 490 (365,000 W)[24]

1999 Acura 3.2 TL sedan: 225 (168,000 W)[25]

2010 Ford Fusion: 175 (130,000 W)[26]

1908 Ford Model T: 22 (16,000 W)[27]

Average home air-conditioning compressor: 5.6 (4,200 W)[28]

Honda Cub motorbike: 4 (3,000 W)[29]

Average lawnmower: 3.5 (2,600 W)[30]

Dyson vacuum cleaner: 1.68 (1,250 W)[31]

Toaster: 1.67 (1,250 W)[32]

Lance Armstrong, pedaling at maximum output: 1.34 (1,000 W)[33]

Coffeemaker: 1.08 (800 W)[34]

Cuisinart: 0.16 (117 W)[35]

Human walking at a brisk pace: 0.14 (106 W)[36]

20-inch iMac computer: 0.11 (80 W)[37]

Ryobi 3/8-inch cordless drill battery charger: 0.07 (49 W)[38]

60-watt lamp: 0.07 (54 W)[39]

Table fan: 0.03 (25 W)[40]

Recharging an Apple iPhone: 0.0013 (1 W)[41]

CHAPTER 4

Wood to Coal to Oil
The Slow Pace of Energy Transitions

GIVEN OUR CURRENT OBSESSION with Big Oil and Big Coal, it's worth noting that the fuel source that has had the longest reign in the American energy business is plain old firewood. Wood's reign as the most important fuel in the United States lasted longer than any other. For 265 years after the Pilgrims founded the Plymouth Colony, and for 109 years after the signing of the Declaration of Independence, wood was the dominant source of energy in America. It wasn't until 1885—the year that Grover Cleveland was first sworn in as president—that coal finally surpassed wood as the largest source of primary energy in the United States.

For the next seventy-five years, coal was king. During the first two decades of the twentieth century, coal was supplying as much as 90 percent of all the primary energy in the United States, fueling factories, heating homes, and providing boiler fuel for essentially all of the nation's electric power plants. But coal's dominance was not to last. Thanks in large part to the booming demand for kerosene for lighting, and more particularly, for gasoline to fuel automobiles, oil began whittling away at coal's market share.

World War II was a turning point. The massive production of airplanes, ships, and motor vehicles during the war years accelerated the demand for oil. And prolific oil fields in Texas and Oklahoma were ready and able to provide nearly all the gasoline and diesel fuel that consumers and industry wanted. Between 1945 and 1950, the number of cars on

FIGURE 3 U.S. Primary Energy Consumption, by Source, 1825 to 2008

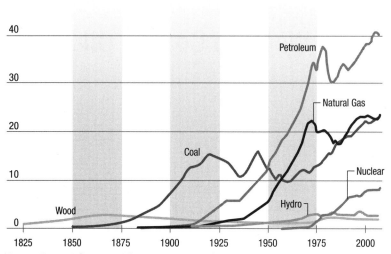

Source: Energy Information Administration, Annual Energy Review 2008, Figure 5, "Primary Energy Consumption by Source, 1635–2008," http://www.eia.doe.gov/emeu/aer/ep/ep _frame.html.

U.S. roads increased by 60 percent. Over the next ten years, the U.S. auto fleet grew by another 50 percent.[1] The increasing mobility of the average American resulted in a huge increase in demand for oil. In 1949, coal accounted for about 37.4 percent of the U.S. primary energy market, with oil trailing close on its heels with a 37.1 percent share. But in 1950, oil hit the tipping point, surpassing coal as the biggest source of U.S. primary energy. And for the past sixty years, oil's primacy has not been challenged. In fact, in 2008, oil's share of the U.S. energy market was at the exact same level as it was back in 1950: 38.4 percent.[2]

Although oil has been the undisputed champion, the jockeying for second place has been ferocious. In 1958, natural gas sped past coal to become the second-largest source of primary energy in the United States. Gas kept its second-place status behind oil for nearly two decades, and by 1971, the United States was consuming nearly twice as much energy in the form of natural gas as it was in the form of coal.[3] But Congress and federal regulators decided that the market couldn't be trusted; thanks to their ham-handed interventions, coal rebounded in a big way. In 1986, coal overtook natural gas to reclaim second place in the U.S. primary energy market. Since then, coal and natural gas have been running neck and neck, with each claiming about 25 percent.

The decades-long jousting for primacy among the various hydrocarbons provides more evidence of just how difficult it will be to replace them. As Vaclav Smil explained in his 2008 book, *Global Catastrophes and Trends*, there's no reason to expect that the transition toward renewable sources such as solar and wind will be completed quickly. In fact, he says to expect the opposite:

> There is no urgency for an accelerated shift to a nonfossil fuel world: the supply of fossil fuels is adequate for generations to come; new energies are not qualitatively superior; and their production will not be substantially cheaper. The plea for an accelerated transition to nonfossil fuels results almost entirely from concerns about global climate change, but we still cannot quantify its magnitude and impact with high confidence.[4]

Furthermore, the longer we use hydrocarbons, the more entrenched they become in our way of life—and the more energy we produce with hydrocarbons, the more energy we are able to produce. That may sound like an exaggeration, but it's a statement that can easily be confirmed by looking back at the history of the coal business. The first railroads were built to haul coal, and the locomotives that hauled the coal also burned coal. As author Jeff Goodell wrote in his book *Big Coal*, the railroads were a key invention that led to more coal production because, "In effect, coal hauled itself."[5] Of course, the railroads were only part of the equation. By perfecting the steam engine, James Watt enabled British mines to produce coal more economically, because his engines pumped water and lifted coal out of the mines.[6]

The idea that hydrocarbons beget more hydrocarbons can also be seen by looking at the Cardinal coal mine in western Kentucky. The mine produces more than 15,000 tons of coal per day. And the essential commodity that facilitates the mine's amazing productivity is electricity. The massive machines that claw the coal from the earth run on electricity provided by power plants on the surface that burn coal. In fact, about 93 percent of Kentucky's electricity is produced from coal.[7] To paraphrase Goodell, at the Cardinal Mine, the coal, in effect, is mining itself.

Hydrocarbons are begetting more hydrocarbons in the oil and gas business. Modern drilling rigs can bore holes that are five, six, or even eight miles long in the quest to tap new reservoirs of oil. And the energy they use to access that oil is . . . oil. Diesel fuel has long been the fuel of choice for drilling rigs around the world. On offshore drilling rigs, the power is often supplied by diesel fuel. But in some cases, the power is provided by natural gas that the rig itself produces. Thus, on those off-shore platforms, the natural gas is, in effect, mining itself.

The transition away from oil, coal, and natural gas will be a decades-long process because the companies that produce those commodities are getting ever better at finding and exploiting them. The oil and gas industry provides a clear example of this. For about a century, analysts have been forecasting an end to the supply of petroleum. And they have consistently been proven wrong. Why? Because the companies that produce oil and gas continue to discover new ways to gain access to previously inaccessible resources.

Though environmental groups and energy analysts eagerly publicize the inventiveness of entrepreneurs working to improve wind- and solar-power technology and other ways to harness alternative sources of energy, they seldom mention the ongoing innovations that are occurring on the hydrocarbon side of the ledger. And in doing so, they frequently forget the sheer size of the industry that is constantly searching for techniques that can get oil and gas out of the ground and do so faster and cheaper than before.

In the United States, there are about 5,000 independent oil and gas companies, every one of which is continually spending money and testing new concepts that will wring yet more petroleum and natural gas out of their leases.[8] In 2007 alone, those companies spent $226 billion drilling and equipping some 54,300 wells.[9] And that doesn't include the money spent on research and technology. All of the money spent on drilling and outfitting those wells, and the investment those companies have made in research and development, helps to assure that the installed fleet of machinery that supplies us with horsepower will continue to be fueled primarily by hydrocarbons.

It was only six decades ago that the oil industry drilled its first off-shore oil well—the Kermac 16—out of the sight of land.[10] And that well was drilled in just 20 feet of water.[11] Today, Anadarko Petroleum is producing natural gas at the Independence Hub in the Gulf of Mexico, where the water depth is 8,000 feet, and that one platform provides enough natural gas to supply about 5 million homes.[12] Moreover, the companies that are drilling for oil around the world are continually pushing into ever-deeper waters. In 2003, Transocean, the world's largest off-shore drilling contractor, announced that it had drilled a well in 10,000 feet of water.[13] Five years later, the firm announced that it had drilled a well off the coast of Qatar with a horizontal section that extended some 6.7 miles. The total measured depth of the well was 40,320 feet (7.6 miles), making it the longest extended-reach well ever drilled.[14] But that record will almost certainly be eclipsed in the next few years, as Houston-based Parker Drilling has recently completed the design and construction of a rig that will be capable of extended-reach wells with lengths of up to 44,000 feet, or 8.3 miles.

Conceiving of an 8-mile-long well boggles the mind, particularly when you learn that the Daisy Bradford No. 3, the well that started the flood of oil development in the East Texas Field, was only 3,500 feet deep.[15] By drilling deeper and faster, and in locations that were previously thought to be uneconomic, the oil and gas industry has continually extended its life expectancy.

In the natural gas sector, recent breakthroughs in shale gas technology have unlocked vast quantities of methane. Over the past five years, U.S. shale gas production has soared, thanks to techniques such as micro-seismic analysis, horizontal drilling, and enhanced well completion. The ever-increasing use of technology in the oil and gas business has resulted in huge improvements in drilling success rates. For instance, the success rate today for "wildcats" (wells drilled in frontier areas) is 50 percent or better. Three decades ago, that success rate was about 10 percent.[16]

While the oil and gas industry continues to improve the techniques that allow companies to drill wells deeper, faster, with greater precision, at ever-lower costs, the coal industry continues to show its resilience. Although oil passed coal as the most important source of U.S. energy back in 1950, coal hasn't gone away. In fact, over the past few years, thanks to soaring global demand for electricity, coal has enjoyed a resurgence. Although we now live in the Age of Oil, the Age of Coal hasn't yet passed. The reason for coal's enduring popularity is that it provides huge quantities of the essential commodity of modernity: electricity.

Over the past two decades, global electricity consumption has grown faster than any other type of energy use, and since 1990 electricity use has increased nearly three times as fast as oil consumption. In their thoughtful 2005 book, *The Bottomless Well*, Peter Huber and Mark Mills declared that "economic growth marches hand in hand with increased consumption of electricity—always, everywhere, without significant exception in the annals of modern industrial history."[17]

Electricity is the energy commodity that separates the developed countries from the rest. Countries that can provide cheap and reliable electric power to their citizens can grow their economies and create wealth. Those that can't, can't. The essentiality of electricity takes us back to coal. Love it or hate it, coal provides the cheapest option for electricity generation in dozens of countries around the world, and in heavily populated developing countries such as China, India, and Indonesia—all of which have large coal deposits—the need for increased capacity for the generation of electricity is acute.

Nearly 130 years ago, Thomas Edison began electrifying the world by burning coal—and in the intervening century, not much has changed.

CHAPTER 5

Coal Hard Facts

PHOTO 3 A cold day at 255–257 Pearl Street, New York City, January 15, 2009

THE BEGINNING OF the modern world can be traced to a single address: 255–257 Pearl Street, New York City.

Modern visitors to that address, located just a block or two west of the entrance to the tourist attractions at South Street Seaport, are unlikely to be impressed. The ground floor of the red brick high-rise building that now occupies the site is dominated by a Duane Reade drugstore. Aside from a bronze plaque mounted on the outside of the building, there's nothing that tells visitors about the importance of that plot of ground.

And yet, on September 4, 1882, Thomas Edison's operations at 255–257 Pearl Street ignited a revolution that transformed the world. On that date, he and his team of workmen began producing commercial quantities of electricity at the world's first central power station.

Edison chose the Pearl Street site because it was affordable and close to office buildings full of likely customers for his product. He paid about $65,000 for the two buildings, then reinforced the interior of the four-story structure at 257 Pearl with iron beams so it could hold the weight of six generators, each of which was named "Jumbo." On the floor below the generators were the boilers. The smaller building at 255 Pearl was used for office space, storage, and sleeping quarters for the workers.[1] The location of the two buildings proved ideal. Within months of starting his business, Edison had 203 customers who were using a total of 3,477 of his incandescent lights. By October 1883, having more than doubled his business, he had 508 customers who were using 10,164 lamps.[2] Edison clearly understood the importance of the Pearl Street endeavor, later calling it "the biggest and most responsible thing I had ever undertaken. . . . Success meant world-wide adoption of our central-station plan."[3]

Edison's technological breakthroughs at Pearl Street—that list includes not only the incandescent bulb, but also the safety fuse, the light socket, the key switch, the generator, and insulated wiring—led to a tsunami of electrification that continues to this day. By 1890, just eight years after Edison launched the beginning of the new world, there were 1,000 central power stations in the United States, and new ones were being added at a frenzied pace.[4]

In retrospect, it's remarkable to note just how small and inefficient Edison's Pearl Street station was. In 1882, Edison's state-of-the-art machinery converted less than 2.5 percent of the heat energy in the coal into electricity.[5] For comparison, some modern coal-fired power plants, using "ultra-supercritical" technology, can convert nearly half of the coal's heat energy into electric power.[6] As for its size, the Pearl Street plant was a midget by modern standards. Edison's first power plant produced 600,000 watts, or the equivalent of about 804 horsepower.[7] That's only a bit more output than a 2009 Ferrari 599 GTB Fiorano, which comes screaming out of the factory with a 620-horsepower engine.[8]

PHOTO 4 The Wizard of Menlo Park next to his original
dynamo at Orange, New Jersey, 1906
Source: Library of Congress, LC-USZ62-93698.

Though the scale of what happened at 255–257 Pearl Street may
seem downright puny, the conveniences and necessities of the modern
world—lights, air conditioning, television, heart monitors, cell phones,
iPods, and a panoply of other gizmos—were all made possible by the
work that Edison pioneered at those two long-gone buildings near the
southern tip of Manhattan. And though much of the world has under-
gone radical change since Edison began selling the juice from that power
plant back in 1882, one crucial element of the electricity-generation
business has not changed at all: coal.

At Pearl Street, Edison's Jumbo generators were driven by coal-fired
boilers. Today, coal continues to play a central role in the global electric

FIGURE 4 Consumption Increases for Various Energy Types, 1990 to 2007

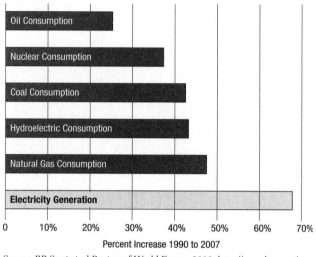

Source: BP Statistical Review of World Energy 2008, http://www.bp.com/liveassets/bp_internet/globalbp/globalbp_uk_english/reports_and_publications/statistical_energy_review_2008/STAGING/local_assets/downloads/pdf/statistical_review_of_world_energy_full_review_2008.pdf.

sector. In fact, no other fuel comes close. In 2006, coal provided 41 percent of the world's total electricity, with the next biggest share of the market belonging to natural gas, with 20.1 percent. From Pearl Street to the present day, coal has always been an essential element in electricity production. And electricity means prosperity.

The world's thirst for prosperity has led to huge increases in electricity demand. Between 1990 and 2007, electricity generation jumped by 67.8 percent. During that same time frame, demand for oil increased by 25.3 percent and coal demand increased by 42.5 percent.[9]

Despite the soaring demand for electricity—and the coal needed to produce it—there is a growing chorus of voices calling for an end to the use of coal. And while that may be a laudable goal, it raises an obvious question: If not coal, then what?

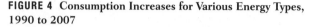

There's an old saw about the garbage business: Everybody wants their trash picked up, but nobody wants it put down. The same thing is true about the coal industry, the red-headed bastard stepchild of the modern energy business.

Everybody wants electricity—and all of the conveniences of modern life that come with it. But few people, particularly in wealthy countries like the United States, want that juice to be manufactured from coal. There's nothing new in that attitude. For the three centuries or so that humans have been using coal in significant quantities, the black fuel has always engendered an intense love-hate relationship. Coal heated people's homes and fueled the factories of eighteenth- and nineteenth-century England, but it also made parts of the country, particularly the smog-ruined cities, nearly uninhabitable. In London in 1812, a combination of coal smoke and fog became so dense that, according to one report, "for the greater part of the day it was impossible to read or write at a window without artificial light. Persons in the streets could scarcely be seen in the forenoon at two yards distance."[10] In 1952, a killer smog in London caused more than 4,000 premature deaths.[11] And the problems with coal-related pollution continue to this day, particularly in China. In Datong, known as the "City of Coal," the air pollution on some winter days is so bad that "even during the daytime, people drive with their lights on."[12] Nor is it just Datong. In 2007, the World Bank reported that sixteen of the twenty most polluted cities in the world are located in China, and much of that air pollution is due to the country's heavy reliance on coal.[13]

Coal is under attack. In September 2008, Al Gore encouraged activists to engage in civil disobedience to keep new coal-fired power plants that did not meet certain standards from being built. "If you're a young person looking at the future of this planet and looking at what is being done right now, and not done, I believe we have reached the stage where it is time for civil disobedience to prevent the construction of new coal plants that do not have carbon capture and sequestration," he said.[14]

A few weeks after Gore's declaration, anti-coal activists got a boost when a huge coal-ash holding pond failed at a power plant operated by the Tennessee Valley Authority. The resulting spill flooded more than 300 acres with coal ash contaminated with a variety of heavy metals, including

FIGURE 5 Global Electricity Generation, by Fuel, 1973, 2006, and Projected to 2030

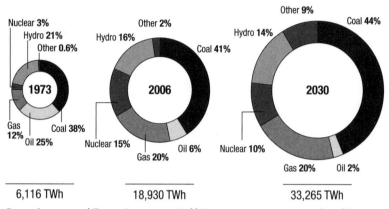

| 6,116 TWh | 18,930 TWh | 33,265 TWh |

Source: International Energy Agency, *Key World Energy Statistics 2008*, 24, and *World Energy Outlook 2008*, 507.

arsenic, lead, barium, chromium, and manganese.[15] In December 2008, James Hansen, the high-profile NASA scientist who is closely aligned with former vice president Al Gore on the issue of global warming, sent an open letter to President-Elect Barack Obama and his wife, Michelle, in which he called coal-fired power plants "factories of death."[16] Two months later, Hansen wrote an opinion piece for Britain's *Guardian* newspaper in which he said that "coal is the single greatest threat to civilization and all life on our planet."[17]

In July 2009, environmentalist Robert F. Kennedy Jr. wrote a piece for the *Financial Times* in which he said that the United States must end its "dependence on deadly, destructive coal." Kennedy pointed out that coal has caused acid rain and that the widespread use of mountaintop-removal mining has "buried 2,000 miles of rivers and streams, and will soon have flattened an area the size of Delaware."[18]

The litany of problems caused by coal mining, coal transport, and coal combustion could fill an entire bookshelf, or maybe even a small library. Acid rain, airborne particulates, water pollution, and air pollution are just a few of the issues. Coal-fired power plants are the largest emitters of mercury in the United States, pumping some 96,000 pounds of

TABLE 1 Top Five Countries with Largest Coal Reserves

Country	Reserves (million tons)	Percentage of world total	Reserves-to-production ratio*
US	238,308	28.9%	224
Russia	157,010	19.0%	481
China	114,500	13.9%	41
Australia	76,200	9.2%	190
India	58,600	7.1%	114

*Indicates the number of years of reserves remaining given current levels of production.
Source: BP Statistical Review of World Energy 2009, http://www.bp.com/liveassets/bp
_internet/globalbp/globalbp_uk_english/reports_and_publications/statistical_energy_review
_2008/STAGING/local_assets/2009_downloads/renewables_section_2009.pdf.

mercury into the air each year.[19] Most humans who encounter the metal released by the power plants do so by eating fish caught from bodies of water that have been affected by airborne mercury. Mercury is a neurotoxin that is particularly harmful when ingested by pregnant women, children, and the elderly.[20] Mercury exposure has been linked to higher risks for autism, impaired cognition, and neurodegenerative disorders such as Alzheimer's disease.[21]

In addition to mercury, U.S. coal plants annually release about 176,000 pounds of lead, 161,000 pounds of chromium, and 100,000 pounds of arsenic—all of which are extremely damaging to humans if they are ingested. In addition, those plants produce some 130 million tons of solid waste, a category that includes both ash and scrubber sludge, the material that is produced by the plants' air pollution control equipment. That volume of material is about three times as much as all of the municipal garbage produced in the United States every year.[22] And in the wake of the massive coal-ash spill in Tennessee, that gargantuan volume of waste has started getting the kind of regulatory scrutiny it deserves.

There's no question that other sources of energy—particularly nuclear and natural gas—can provide large amounts of electric power without putting pollutants into the atmosphere. The problem with replacing coal with something else—anything else—is, once again, an issue of scale. On an average day, the world consumes about 66.3 million barrels of oil equivalent in the form of coal. Though coal surely deserves much of the criticism that it gets, it has become the de facto standard for electricity generation, particularly in the world's most populous countries. Finally, production of the world's coal is not controlled by the Organization of the Petroleum Exporting Countries (OPEC) or any similar entity.

Those attributes, combined with the world's insatiable demand for electricity, are driving demand. Between 2007 and 2008, global coal use increased by about 800 million barrels of oil equivalent.[23] That increase—which works out to some 2.2 million barrels of oil equivalent per day—is about twenty-five times as much energy as that produced by all the solar panels and wind turbines in the United States in 2008.

The world's developing countries are using their coal for electricity generation, and that electricity is propelling economic growth around the world, particularly in rapidly developing countries such as China, Indonesia, and Malaysia. Between 1990 and 2008, electricity generation in those three countries jumped by more than 300 percent. The five countries with the biggest increases in electricity generation during that time included China (452 percent increase), Indonesia (353 percent), United Arab Emirates (352 percent), Malaysia (321 percent), and Qatar (307 percent).[24] Those are astounding rates of increase, particularly when you consider that between 1990 and 2008, global electricity generation increased by just 70 percent and U.S. electric output rose by 35.5 percent.[25]

The close correlation between electricity use and economic growth has become so obvious that it is accepted as fact. Investment bankers in the United States and elsewhere use China's electricity-production data as a barometer of that country's industrial output.[26] The five countries with the highest per-capita rates of electricity consumption are Iceland, Norway, Finland, Canada, and Qatar—all of which are among the world's wealthiest countries on a per-capita basis. Conversely, the countries and territories with the lowest electricity consumption—Gaza, Chad, Burundi, Central African Republic, and Rwanda—are among the poorest.[27]

In 2000, Alan Pasternak of Lawrence Livermore National Laboratory did a systematic analysis of electricity use and wealth. He used the United Nations Human Development Index—which ranks countries based on measures such as life expectancy, nutrition, health, mortality, poverty, education, and access to safe water and sanitation—as his baseline ranking system. He then compared each country's rank in the Human Development Index with its electricity consumption. Pasternak's

FIGURE 6 Electricity Consumption and the Human Development Index: A Near-Perfect Correlation

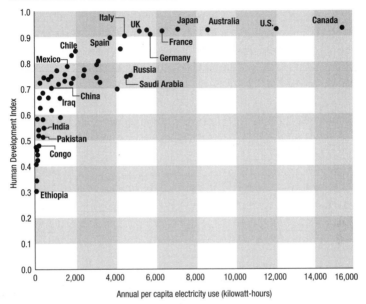

Source: Alan D. Pasternak, "Global Energy Futures and Human Development: A Framework for Analysis," Lawrence Livermore National Laboratory, October 2000, https://e-reports-ext.llnl.gov/pdf/239193.pdf, 5.

conclusions were unequivocal: "Neither the Human Development Index nor the Gross Domestic Product of developing countries will increase without an increase in electricity use," he wrote. Pasternak went on, saying: "The estimates of electricity use associated with high levels of human development presented in this analysis argue for substantially increased energy and electricity supplies in the developing countries and the formulation of supply scenarios that can deliver the needed energy within resource, capital, and environmental constraints."[28] In other words, if we want to help developing countries bring more people out of poverty, we need to help them increase the amount of electricity they generate and distribute.

Pasternak's work has since been cited by many other researchers who have looked into the correlation between electricity and human development. Moreover, Pasternak found that providing modest amounts of electricity per capita was not enough to assure good results. He determined that per-capita electricity consumption needed to be at least 4,000 kilowatt-hours per person per year to assure that countries had a Human Development Index of 0.9 or greater. That's an important distinction, because Pasternak found that only "14.6 percent of the sample global population enjoyed an HDI [Human Development Index score] of 0.9 or greater in 1997."

The essentiality of electricity can be demonstrated by looking at Africa. In 2007, the entire continent of Africa, a region with a population of some 955 million people, representing about 14 percent of the world's population, used just 3 percent of the world's electric power. In 2007, the entire population of Africa consumed about 612.6 terawatt-hours of electricity.[29] That's about as much power as was consumed by Canada, a country with 33.3 million residents.[30] Put another way, the average Canadian uses about twenty-eight times as much electricity as the average African.

In fact, the "dark continent," is just that—dark. Satellite photos of the Earth's surface at night show that the majority of the development in Africa has occurred on the continent's northernmost coastal areas and in South Africa. Meanwhile, there are only a handful of lighted areas on the southern coasts and in the interior of the continent.[31] Africa provides a case study in the correlation between electricity and economic development. Of the 1.6 billion people on the planet who live without electricity, about one-third of them, some 547 million people, live in Africa.[32] And the paucity of electricity correlates with some truly awful quality-of-life statistics. Of the 15 countries with the highest death rates, 14 of them are in Africa. (The outlier on that list is Afghanistan.)[33] Of the 22 countries with the highest infant mortality rates, 21 of them are in Africa. (Afghanistan, again, makes the list.)[34]

Addressing the lack of electricity in Africa will require billions of dollars in new investment. And the continent has shown some remarkable progress in recent years. Between 1990 and 2007, electricity generation in Africa nearly doubled.[35] But much of that new power was generated with coal. South Africa, one of the continent's most prosperous countries, relies heavily on coal for its electric power, deriving some 76 percent of its primary energy from the black hydrocarbon.[36]

The paucity of electricity in the developing world and the absolute need for more electric power to help bring people out of poverty brings us to two key points: First, much of the new generation capacity that will be installed in the developing world is going to come from coal-fired power plants because that is often the most-affordable option. Second, the countries building these plants are far more concerned about raising the living standards of their citizens than they are about the amount of carbon dioxide they are producing.

Consider the world's two most populous countries: China and India. Both have huge deposits of coal, and both have made it clear that they will use their coal to make electricity. At the end of 2008, China had about 800,000 megawatts of electricity generation capacity.[37] (For comparison, the United States has about 1 million megawatts of capacity.)[38] About 80 percent of the electricity China generates comes from coal-fired power plants.[39] With about 114 billion tons of reserves, China has enough coal to last about forty-one years at current levels of production.

India, with 1.1 billion people, sits atop 58 billion tons of coal reserves.[40] Between 1990 and 2008, India's coal output jumped by 129 percent, to some 512 million tons. During that period, India had the third-largest increase in coal production among the world's major economies, trailing only Indonesia (2,038 percent increase) and China (157 percent increase).[41] That coal is fueling frenzied expansion of coal-fired generation capacity. Between 1990 and 2009, India's electricity production nearly tripled, reaching 834.3 terawatt-hours in 2008, and about 68 percent of that power generation now comes from the burning of coal.[42] But even with the huge increases in power production, 40 percent of Indian homes still don't have electricity and 60 percent of Indian industrial firms rely on alternate forms of generation because the power grid isn't reliable.[43]

India is tired of lagging behind the rest of the world. That message was made clear by none other than Rajendra Pachauri, the Indian academic who chairs the UN's Intergovernmental Panel on Climate Change (IPCC). In July 2009, Pachauri asked reporters, "Can you imagine 400 million people who do not have a light bulb in their homes?" And he went on to explain where India was going to be getting its future power: "You cannot, in a democracy, ignore some of these realities and as it happens with the resources of coal that India has, we really don't have any choice but to use coal."[44]

The necessity of coal in developing countries was made clear in October 2009 by none other than U.S. Secretary of State Hillary Clinton. During a visit to Pakistan, Clinton advised the Pakistanis that they should be burning more coal in order to produce more electricity and attract more foreign investment. "Now, obviously, that is not the best thing for the climate, but everybody knows that," she said during a meeting with business leaders in Lahore. "Many of your neighbors are producing coal faster than they can even talk about it. It's unfortunate, but it's a fact that coal is going to remain a part of the energy load until we can transition to cleaner forms of energy." Then, speaking of Pakistan's coal deposits, she said, "You have these kinds of reserves, you should see the best and cleanest technology for their extraction and their use going forward."[45]

Though India and Pakistan are burning more coal, it is unlikely that they will ever match the consumption of the United States, which now generates about 48 percent of its electricity with coal.[46] The United States is the world's second-largest coal consumer, with daily consumption equal to about 11.3 million barrels of oil.

America's hunger for electricity—and therefore, for the coal needed to keep the lights on—can be illustrated by looking at the period from the beginning of 1994 to the end of 2008. During those fifteen years, the United States added three Spains to its electric grid.

Let me explain. During that time period, America's electric power generators increased their output from 3,247,000,000 megawatt-hours per year to 4,110,000,000 megawatt-hours. That's an increase of 863,000,000 megawatt-hours.[47] The latest data shows that Spain generates about 294,000,000 megawatt-hours of electricity per year.[48] Thus, the net increase in U.S. electricity generation over that decade and a half was nearly three times the annual electricity output of Spain. And one of the "Spains" that was added to the U.S. power grid was supplied by an increase in coal-fired generation. (The bulk of the remainder came from increases in natural gas and nuclear-fired production.)[49]

Or think of it this way: In 2008, the amount of energy America used in the form of coal nearly equaled the total energy consumption—from

FIGURE 7 Increases in U.S. Electricity Production from Solar, Wind, and Coal, 1995 to 2008

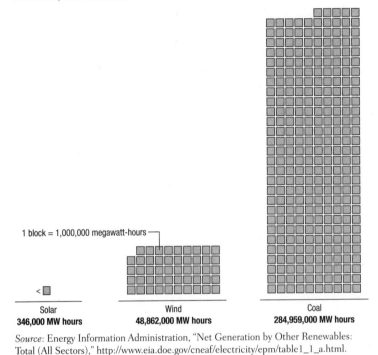

1 block = 1,000,000 megawatt-hours

< ▩

Solar	Wind	Coal
346,000 MW hours	48,862,000 MW hours	284,959,000 MW hours

Source: Energy Information Administration, "Net Generation by Other Renewables: Total (All Sectors)," http://www.eia.doe.gov/cneaf/electricity/epm/table1_1_a.html.

all sources, coal, oil, natural gas, hydro, and nuclear—of *all* of the countries of Central and South America *combined*.[50]

And the United States continues to add new coal-fired capacity at a rate that is far faster than the rates for new wind and solar capacity. Between 1995 and 2008, U.S. wind output increased dramatically, going from 3,164,000 megawatt-hours per year to 52,026,000 megawatt-hours per year, for a total increase of 48,862,000 megawatt-hours per year. That's an increase of about 1,500 percent. During that same time period, solar power production (which includes thermal solar and photovoltaics) increased by 69 percent, going from 497,000 megawatt-hours to 843,000 megawatt-hours, for an increase of 346,000 megawatt-hours.[51]

Meanwhile, coal-fired generation increased by a much more modest percentage: just 16.7 percent. But, once again, it's all about scale. And the amount of energy in that 16.7 percent increase is enormous. In 1995, coal-fired power plants delivered 1,709,426,000 megawatt-hours of electricity. By 2008, coal plants were delivering 1,994,385,000 megawatt-hours per year, an increase of 284,959,000 megawatt-hours.[52]

In other words, the absolute increase in total electricity produced by coal was about 5.8 times as great as the increase in output from wind and 823 times as great as the increase from solar. And yet, over the past decade, citizens in the United States have been bombarded with the notion that wind and solar power are the resources of the future.

The reality is that coal—even with its many negative attributes, continues to be the fuel of choice for creating electricity for a simple reason: cost. And though critics contend that coal imposes many costs that are not paid in the final price of electricity—such as air pollution, ecosystem destruction, miner deaths, and heavy metals contamination, to name just a few—the reality is that those "external" costs, large though they may be, have become an accepted part of the tradeoff. The always-on, super-clean, super-abundant horsepower that electricity provides has so much value that citizens around the world are willing to ignore the heavy costs exacted by mining and burning coal. That was true for Edison on Pearl Street and it's still true today. And coal's persistence gives us a good indicator of what lies ahead with regard to oil.

Coal's reign as the most important source of primary energy in the United States lasted for seventy-five years. Oil's reign as the most important source of primary energy in the United States has—so far—lasted six decades. But oil is superior to coal in nearly every respect. It has higher energy density. It is also cleaner, easier to transport, and far more flexible than coal. Add all of those factors together, and it means that oil, like coal, is here to stay.

TABLE 2 Top Twenty Countries Ranked by Gross Domestic Product and Total Electricity Generation, 2008

Wealth and electricity generation travel hand-in-hand. The latest available data for electricity generation and GDP in the top 20 countries in each category reveals the correlation.

Rank	GDP	Electricity generation
1	U.S.	U.S.
2	China	China
3	Japan	Japan
4	India	Russia
5	Germany	India
6	UK	Germany
7	Russia	Canada
8	France	France
9	Brazil	S. Korea
10	Italy	Brazil
11	Mexico	UK
12	Spain	Italy
13	Canada	Spain
14	S. Korea	S. Africa
15	Indonesia	Australia
16	Turkey	Mexico
17	Iran	Taiwan
18	Australia	Iran
19	Taiwan	Turkey
20	Netherlands	Saudi Arabia

Sources: BP Statistical Review of World Energy 2009, http://www.bp.com/liveassets/bp_internet/globalbp/globalbp_uk_english/reports_and_publications/statistical_energy_review_2008/STAGING/local_assets/2009_downloads/renewables_section_2009.pdf; Central Intelligence Agency, World Factbook (data retrieved via Wikipedia, http://en.wikipedia.org/wiki/List_of_countries_by_GDP_(PPP).

From Pearl Street to EveryGenerator.com: A Story of Rising Power Density and Falling Costs

Electricity and electricity generation have become so commonplace that we forget just how cheap electricity has become. But a comparison of the hardware used by Edison with today's generators brings the enormous improvements made over the past century into focus.

At his Pearl Street facility, Thomas Edison used six of his Jumbo dynamos, each of which had a capacity of 100,000 watts and weighed about 54,000 pounds.[53] Thus the Jumbo was capable of generating about 1.85 watts of electricity per pound—and that figure doesn't include the weight of the engines or the boilers needed to feed the dynamos. Compared with modern, off-the-shelf generators, those numbers are almost cartoonish. Today, consumers can buy generators that are far cheaper and have power-to-weight ratios about which Edison could only dream. For instance, EveryGenerator.com sells a 10,000-watt gasoline-powered unit made by Briggs and Stratton that weighs 288 pounds, which computes to about 34.7 watts per pound.

The result: The Briggs and Stratton generator provides an eighteen-fold improvement in power density over the hulking machines that Edison used.

Edison's total investment in the Pearl Street station was about $600,000. (That figure included the cost of the real estate plus the cost of the wires, conduits, and machinery.)[54] That's about $12.7 million in 2007 dollars. Thus, to generate 600,000 watts back in 1882, Edison had to spend—in current dollars—about $21 per watt. If Edison wanted to supply that same amount of power using the Briggs and Stratton generators, he could simply buy sixty of them. His total cost for that power capacity (in late 2009, the Briggs and Stratton units cost $1,999.99 each) would be about $120,000.[55] That works out to about $0.20 per watt. The result: a 105-fold improvement over the costs Edison faced when he built the Pearl Street station.

Indeed, if the Wizard of Menlo Park were still around, he could buy all the cheap generating capacity he wanted. And with an Internet connection and a credit card, he could even get free shipping.

CHAPTER 6

If Oil Didn't Exist,
We'd Have to Invent It

AMIDST ALL THE RHETORIC about the evils of oil, the evils of OPEC, the claims that we are "addicted" to oil, that oil fosters terrorism, that we can "win the oil endgame," or that oil is killing the planet, the simple, unavoidable truth is that using oil makes us rich. In fact, if oil didn't exist, we'd have to invent it.

Of course, many people would argue that point, but there's simply no denying that as oil consumption increases, so does prosperity. And the correlation is so clear as to be undeniable. In mid-2009, the International Energy Agency published the graph reproduced in Figure 8, which shows oil demand per capita for the members of the Organization for Economic Cooperation and Development (OECD) versus the rest of the world. In the countries where oil demand is more than 12 barrels per capita per year, GDP is at least two times as high as those where oil demand is 6 barrels or lower.

None of that is to discount the myriad problems that oil creates. The cofounder of OPEC, Venezuela's Juan Pablo Pérez Alfonso, famously called oil "the devil's excrement."[1] The pursuit of oil and the riches that come with it have ruined countries and created generational corruption that persists to this day. Average residents of countries such as Nigeria and Angola, both of which sit atop massive deposits of oil and gas, have gained little from the exploitation of the mineral wealth beneath their

FIGURE 8 Gross Domestic Product and Oil Demand Per Capita, 2008

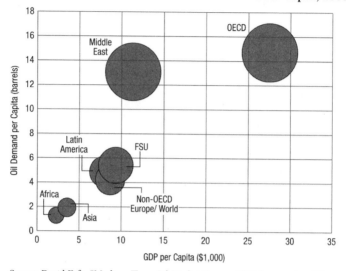

Source: David Fyfe, "Medium-Term Oil Market Report, 2009," supporting slides, International Energy Agency, June 29, 2009, http://www.iea.org/textbase/speech/2009/Fyfe_mtomr2009_launch.pdf, 6.

feet. Frequent wars sprout in the Middle East as various countries vie to control the flow of oil from the region, with the Second Iraq War being the most recent example, and it has become one of the most militarized areas on the planet.

And yet, for all of the problems that oil creates, it also provides us with unprecedented mobility, comfort, and convenience. Although we think of oil primarily as a transportation fuel, it's also a nearly perfect fuel for heating, can be used to generate electricity, and, when refined, can be turned into an array of products, from cosmetics to shoelaces and bowling balls to milk jugs.

In short, oil may be the single most flexible substance ever discovered. The consumption of oil has so radically changed human society over the course of the past century that this entire book could be focused on that one topic. More than any other substance, oil helped to shrink the world. Indeed, thanks to its high energy density, oil is a nearly perfect fuel for use in all types of vehicles, from boats and planes to cars and motorcycles. Whether measured by weight or by volume, refined oil products provide more energy than practically any other commonly available substance, and they provide it in a form that's easy to handle, relatively cheap, and relatively clean.[2] Furthermore, oil provides the fuel for the two prime movers that have done more for the cause of globalization than any other: the diesel engine and the jet turbine.

That is not to downplay the significance of gasoline-fueled engines, which have brought mobility and useful power of all types (generators, motorcycles, weed whackers, and so on) to hundreds of millions of people. But since World War II, the diesel engine and the jet turbine have fundamentally changed the world. Since their use became widespread in the 1950s and 1960s, those two prime movers have had a greater impact on the global economy than any corporate marketing effort or international trade agreement.

A decade ago, a colorful railroad lawyer named Don Cheatham told me something that stuck in my head: "Without transportation" he declared, "there is no commerce." Cheatham's point is clearly true. But it leads to a corollary point, which perhaps can be called Bryce's Hypothesis: If it is true that without transportation there is no commerce, then without oil there is no commerce. Proving Bryce's Hypothesis is not overly difficult. The global transportation system depends almost exclusively on oil. No other substance provides such high energy density with such incredible versatility.

The diesel engine and the jet turbine effectively reduced the size of the Earth. By offering greater reliability and range than engines powered by gasoline, they cut the amount of time required to traverse the distances between countries and thereby fostered unprecedented volumes of trade. Thanks to the characteristics of the fuels they use and their more efficient use of heat energy, diesel engines and jet turbines offer about 12 percent more range than comparable gasoline-fueled engines, and they do it with greater reliability.[3]

The pivotal role of diesel engines and jet turbines in the global economy underscores the essentiality of oil. Why? The fuels that drive those machines cannot be effectively replaced. A number of alternatives can be used to substitute for gasoline in the light-duty vehicle market, including electricity, natural gas, and ethanol, but none of those alternatives can be used to substitute for diesel fuel or jet fuel. Airlines are not going to be flying Boeing 737s from Tulsa to Tacoma by filling them with compressed natural gas or huge banks of batteries. Container ships that ferry consumer goods from Singapore to Rotterdam don't run on corn ethanol.

Of course, there is growing interest in biodiesel made from soybeans and artificial jet fuel made from various substances. But none of those alternatives can provide anything close to the scale of production that would be needed to keep the world's fleet of diesel engines and jet turbines on the move. For instance, even if the United States converted all of the soybeans it produces in an average year into biodiesel, doing so would provide less than 10 percent of America's total diesel-fuel needs.[4] Now suppose an inventor found a way to convert soybeans into jet fuel. Even with that invention, the conversion of all of America's yearly soybean production into jet fuel would only provide about 20 percent of U.S. jet-fuel demand.[5]

When it comes to the global commercial transportation market, there simply is no substitute for oil. The centrality of diesel engines—and the diesel fuel needed to power them—can be demonstrated by this one fact: Ninety-four percent of the goods shipped in the United States are transported on diesel-powered vehicles.[6] The same percentage likely holds true for goods shipped internationally. Sony and Samsung may be producing fancy big-screen televisions for sale at the nearest Costco, but those TVs would likely still be in China or somewhere else in the Far East if there were no giant diesel engines to propel the container ships that bring those TVs to the United States.

While diesels are driving surface-based trade, jet turbines (and thus, jet fuel) have made global air travel into a routine experience. Six decades ago, passenger airliners relied heavily on piston-driven engines that used high-octane gasoline, but by the mid-1960s, the era of piston-driven engines gave way to the jet age. Jet aircraft became dominant because they can fly about three times as fast and two times as high as their gasoline-powered cousins. That means that passengers can save huge amounts of time and do so while flying in the upper reaches of the troposphere, which is usually above the levels where weather and air turbulence can present a problem.[7] The astounding success of the jet turbine can be seen by the growth in air travel. In 1950, the total volume of air travel—

PHOTO 5 General Electric's GE90-115B is the world's most powerful jet turbine. The turbine's 4-foot-long blades—made from titanium, carbon fiber, and epoxy—are designed to move large amounts of air quietly. One of the blades was recently displayed at the Museum of Modern Art in New York City.

Source: General Electric, "GE90-115B Aircraft Engine," n.d., http://ge.ecomagination.com/ site/water/products/ge90.html. On technical specifications, see GE press release, "It's Great Design, Too: World's Biggest Jet Engine Fan Blade at the Museum of Modern Art," November 16, 2004, http://www.geae.com/aboutgeae/presscenter/ge90/ge90_20041116.html. For more technical data, see Museum of Modern Art, "Jet Engine Fan Blade," n.d., http://www.moma.org/collection/object.php?object_id=93637.

measured in passenger-kilometers—was 28 billion. By 2005, that quantity of air travel had increased to some 3.7 trillion—a 130-fold increase.[8]

In the late nineteenth century, Jules Verne's character Phileas Fogg became famous thanks to Verne's novel *Around the World in 80 Days*. Today, thanks to high-speed jet airliners, Fogg could fly to almost any modern airport on the planet in thirty-six hours or less. In fact, if he were so inclined, Fogg could probably fly all the way around the world and be back at the Reform Club in London with a dry martini in his hand within seventy-two hours of his departure—if, of course, he kept his seatbelt fastened and his tray table in its upright and locked position.

Even if the world's leading politicians wanted to quit using oil, their ability to do so would be stymied, because global commerce depends on the use of diesel engines and jet turbines. In fact, the ongoing improvements to those machines are likely to increase their dominance over the coming decades. Ever since Rudolf Diesel first patented the engine that carries his name in 1892, his design has been undergoing continual improvement.[9] Dramatic efficiency improvements are also being made to jet turbines—first flight-tested in the late 1930s—to make them quieter, more powerful, and more efficient. In 2008, General Electric, the world's largest producer of jet turbines, announced it was developing a new design, the Leap-X, which could cut fuel consumption by 16 percent compared to existing models.[10]

The improving efficiency of the diesel engine and jet turbine—and their increasing popularity—provides yet more evidence of our centuries-long quest for horsepower. And the central role that oil plays in fueling those machines provides evidence that petroleum, and the many products and services that we derive from it, will remain irreplaceable for years to come.

Thus far, much of the discussion has focused on power density and energy density, while the issue of scale has largely been ignored. In the next chapter, the final chapter of Part 1, I will show just how daunting the challenge of replacing hydrocarbons will be. And in keeping with the themes of this book, I will provide those scale comparisons in both energy equivalents and power equivalents.

CHAPTER 7

Twenty-Seven Saudi Arabias Per Day

THE GARGANTUAN SCALE of our energy consumption is almost impossible to comprehend. The BP Statistical Review of World Energy estimates daily global commercial energy use at about 226 million barrels of oil equivalent. Of that quantity, about 79 million barrels comes from oil, 66 million from coal, 55 million from gas, 12 million from nuclear, and 14 million from hydropower. Obviously, hydrocarbons are the biggest portion of the global energy mix, accounting for about 200 million barrels of oil equivalent per day. But how can we even imagine what those quantities of energy represent?

What is 226 million barrels of oil equivalent? Well, try thinking of it this way: It's approximately equal to the total daily oil output of twenty-seven Saudi Arabias. Since the 1973 Arab Oil Embargo, Saudi Arabia's oil production has averaged about 8.5 million barrels per day.[1]

Over the past few years, we have repeatedly been told that we should quit using hydrocarbons. Fine. Global daily hydrocarbon use is about 200 million barrels of oil equivalent, or about 23.5 Saudi Arabias per day. Thus, if the world's policymakers really want to quit using carbon-based fuels, then we will need to find the energy equivalent of 23.5 Saudi Arabias every day, and all of that energy must be carbon-free.

While Saudi Arabia provides an easily understandable metric for global energy use, this book focuses on our desire for power. So let's

FIGURE 9 World Power Consumption, by Primary Energy
Source, in Horsepower (and Watts)

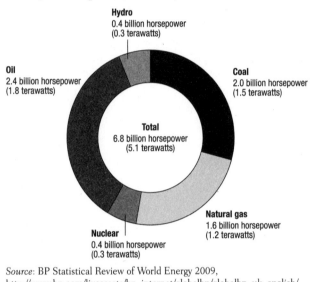

Hydro
0.4 billion horsepower
(0.3 terawatts)

Oil
2.4 billion horsepower
(1.8 terawatts)

Coal
2.0 billion horsepower
(1.5 terawatts)

Total
6.8 billion horsepower
(5.1 terawatts)

Natural gas
1.6 billion horsepower
(1.2 terawatts)

Nuclear
0.4 billion horsepower
(0.3 terawatts)

Source: BP Statistical Review of World Energy 2009,
http://www.bp.com/liveassets/bp_internet/globalbp/globalbp_uk_english/
reports_and_publications/statistical_energy_review_2008/STAGING/local
_assets/2009_downloads/renewables_section_2009.pdf.

convert those global energy consumption numbers into power terms.
That will be easy, as they are provided in barrels of oil equivalent *per day*,
and that means they are readily converted into our now-familiar power
metrics: watts and horsepower.

SI provides the easiest way to compute power. A barrel of oil contains
5.8 million Btu. That's equal to about 5.8 billion joules (5.8 GJ). To obtain watts, we must divide those joules by seconds. (Remember that
power = energy/time.) We must therefore divide our 5.8 gigajoules by
86,400 seconds, which is the number of seconds in 24 hours. We must
also account for the heat lost during the conversion of that heat energy
into useful power. The result: Each barrel of oil equivalent produces
about 22,152 watts, or about 29.7 horsepower. For simplicity, let's call it
30 horsepower per barrel of oil equivalent per day.[2]

Multiplying global energy use (226 million barrels of oil equivalent
in primary energy each day) by horsepower per barrel (30), we find that

FIGURE 10 Per-Capita Power Consumption in the Six Most Populous Countries, in Watts

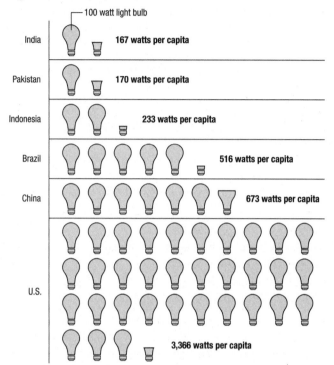

Sources: BP Statistical Review of World Energy 2009, http://www.bp.com/liveassets/bp_internet/globalbp/globalbp_uk_english/reports _and_publications/statistical_energy_review_2008/STAGING/local_assets/ 2009_downloads/renewables_section_2009.pdf; Central Intelligence Agency, World Factbook, https://www.cia.gov/library/publications/the-world-factbook/.

the world consumes about 6.8 billion horsepower—all day, every day. Therefore, roughly speaking, the world consumes about 1 horsepower per person. Of course, this power availability is not spread evenly across the globe. Americans use about 4.5 horsepower per capita, while their counterparts in Pakistan and India use less than 0.25.

Although those numbers are telling, they are somewhat unwieldy. Thus, it makes sense to also look at global power consumption in watts. Using the numbers cited above, we find that global power consumption is about 5.1 trillion watts, or 5.1 terawatts. Using watts allows us to see the disparity in power consumption much more clearly. For instance, the average resident of India consumes about 167 watts, while the average Brazilian uses 516 watts. Meanwhile, the average resident of the United States consumes 3,366 watts.

The wealth of power in the United States provides an obvious explanation for America's incredible economic success. U.S. residents have enormous amounts of power at their disposal that can be used for whatever bit of work they choose to do: running a spreadsheet, recharging an electric drill, mixing cookie dough, manufacturing computers, or running the air compressor in the garage in order to inflate the tires on the car.

In the never-ending quest for horsepower, the residents of the United States are leading the world, and they are doing so because America leads the world in the production of high-quality energy. The United States ranks first in the world in the production of electricity from nuclear reactors (ahead of France). It ranks second in coal production (behind China), second in natural gas production (behind Russia), third in oil production (behind Saudi Arabia and Russia), and fourth in hydro production (behind China, Canada, and Brazil).[3] In all, the United States produces about 74 percent of the primary energy it consumes, a fact seldom mentioned by the many neoconservatives and energy posers who have been sounding the alarm about the evils of foreign energy.[4] America's enormously productive energy sector—combined with significant imports of oil—allows the United States to provide colossal quantities of power to its citizens. And it's that power availability that has turbocharged the American economy and made it into a powerhouse.

Furthermore, the United States has more hydrocarbon reserves than any other country. In October 2009, the Congressional Research Service reported that the proved hydrocarbon reserves of the United States totaled nearly 970 billion barrels of oil equivalent. The vast majority of that total (about 906 billion barrels of oil equivalent) is in the form of coal. Running second behind the United States in total hydrocarbon reserves is Russia, which has about 955 billion barrels of oil equivalent, followed by China with 466 billion barrels of oil equivalent.[5]

Given America's enormous energy production and energy reserves, why are so many Americans willing to believe that they should trade reliable resources, such as nuclear energy, coal, oil, and natural gas—all of which have high power density—for unreliable, low-power-density sources, such as solar energy and wind power? The answer is that many Americans are too willing to believe the hype. They haven't bothered to investigate the claims or do the calculations that would allow them to see through the hype. Or perhaps the self-satisfaction they get from aligning themselves with such grand ideals is so alluring that they don't even want to try.

In the next section, I will expose many of the myths of "green" energy. In doing so, I will set the stage for Part 3, where I will explain why the most logical, or rather, *the inevitable*, energy policy for the future is N2N: natural gas to nuclear.

PART II
THE MYTHS OF "GREEN" ENERGY

CHAPTER 8

Wind and Solar Are "Green"

> Density is green.
> WITOLD RYBCZYNSKI[1]

THE ESSENCE OF PROTECTING the environment can be distilled to a single phrase: Small is beautiful.

That phrase gained widespread traction in the early 1970s when British economist E. F. Schumacher published a collection of essays in a book that carried that title. As Schumacher made clear, when it comes to environmental protection, manmade disturbances of the natural world should be kept to a minimum. His maxim applies most particularly to energy production: The best energy sources have the highest power densities, that is, they generate lots of power from small pieces of real estate.[2] From a "deep green" perspective, that's ideal: small footprints and big power outputs.

Of course, every source of energy production takes a toll on the environment. The goal should be to minimize those costs, which, for hydrocarbons and nuclear power, include, but are not limited to, air pollutants, long-lived waste issues, oil spills, carbon dioxide emissions, the effects of drilling operations and pipelines, and the potential for catastrophic accidents. Those downsides are well known and have been accepted as a cost of doing business for many decades. And though these costs are significant, hydrocarbons and nuclear power are prevailing in the modern energy diet because they satisfy the Four Imperatives.

The essential problem with renewables is that they fail the first test of the Four Imperatives: power density. The weak power density of renewables has become so apparent that the Nature Conservancy, one of the biggest and most conservative of the U.S. environmental groups, recently coined the term "energy sprawl," a reference to the vast stretches of land that are needed for the production and transportation of energy from wind and solar installations.

The energy sprawl of renewables can easily be illustrated by comparing the footprint of a typical U.S. nuclear power plant, in this case, the South Texas Project, with that of wind and solar. Using conservative calculations—which means counting all 12,000 acres of the South Texas Project's land area as part of the two-reactor plant's footprint—yields a power density of about 300 horsepower per acre (56 watts per square meter). Compare that with wind power, which produces about 6.4 horsepower per acre (1.2 watts per square meter).[3] Or look at solar photovoltaic, which produces about 36 horsepower per acre (6.7 watts per square meter).[4] The results: Wind power requires about 45 times as much land to produce a comparable amount of power as nuclear, and solar photovoltaic power requires about 8 times as much land as nuclear. The corn ethanol scam is even worse, requiring about 1,150 times as much land as nuclear.[5]

Of course, estimates of power density vary. The numbers in the previous paragraph come from a variety of sources, including my own calculations. But those numbers are quite similar to the ones used by the Nature Conservancy in a 2009 study called "Energy Sprawl or Energy Efficiency." That study estimates that when considering all land-use impacts, corn ethanol requires about 144 times as much land as nuclear, wind power requires about 30 times as much, and solar photovoltaic requires about 15 times as much. The same study found that wind power generation requires nearly 4 times as much land as natural gas and about 7 times as much as coal.[6]

"Deep green" means—or at least, should mean—not paving any land unless it is essential to do so. Renewables such as wind and solar power require huge swaths of land—which often becomes unusable for other purposes. Humans cannot live close to wind farms because of the low-level noise caused by the massive blades. That noise, say neighbors and critics, disturbs sleep patterns and can cause headaches, dizziness, and other health problems.

In January 2010, I interviewed Charlie Porter, a successful horse trainer who lives in northwestern Missouri. Porter said that in 2007, shortly after a wind farm was completed within a half mile of his twenty-acre farm, he, his wife, and daughter began having trouble sleeping due to the rumbling from the turbines. Porter compared the noise to having "a hat on that's way too small. It just makes your world tiny." In late 2009, exhausted by the turbine noise, the Porters purchased a house in nearby King City and moved off of their farm. The turbines, Porter told me, "drove us out of our home. They just ruined life out in the country."[7]

Porter's story is not unique. Dr. Nina Pierpont, a pediatrician who lives and works in rural upstate New York, has documented dozens of cases of what she calls "wind turbine syndrome," and in late 2009, she published her findings in a book.[8] Pierpont recommends that wind turbines have set-backs of at least 1.25 miles from any human habitation. Pierpont's work clearly worries the American Wind Energy Association (2007 budget: $14 million) which published a report in December 2009 which claimed that "there is no evidence that the audible or sub-audible sounds emitted by wind turbines have any direct adverse physiological effects" and that the vibrations from the turbines are "too weak to be detected by, or to affect, humans."[9] But the scientific literature shows that the problems associated with low-level turbine noise have been known for years. And people living near wind farms in Texas, Oregon, New York, and Minnesota, as well as in numerous foreign countries, including England, New Zealand, Canada, France, and Australia, have complained about the noise and cited some of the same problems that Porter named, including sleep deprivation.[10]

Solar farms require huge arrays of panels or mirrors that cover nearly every square meter of their property. In addition to the land used by the actual wind and solar farms, those same sources usually require the con-struction of many miles of new high-voltage transmission lines, and those lines will zig-zag across huge swaths of the United States.

The need for more high-voltage transmission lines is perhaps the most controversial aspect of renewable energy. Across the country, citizen and

FIGURE 11 The 2,700-Megawatt Challenge: Comparing the Power Densities of Various Fuels

The two fission reactors at the South Texas Project produce 2,700 megawatts of power. How many acres of corn ethanol would it take to produce that much power? What about wind turbines? Here are the power densities of those sources as well as the footprints they would require to produce 2,700 megawatts.

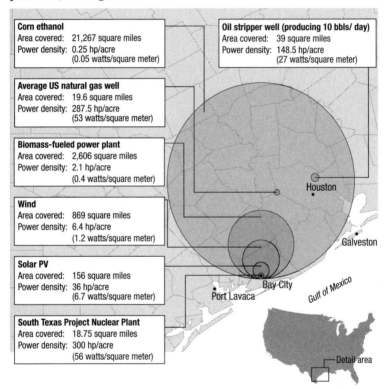

Sources: The calculations for the energy densities of the renewable sources are from Jesse Ausubel, "The Future Environment for the Energy Business," *APPEA Journal* (2007), http://phe.rockefeller.edu/docs/ausubelappea.pdf, 8. Other sources include Energy Information Administration data. For the wind calculation, the math is as follows: 2.7 billion watts /1.2 watts per square meter = 2.25 billion square meters.

environmental groups are fighting the construction of new power lines—many of which are needed to transport electricity from new wind and solar facilities to distant towns and cities. Some 40,000 miles of new lines will be needed by the wind sector alone.[11] If we assume that each of these transmission lines requires a 100-foot-wide swath of right-of-way (this is a conservative estimate, the actual right-of-way may be much wider, particularly for high-voltage lines), then those 40,000 miles of transmission lines will cover about 750 square miles of territory, which is about half the size of the state of Rhode Island.[12]

The prolonged fight over the Sunrise Powerlink project provides a good example of the controversy over transmission lines. In 2005, San Diego Gas and Electric announced plans for a new high-voltage transmission line that would carry electricity from the Imperial Valley to customers in and around San Diego. The line would help bring solar power from the desert to consumers in the San Diego area. The utility claims the line is essential if it is to meet California's mandates on renewable energy. But the 123-mile, $1.9 billion transmission project is opposed by several environmental groups, including the San Diego–based Desert Protective Council and the Sierra Club.[13]

The Desert Protective Council opposes "big solar" and says the "land rush" of solar companies could "potentially cover over 600,000 acres in California alone." "While these projects are touted as solving the global warming crisis," the council website says, "they also have serious environmental impacts that need to be carefully considered."[14] In late 2008, despite the opposition, the project was approved for construction by the California Public Utility Commission, but it may yet face a legal battle in the courts.[15]

While the battle over Sunrise Powerlink may be the best known of the transmission-line skirmishes, many other projects are being developed across the United States, and just as with Sunrise Powerlink, nearly all of them are generating controversy and opposition.

To be certain, battles between landowners and the electric utilities over transmission lines are nothing new. Consider the case of former U.S. senator Phil Gramm, the man who did more to enable Enron and the Wall Street pirates than perhaps any other member of the U.S. Congress. You remember Gramm. He was the senator who got more campaign contributions from the now-defunct accounting firm Arthur Andersen than any other. As head of the Senate Banking Committee, he engineered the 1999 financial services bill that repealed the Glass-Steagall Act, the Depression-era set of rules that helped to slow the trend toward gigantism in the banking, insurance, and securities business. Gramm led congressional efforts to put a leash on both the Securities and Exchange Commission and the Commodity Futures Trading Commission. He was also the author of the "Enron exemption," the provision that he slipped into a bill that furthered Enron's ability to escape federal regulation of EnronOnline, the company's massive trading operation. Gramm declared that his legislation would "protect financial institutions from overregulation." He went on, saying that it guaranteed that the United States would "maintain its global dominance of financial markets."[16]

A decade later, a grateful nation may look to Gramm as not only a key enabler and architect of the worst financial collapse in modern history, but also a true American NIMBY. In 2003, just a few months after Gramm quit the Senate (before his term expired), he and his wife, former Enron board member Wendy Gramm, joined a group of landowners opposing a high-voltage transmission line that was to be installed close to their ranch northwest of San Antonio.[17] Instead of going across their property, the Gramms wanted San Antonio's City Public Service to route the transmission lines across Government Canyon State Natural Area, a park that had recently been established by the Texas Parks and Wildlife Department.[18] The Gramms failed to get the power lines moved onto the park. But their efforts illustrate that NIMBY is alive and well, and living, well, just about everywhere.

NIMBY issues are not unique to transmission lines. Landowners and neighborhood groups all across the United States are always lining up to fight proposals for various development projects, whether they are shopping malls, confined animal feeding operations, slaughterhouses, landfills, nuclear reactors, or coal-fired power plants. Those types of projects have always faced opposition and likely always will. But the emergence of transmission lines as a pivotal land-use issue for wind and solar power has been one of the biggest surprises of the push for more "green" energy. And it has created an entirely new set of opponents.

In upstate New York, a privately held investment group called New York Regional Interconnect has been trying for years to build some 200 miles of transmission lines, with towers thirteen stories high, that would carry electricity from the northern part of the state, where the wind resources are, to customers further south. But the line is opposed by a number of local groups who don't want the lines to cross through their communities.[19]

In July 2009, a coalition of environmental groups, including the Sierra Club, the Natural Resources Defense Council, and the Wilderness Society, filed a lawsuit against several federal agencies in an effort to force them to change the routes of a number of planned transmission lines. The suit, which names the Departments of Interior, Energy, and Agriculture, as well as the Bureau of Land Management and the Forest Service, as defendants, claims that the federal government has created 6,000 miles of rights-of-way in the western states without considering all of the environmental impacts of the transmission corridors.[20] The suit invokes a variety of federal laws, including the Endangered Species Act, and asks the court to "declare unlawful and set aside" the power transmission corridors laid out by the federal government.[21]

In mid-2009, the Lower Colorado River Authority announced plans for 600 miles of transmission lines to carry wind power from western Texas to customers in the central part of the state.[22] The project quickly garnered opposition from landowners and local citizens who said the lines would damage their property values. In a story on the controversy, Asher Price, a reporter for the *Austin American-Statesman*, discussed one property owner, Bill Neiman, the owner of a seed farm near Junction, who claimed that "his property and livelihood would be ruined by transmission lines." A spokesperson for the river authority neatly summed up the situation by saying that most people understand the need for more transmission lines, "they just don't want them on their property or within view."[23]

While transmission lines are a key limiter for renewable energy schemes, the environmental impact of renewables extends beyond the problem of energy sprawl. As discussed above, the low power density of wind and solar means that they need lots of land. But that same low power density also requires large resource inputs, specifically, huge quantities of steel and concrete.

Of course, every method of large-scale electricity production requires significant quantities of such materials. But the resource requirements of wind are several times higher than those of natural gas and nuclear. And those higher inputs mean higher relative costs per unit of power delivered. Put another way, power-generation systems such as natural gas and nuclear power plants are far more efficient users of steel and concrete than are wind power systems.

Consider the Milford Wind Corridor, a 300-megawatt wind project that was built in Utah in 2009. The project was the first to be approved under the Bureau of Land Management's new wind program for the western United States.[24] To construct the wind farm, which uses 139 turbines spread over 40 square miles, the owners of the project installed a concrete batch plant that ran six days a week, twelve hours per day, for six months. During that time, the plant consumed about 14.3 million gallons of water to produce 44,344 cubic meters of concrete. Thus, each megawatt of installed wind capacity consumed about 319 cubic meters of concrete.[25]

But those numbers must be adjusted to account for wind's capacity factor—the percentage of time the generator is running at 100 percent of its designed capacity. Given that wind generally has a capacity factor of 33 percent or less, the deployment of 1 megawatt of reliable electric-generation capacity at Milford actually required about 956 cubic meters of concrete. The concrete numbers on the Milford project are similar to those described in a report delivered by Per Peterson to the President's Council of Advisors on Science and Technology in September 2008. Peterson, a professor in the nuclear engineering department at the University of California at Berkeley, reported that when accounting for capacity factor, each megawatt of wind power capacity requires about 870 cubic meters of concrete and 460 tons of steel.

For comparison, each megawatt of power capacity in a combined-cycle gas turbine power plant (the most efficient type of gas-fired electricity production) requires about 27 cubic meters of concrete and 3.3 tons of steel. In other words, a typical megawatt of reliable wind power capacity requires about 32 times as much concrete and 139 times as much steel as a typical natural gas-fired power plant.

To be fair, the concrete and steel requirements of a gas-fired power plant are only part of the electricity equation. The miles of steel pipelines required to move gas from the wellhead to the turbines, and the steel and concrete used to line each gas well, must also be included in any

FIGURE 12 Resource Intensity of Electric Power Generation Capacity: Comparing Wind with Natural Gas, Nuclear, and Coal

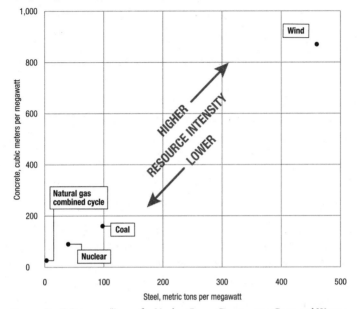

Source: Per F. Peterson, "Issues for Nuclear Power Construction Costs and Waste Management," September 16, 2008, http://www.ostp.gov/galleries/PCAST/PCAST% 20Sep.%202008%20Peterson%20slides.pdf, 4.

rigorous materials-intensity analysis of the life cycle of natural gas as it relates to electricity production. That said, the resource intensity of wind is also far higher than that of our other favorite fuel: nuclear.

Peterson's report shows that the concrete and steel requirements for wind are 9.6 times greater and 11.5 times greater, respectively, than those needed for a nuclear power plant. Each megawatt of power capacity in a nuclear power plant requires about 90 cubic meters of concrete and 40 tons of steel.[26]

Even though wind power has low power density and a huge appetite for steel and concrete, it appears that it will continue to be pursued in the years ahead, with wind turbines being located across the country. The entire wind industry will nevertheless be dogged by the fundamental problem of power density, because the math makes it unavoidable. And the power-density challenge is fundamentally about ethics and aesthetics.

Energy sources with high power densities have the least deleterious effect on open space. They allow us to enjoy mountains, plains, and deserts without having our views obstructed or disturbed by spinning wind turbines, sprawling solar arrays, towering transmission lines, or miles of monocultured crops. As the architect Witold Rybczynski wrote in *Atlantic Monthly* in an essay expounding the environmental benefits of cities, "density is green."[27]

Rybczynski's endorsement of cities echoes that of Stewart Brand, who, in his latest book, the *Whole Earth Discipline*, argues that cities, and even densely populated slums, provide a path out of poverty for millions of people. Brand says that "cities are probably the greenest things that humans do."[28]

Embracing the density of cities make sense. And to properly fuel them, we need energy sources with the highest possible densities. Energy projects with small footprints are not only green, they reduce the potential for NIMBY objections. And that not-in-my-backyard attitude is particularly virulent among people like Phil and Wendy Gramm, who can afford to fight power lines and other infrastructure projects. In fact, NIMBYism even occurs among the most ardent supporters of wind power. T. Boone Pickens wants to carpet the Great Plains with thousands upon thousands of wind turbines and endless rivers of transmission lines. But the Dallas-based billionaire wants those turbines and transmission lines on other people's land, not his. In May 2008, Pickens declared that his 68,000-acre ranch, located in the Texas Panhandle, one of America's windiest regions, will not sport a single turbine.

"I'm not going to have the windmills on my ranch," quoth he. "They're ugly."[29]

Ugly or not, wind power has become the most-hyped segment of the "green" energy alternatives. And few aspects of the wind-power hype have gotten more attention than the claim that adding wind turbines will mean big reductions in carbon dioxide emissions. There's only one problem with that claim: It's not true.

All About Power Density: A Comparison of Various Energy Sources in Horsepower (and Watts)

Nuclear: 300 hp/acre* (56 W/square meter)[30]

Average U.S. natural gas well, producing 115,000 cubic feet per day: 287.5 hp/acre† (53 W/square meter)[31]

Gas stripper well, producing 60,000 cubic feet per day: 153.5 hp/acre† (28 W/square meter)[32]

Oil stripper well, producing 10 barrels per day: 150 hp/acre† (27 W/square meter)[33]

Solar PV: 36 hp/acre (6.7 W/square meter)[34]

Oil stripper well, producing 2 barrels per day: 30 hp/acre† (5.5 W/square meter)

Wind turbines: 6.4 hp/acre (1.2 W/square meter)[35]

Biomass-fueled power plant: 2.1 hp/acre (0.4 W/square meter)[36]

Corn ethanol: 0.26 hp/acre (0.05 W/square meter)[37]

Note: Calculations may not be exact because of rounding. Assumptions: 1 Btu equals 1,000 joules, and 1 acre equals 4,000 square meters.
* Calculation uses entire 12,000 acres of the South Texas Project.
† Assumes well site is 2 acres.

CHAPTER 9

Wind Power Reduces CO$_2$ Emissions

GIVEN THE HYPE about wind power, it would be logical to assume that wind-power advocates have multiple studies on their shelves to prove that wind power cuts carbon dioxide emissions. The problem: They don't have a single study to support that claim. Yes, they have reports based on models that look at various scenarios of wind-power use, and those models provide projections about what the carbon dioxide reductions might be.

But the wind-power boosters do not have a single study—based on actual data collected from the world's existing fleet of wind turbines and conventional electricity-generation plants—showing that wind power actually reduces carbon dioxide emissions. That's remarkable, given that the Global Wind Energy Council has declared that "a reduction in the levels of carbon dioxide being emitted into the global atmosphere is the most important environmental benefit from wind power generation."[1]

In 2009, when I asked the American Wind Energy Association for studies proving that wind power reduced carbon emissions, association officials pointed to two reports relying on models that assume certain levels of future reductions.[2] The Global Wind Energy Council could only provide its annual report, "Global Wind Energy Outlook 2008," which, like the reports that used models, claimed that wind would, sometime in the future, decrease the use of hydrocarbons for electricity production.[3]

But that claim ignores the fact that all wind-power installations must be backed up with large amounts of dispatchable electric generation capacity. In Denmark's case, that has meant having large quantities of available hydropower resources in Norway and Sweden that can be called upon when needed. But even with a perfect zero-carbon backup system, the Danes haven't seen a reduction in carbon dioxide emissions. (Denmark's wind sector is discussed at length in the next chapter.) And that bodes ill for countries that don't have the access to hydropower that Denmark has. Nearly every country that installs wind power must back up its wind turbines with gas-fired generators.

This reality was explained in a 2008 report by Cambridge Energy Research Associates (CERA). In the 23-page report, the firm concluded that wind power "is more expensive than conventional power generation, in part because wind's intermittent production patterns need to be augmented with dispatchable generators to match power demand."[4] The CERA report goes on to explain that wind turbines may be good for producing electricity during fall and spring, but wind power "has limited capability as a capacity resource as its production patterns generally do not correlate well with peak summer demand. Consequently, the capacity provided by wind projects is typically valued at 10% to 20% of their maximum rated capacity."[5]

In the electric power business, generating plants are rated by their "capacity factor," which is based on the amount of time they will produce power at 100 percent of their maximum output. As the CERA report makes clear, many wind projects have a capacity factor of 10, 20, or 30 percent. But some grid operators are using capacity factors that are far lower than the estimates from CERA. For proof of that, look no further than the Lone Star State.

Texas has repeatedly been lauded as a leader in wind power development. In 2008, the state installed nearly 2,700 megawatts of new wind capacity, and by early 2009, if Texas were an independent country, it would have ranked sixth in the world in terms of total wind-power production capacity.[6] Republican governor Rick Perry, among the state's most ardent supporters of wind power, declared a few years ago that "no state is more committed to developing renewable sources of energy." He went on to say that by "harnessing the energy potential of wind, we can provide Texans a form of energy that is green, clean and easily renewable."[7] The Lone Star Chapter of the Sierra Club has also repeatedly trumpeted wind-power development, saying that it "means more jobs for Texas, less global warming from coal plants and less radioactivity from nuclear plants." The group says that wind power in the state "has exceeded all expectations," bringing in "an estimated $6 billion in investments and 15,000 new jobs" for the state.[8]

In June 2009, shortly before the U.S. House of Representatives was to vote on a major cap and trade bill aimed at increasing use of renewable energy, President Obama reminded reporters that Texas had one of the "strongest renewable energy standards in the country. . . . And its wind energy has just taken off and been a huge economic boon to the state."[9]

Alas, the hype doesn't match the reality. The Electric Reliability Council of Texas (ERCOT), which manages 85 percent of the state's electric load, pegs wind's capacity factor at less than 9 percent.[10] In a 2007 report, the grid operator determined that just "8.7% of the installed wind capability can be counted on as dependable capacity during the peak demand period for the next year." It added that "conventional generation must be available to provide the remaining capacity needed to meet forecast load and reserve requirements."[11] In 2009, the grid operator reaffirmed its decision to use the 8.7 percent capacity factor.[12]

By mid-2009, Texas had 8,203 megawatts of installed wind-power capacity.[13] But ERCOT, in its forecasts for that summer's demand periods, when electricity use is the highest, was estimating that just 708 megawatts of the state's wind-generation capacity could actually be counted on as reliable. With total summer generation needs of 72,648 megawatts, the vast majority of which comes from gas-fired generation, wind power was providing just 1 percent of Texas's total reliable generation portfolio. ERCOT's projections show that wind will remain a nearly insignificant player in terms of reliable capacity through at least 2014, when the grid operator expects wind to provide about 1.2 percent of the state's needed generation.[14]

Given the data from the Global Wind Energy Council and ERCOT, it's clear that wind power cannot be counted on as a stand-alone source of electricity but must always be backed up by conventional sources of electricity generation. In short, wind power does not reduce the need for conventional power plants, a point that was underscored in early January

FIGURE 13 Reliable Summer Generation Capacity in Texas, by Fuel Type, 2009 and 2014

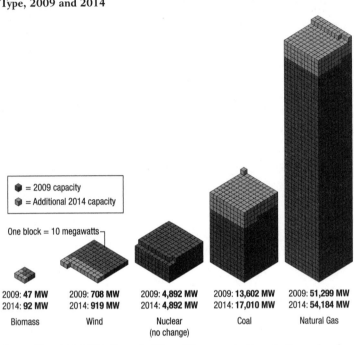

● = 2009 capacity
◉ = Additional 2014 capacity

One block = 10 megawatts

2009: **47 MW**	2009: **708 MW**	2009: **4,892 MW**	2009: **13,602 MW**	2009: **51,299 MW**
2014: **92 MW**	2014: **919 MW**	2014: **4,892 MW**	2014: **17,010 MW**	2014: **54,184 MW**
Biomass	Wind	Nuclear (no change)	Coal	Natural Gas

Source: Electric Reliability Council of Texas, "Report on the Capacity, Demand, and Reserves in the ERCOT Region," May 2009, http://www.ercot.com/content/news/presentations/2009/2009%20ERCOT%20Capacity,%20Demand%20and%20Reserves%20Report.pdf, 13.

2010 when Britain was hit by a record-setting cold snap. At the same time that energy demand soared due to the cold weather, Britain's wind farms produced practically no electricity.[15]

In its 2008 report, CERA determined that "in order to provide reliable capacity throughout the year, every megawatt of wind capacity needs to be matched up with a megawatt of dispatchable capacity."[16] Those findings were affirmed in early 2009 by Peter Lang, an engineer with forty years of experience in the energy business who is based in Canberra, Australia. In a report called "Cost and Quantity of Greenhouse Gas Emissions Avoided by Wind Generation," Lang concluded: "Because wind cannot be called up on demand, especially at the time of peak demand, installed wind generation capacity does not reduce the amount of installed conventional generating capacity required. So wind cannot contribute to reducing the capital investment in generating plants. Wind is simply an additional capital investment."[17]

In other words, thanks to its variability and intermittancy, wind power does not, and cannot, displace power plants, it only adds to them. The conclusions reached by Lang, and the analysts at CERA, are similar to those reached by British consultant James Oswald, who studied the potential effects of increased wind-power consumption in Britain. In a 2008 article published in the journal *Energy Policy*, Oswald and his two coauthors concluded that increased use of wind would likely cause utilities to invest in lower-efficiency gas-fired generators that would be switched on and off frequently, a move that cuts their energy efficiency and increases their emissions. Upon publication of the study, Oswald said that carbon dioxide savings from wind power "will be less than expected, because cheaper, less efficient [gas-fired] plant[s] will be used to support these wind power fluctuations. Neither these extra costs nor the increased carbon production are being taken into account in the government figures for wind power."[18]

China provides another example of the limited role that wind power will play in cutting carbon dioxide emissions. In September 2009, Jing Yang of the *Wall Street Journal* reported that "China's ambition to create 'green cities' powered by huge wind farms comes with a dirty little secret: Dozens of new coal-fired power plants need to be installed as well." Chinese officials are installing about 12,700 megawatts of new wind turbines in the northwestern province of Gansu. But along with those turbines, the government will install 9,200 megawatts of new coal-fired generating capacity in Gansu, "for use when the winds aren't favorable." That quantity of coal-fired capacity, Jing noted, is "equivalent to the entire generating capacity of Hungary."[19]

The obvious problem with the Chinese plan is that coal-fired plants are designed to provide continuous, baseload power. They cannot be turned on and off quickly. That likely means that all of the new coal plants being built in Gansu province to back up the new wind turbines will be run continuously in order to assure that the regional power grid doesn't go dark.

One other analysis of the wind–carbon dioxide question deserves mention: In November 2009, Kent Hawkins, a Canadian electrical engineer, published a detailed analysis on the frequency with which gas-fired generators must be cycled on and off in order to back up wind power. Hawkins' findings: The frequent switching on and off results in more gas consumption than if there were no wind turbines at all. His analysis suggests that it would be more efficient in terms of carbon dioxide emissions to simply run combined-cycle gas turbines on a continuous basis than to use wind turbines backed up by gas-fired generators that are constantly being turned on and off. Hawkins concluded that wind power is not an "effective CO_2 mitigation" strategy "because of inefficiencies introduced by fast-ramping (inefficient) operation of gas turbines."[20]

During an interview, Hawkins told me that he had been studying the wind-power sector for years and had been motivated to do his analysis because "nobody has done a comprehensive study." He said the wind industry has no interest in trying to produce proof that wind turbines cut carbon dioxide emissions. "Why do they have to prove anything with regard to CO_2 emissions? The industry already has all the political support, and the media, behind it."[21]

Given the work by Lang, Oswald, and Hawkins suggesting that wind power doesn't reduce carbon dioxide emissions, what does the wind industry claim? In its 2008 outlook, the Global Wind Energy Council put forward a "reference scenario." The trade group projects that by 2030, global wind-power capacity will be nearly 500,000 megawatts, a five-fold increase over 2007, when installed capacity totaled about 94,000 megawatts.[22] If that massive expansion of wind power occurs, the group expects global annual carbon dioxide emissions to be reduced by 731 million tons by 2030. That sounds significant.

But the council warned that "under this scenario, *carbon dioxide savings under wind would be negligible*, compared with the 18,708 million tons of carbon dioxide that the IEA expects the global power sector will emit every year by 2030" (emphasis added).[23] Put another way, at the same time that wind promoters are claiming that carbon dioxide reductions are a key benefit of adding new wind-power capacity, their own projections reveal that even if the wind-power sector continues to experience rapid growth, it will only reduce electricity-related carbon dioxide emissions by about 4 percent by 2030.[24] And given that the electric-generation sector represents about 40 percent of total global carbon dioxide emissions, that 4 percent reduction from wind—if it occurs—will be almost insignificant, amounting to a reduction of perhaps 1.5 percent of the total annual volume of anthropogenic carbon dioxide emissions.

In short, the wind-power promoters promise major carbon dioxide benefits, even claiming that carbon dioxide reductions are the industry's "most important environmental benefit." But the wind industry has no proof that it can achieve that claim. And even if you believe the industry's data for optimum investment in additional wind capacity, the resulting reductions in carbon dioxide emissions will barely be noticeable.

In fact, even in Denmark—the country that has taken the wind-power experiment further than any other—embracing wind has not reduced carbon dioxide emissions or cut hydrocarbon consumption.

CHAPTER 10

MYTH

Denmark Provides an Energy Model for the United States

> Today, America produces less than 3% of our electricity through renewable sources like wind and solar—less than 3%. Now, in comparison, Denmark produces almost 20% of their electricity through wind power. . . . When it comes to renewable energy, I don't think we should be followers, I think it's time for us to lead.
>
> BARACK OBAMA,
> Earth Day Speech, April 22, 2009[1]

ADVOCATES OF RENEWABLE energy love Denmark. And why shouldn't they? The Danes love themselves.

Surveys in 2006 and 2008 found that the Danes are the happiest people on the planet.[2] The surveys found strong correlations for satisfaction with availability of health care, higher personal income, and access to education. (In the 2006 study, the United States ranked twenty-third in the happy ratings.)[3] Although the studies don't mention renewable energy as an element of the happiness quotient, if the wind-power promoters are to be believed, then Denmark's happiness surely must correlate with the number of wind turbines that are installed in the country.

America's leading energy cheerleaders love to cite Denmark as the model to be copied. For instance, in a 2006 interview with *Discover* magazine, Amory Lovins enthused about wind power, saying that there is enough available wind energy in South Dakota and North Dakota "to meet the United States' electricity needs." He went on, saying that "Denmark is now one-fifth wind powered."[4] A few months after Lovins proclaimed Denmark's virtues, Fox News, the conservative news outlet, followed suit, saying that Denmark "has become a leader in the field of renewable energy" and that "renewable sources account for a greater share of the nation's energy consumption with each passing year."[5]

In August 2008, *New York Times* columnist Thomas Friedman held up the Danes as the model for the United States. In the wake of the 1973 Oil Embargo, Friedman claimed, Demark "responded . . . in such a sustained, focused and systematic way that today it is energy independent." Friedman went on to lament America's situation, writing that if "only we could be as energy smart as Denmark!"[6]

In mid-2009, Joshua Green, a senior editor at *Atlantic Monthly*, wrote a long article about renewable energy bemoaning the fact that the United States had not done more to embrace renewable energy. He said the election of Obama and a Democratic Congress had made "a significant shift in the nation's energy policy a real possibility for the first time in years."[7] Green then lauded Denmark: "Europe offers a model of how governments can lead the transition to clean energy and thereby reduce demand for fossil fuels. Denmark, which also suffered the shocks of the 1970s, no longer needs to import oil."[8]

While all of the wind power and happiness in Denmark makes me want to fly to Copenhagen for a cup of coffee and a hug, a close look at Denmark's energy sector shows that its embrace of wind power has not resulted in "energy independence"; nor has it made a major difference in the country's carbon dioxide emissions, coal consumption, or oil use. Despite massive subsidies for the wind industry and years of hype about the wonders of Denmark's energy policies, the Danes now have some of the world's most expensive electricity and most expensive motor fuel. And in 2007, their carbon dioxide emissions were at about the same level as they were two decades ago.

Thomas Friedman may like the idea of energy independence, but the data shows that Denmark is not energy independent—it's not even close. The Danes import all of their coal. I repeat, Denmark imports *all* of its coal.[9] Those coal imports—and coal consumption among the Danes—show little sign of declining, even though Denmark's wind power production capacity is increasing. And Denmark is even more dependent on coal than the United States.[10]

Green's claim that Denmark doesn't import any oil is true. Denmark is an oil exporter. But that reality has nothing to do with Denmark's embrace of renewables. Instead, it's a result of the country's decades-old decision to pursue aggressive offshore oil development in the North Sea.[11] Perhaps if the United States were as sensible about offshore drilling as the Danes, the United States wouldn't need to import oil, either.

Alas, the hard facts about Denmark appear to matter little to America's leading purveyors of energy happy-talk. But the facts, particularly when it comes to wind, provide plenty of reason to be skeptical about Denmark's energy policy.

Once again, we must look at the numbers. Between 1999 and 2007, according to data from the Danish Energy Agency, the amount of electricity produced from the country's wind turbines grew by about 136 percent, going from 3 billion kilowatt-hours to some 7.1 billion kilowatt-hours.[12] By the beginning of 2007, wind power was accounting for about 13.4 percent of all the electricity generated in Denmark.[13] And yet, over that same time period, coal consumption didn't change at all. In 1999, Denmark's daily coal consumption was the equivalent of about 94,400 barrels of oil per day.[14] By 2007, Denmark's coal consumption was exactly the same as it was back in 1999.[15] In fact, Denmark's coal consumption in both 2007 and 1999 was nearly the same as it was back in 1981.[16]

Denmark may be leading the world in wind-power installation and production, but the variability of wind assures that Denmark's production of coal-generated electricity will continue to rise and fall depending on the weather. The gyrations in the country's power sector can be seen by considering this fact: In 2006, the Danish grid consumed 50 percent more coal-fired electricity than it did in 2005.[17] The basic problem with Denmark's wind-power sector is the same as it is everywhere else: It must be backed up by conventional sources of generation. For Denmark, that means using coal as well as the hydropower resources of its neighbors. As much as two-thirds of Denmark's total wind power production is exported to its neighbors in Germany, Sweden, and Norway.[18] In 2003, 84 percent of the wind power generated in western Denmark was exported, much of it at below-market rates.[19]

As Hugh Sharman, a British-born energy analyst who lives in Denmark, put it, the Danes are providing an electricity subsidy to their neighbors. And they are doing so because Denmark cannot use all of the wind-generated electricity it produces. The intermittency of the wind resources in western Denmark—located far from the main population center in Copenhagen—means that the country must rely on its existing coal-fired power plants. When excess electricity comes onstream from the country's wind turbines, the Danes ship it abroad, particularly to Sweden and Norway, because those countries have large amounts of hydropower resources that Denmark then uses to balance its own electric grid.

Put another way, Denmark's wind turbines often produce surplus electricity that the country cannot use; thus, that power must be exported. Its neighbors' hydropower resources "are effectively batteries for Danish wind energy," said Sharman.[20] (Norway utilizes more hydropower as a percentage of primary energy than any other country, getting some 68 percent of its total energy needs from water. Sweden ranks seventh in hydropower utilization, getting nearly 30 percent of its primary energy needs from water. For more on this, see Appendix D.)[21]

In September 2009, the Danish Center for Political Studies, known as CEPOS, a Copenhagen-based think tank, came to the same conclusion as Sharman, declaring that this "exported wind power, paid for by Danish householders, brings material benefits in the form of cheap electricity and delayed investment in new generation equipment for consumers in Sweden and Norway but nothing for Danish consumers."[22]

None of this is aimed at belittling Denmark. The country has had remarkable success at keeping energy demand down. It is one of the few countries where energy demand growth is essentially flat. In 2007, the country's total primary energy use, about 363,000 barrels of oil equivalent per day, was roughly the same as it was in 1981.[23] Denmark's ability to keep energy consumption growth flat over such a long period is anomalous. But let's be clear: That near-zero growth in energy consumption has been achieved in part by imposing exorbitant energy taxes and by maintaining near-zero growth in population.

Thanks to their government's exorbitant tax rates, the Danes pay some of the highest electricity rates in the world. In 2006, the Energy Information Administration looked at residential electricity rates in sixty-five countries and found that Denmark's rates were the highest by far, amounting to some $0.32 per kilowatt-hour. That was about 25 percent higher than the electricity costs in the Netherlands, which had the next-highest rates in the survey at $0.25 per kilowatt-hour. And that's not a new phenomenon. From 1999 through 2006, Denmark had either the highest—or the next-highest—electricity rates of the countries surveyed by the EIA. (In 1999 and 2000, Japan's electricity rates were slightly higher than those in Denmark.)[24] Furthermore, Denmark's electricity rates are the highest in Europe—and no other country comes close.[25]

In 2008, electricity rates were even higher, with Danish residential customers paying $0.38 per kilowatt-hour—or nearly four times as much as U.S. residential customers, who were paying about $0.10 per kilowatt-hour.[26] And the Danes were paying more than twice as much as their counterparts in nuclear-heavy France, where residential electricity costs were $0.17 per kilowatt-hour.[27]

While Danish homeowners are getting spanked by expensive electricity, Danish motorists are getting absolutely mugged at the service station. In late 2008, Danish drivers were paying an average of $1.54 per liter for gasoline (the equivalent of $5.83 per gallon), while drivers in the United Kingdom were paying $1.44 per liter ($5.45 per gallon) and U.S. motorists were paying $0.56 per liter ($2.12 per gallon). According to German Technical Cooperation (GTZ), an agency of the German government, only a handful of countries have more expensive fuel than Denmark, a list that includes Italy, Norway, Turkey, and Germany.[28]

Although Friedman, Lovins, and Green want to hold up Denmark as a model to be emulated, the brutal facts show that Denmark is even more reliant on oil—as a percentage of primary energy—than the United States is. In fact, the Danes are among the most oil-reliant people on Earth. In 2007, Denmark got about 51 percent of its primary energy from oil. That's far higher than the percentage in the United States (40 percent) and

FIGURE 14 What Price Wind? Denmark's Residential Electricity
Prices Compared to Those of Other Countries

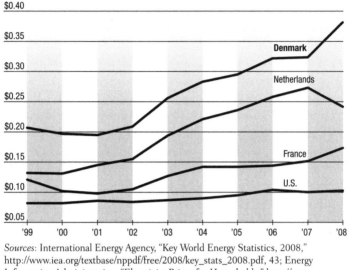

Sources: International Energy Agency, "Key World Energy Statistics, 2008,"
http://www.iea.org/textbase/nppdf/free/2008/key_stats_2008.pdf, 43; Energy
Information Administration, "Electricity Prices for Households," http://www
.eia.doe.gov/emeu/international/elecprih.html.

significantly higher than the world average of 35.6 percent. As stated
above, Denmark is more coal dependent than the United States, getting
about 26 percent of its primary energy from coal while America gets about
24 percent of its primary energy from the carbon-heavy fuel.[29]

But given the global worries about climate change and carbon dioxide,
the key issue for Denmark is this: Has wind power made a difference in
its carbon dioxide emissions? The answer, based on the country's own
data, appears to be no. Although Denmark has more than doubled its
wind power production over the past decade or so, it has not seen major
reductions in its carbon dioxide emissions or its coal consumption.

In 2008, the European Environment Agency released a report that
analyzed the progress (or lack thereof) that European countries are making toward their carbon emission reduction targets under the Kyoto Protocol. The report found that Denmark was falling short of its 2010
reduction targets by 18.8 percent.[30] It also found that between 1990 and
2006, Denmark's overall greenhouse gas emissions increased by 2.1 per-

cent.[31] Although that's a much better emissions profile than that of Spain, where emissions jumped by 50.6 percent over that same time period, it's obvious that Denmark's emissions have not declined despite its big increase in wind power production. Between 1990 and 2006, according to the European Environment Agency, Denmark's annual per-capita emissions of greenhouse gases stayed essentially the same—at about 13 tons of carbon dioxide.[32]

If Denmark's huge wind-power sector were reducing carbon dioxide emissions, you'd expect the Danes to be bragging about it, right? Well, guess what? They're not.

In early 2009, Energinet.dk, the operator of Denmark's electricity and natural gas grids, published a handsome brochure available on its website entitled, "Wind Power to Combat Climate Change: How to Integrate Wind Energy into the Power System."[33] But the 56-page report does not contain any evidence that wind power has done anything to cut carbon dioxide emissions.[34] The only claims it makes are about the role that wind *might* play in cutting emissions in the future. The report quotes Henriette Hindrichsen, a senior consultant at Energinet.dk, who declared that given enough wind power, Denmark "could" cut its use of hydrocarbons. "We could save fossil fuels," she helpfully explained, "and thus reduce carbon dioxide emissions by moving some of our consumption away from peak-load periods to times of considerable wind and low consumption."[35]

It's critical to note that Energinet.dk does not say that wind *will* cut carbon emissions, only that it *could*. The report—which claims to be a "carbon dioxide–neutral product" because the gases produced during its production were "neutralized by buying carbon dioxide quotas from Climate Friendly via the WWF"—also discusses the potential for using wind power to charge electric vehicles in Denmark. It does so by breezily noting that "Danes have never really taken a fancy to electric cars even though they are tax-free. For this reason, only about 300 electric cars buzz along the Danish roads."[36]

Despite all this, Energinet.dk is certain that someday soon—real soon—Danes will be driving electric cars, and when they do so, those cars will "absorb surplus wind power, significantly reduce transport sector carbon dioxide emissions and help to balance the power system if the batteries are charged or drained as required."[37] But of course, that will only happen when, or if, Danes "take a fancy" to electric cars. In late 2009, the Danish government announced that it would offer huge incentives— a $40,000 tax break on each electric car, as well as free parking in downtown Copenhagen—to customers who bought all-electric vehicles. The government also announced plans to set up thousands of charging stations, as part of a deal with Better Place, a California-based company that is promoting electric vehicles.[38]

Though the Danish government is pushing electric cars, Energinet.dk is not making any claims about the ability of wind power to cut carbon dioxide emissions. The grid operator's annual reports from 2006 and 2007 contain no references to reduced carbon dioxide output due to wind.[39] More important is this: Energinet.dk's 2008 environmental report contains a graph showing that in 2007, carbon dioxide levels from electricity generation totaled about 23 million tons, about the same amount as in 1990, before the country began its frenzied construction of wind turbines.[40] And the glossy 44-page document forecasts no decreases in carbon dioxide emissions through 2017.[41] In other words, even though Denmark uses hydropower—a zero-carbon-dioxide source of backup power for its wind turbines—it still hasn't seen a reduction in its carbon dioxide emissions from its electric generation sector.

Nor has Denmark seen a significant decrease in its total carbon dioxide emissions from all sources of electricity generation, transportation, and the like over the past two decades. According to the IEA, in 1990, Denmark emitted a total of 50.7 million tons of carbon dioxide. In 2007, the country's emissions totaled 50.6 million tons, a reduction of just 0.1 percent.[42]

Denmark saw a significant increase in its overall electricity use during that same time period, with consumption rising from 28.8 billion kilowatt-hours in 1990 to about 35 billion kilowatt-hours in 2007.[43] That's an increase of 21.5 percent. Thus, Denmark has, it appears, been able to hold its carbon dioxide emissions flat while increasing electricity production by about one-fifth. That's a laudable achievement. But part of the ability to control carbon dioxide emissions comes from demographics. In 2008, Denmark had a population of 5.5 million, and its population growth rate was just slightly above zero.[44] In fact, Denmark has one of

FIGURE 15 Emissions from Denmark's Electric Power and Combined Heat and Power Sectors, 1990 to 2007

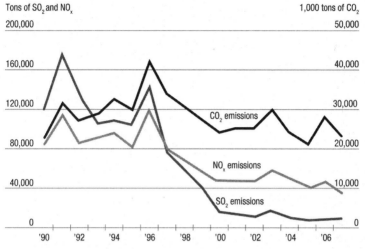

Source: Energinet.dk, "Environmental Report 2008," n.d., http://www.energinet.dk/ NR/rdonlyres/EC3E484D-08D5-4179-9D85-7B9A9DBD3E08/0/ Environmentalreport2008.pdf.

the slowest-growing populations in Europe. Between 1998 and 2008, the population grew by just 200,000 people.[45] During that same time period, the U.S. population jumped by about 33 million people.[46]

Given that population data, several key interrelated questions emerge: Has Denmark been able to keep its carbon dioxide emissions flat because of its slow-growing population? Is the emissions rate holding steady because of the exorbitant electricity and motor fuel taxes? Or is the increased use of domestic wind production and imported hydropower holding the emissions rate down?

There are no clear answers to those questions. But Energinet.dk's environmental report holds some clues. It says that carbon dioxide emissions from power generation declined slightly in 2007, but that this decline was not due to wind power. Instead, Energinet.dk says, the decline was a result of the fact that 2007 "was a wet year in the Nordic region with massive hydropower production in Norway and Sweden." It concluded that "the emissions level for 2007 followed the general trend of low emissions following wet years."[47] The environmental report from Energinet.dk explains that Denmark sees higher levels of carbon dioxide emissions from the electric sector during dry years "because the generation at coal-fired power stations in particular is higher during dry years. The Danish emissions level is thus affected by whether it is a wet year or a dry year."[48]

Nevertheless, give the Danes their due: They are doing some things right. Between 1999 and 2006 (the latest year for which data is available), Denmark's carbon intensity (that is, the amount of carbon dioxide emitted per unit of GDP) declined by nearly 11 percent.[49] That's a pretty good result. Denmark is also reducing the overall energy intensity of its economy. According to the Danish Energy Agency, between 1990 and 2008 the Danish economy grew by more than 42 percent, but energy use increased by less than 5 percent. According to the agency, "This means that today Denmark uses 3 units of energy to produce the same goods as required 4 units of energy in 1990."[50] And the average Dane uses far less energy than the average American. Denmark's per-capita energy use is about 3.85 tons of oil equivalent per year, while the U.S. average is about 7.74 tons.[51]

But the reality is that even amidst all the praise for Denmark's efforts, the United States is doing as well as—and in some cases, better than—Denmark in terms of carbon. Between 1999 and 2006, the carbon intensity of the U.S. economy decreased by nearly 13 percent—a rate about 2 percent better than the reduction rate seen in Denmark over that time period. Furthermore, in a longer time frame—1980 through 2006—America's carbon intensity declined by 43.7 percent; Denmark managed just slightly better, with a 47 percent decrease.[52] Thus, the United States is largely keeping up with Denmark even though it hasn't made the same kind of massive investments in wind or made electricity exorbitantly expensive through punitive tax measures.

Although foreign journalists love to talk about Denmark's wind sector, they seldom look at the country's oil and gas business. If they were to do so, they'd be surprised by its size and by how reliant the Danes are on crude oil and natural gas.

Denmark has become largely self-sufficient in oil and gas, not because it's more virtuous or because it's using more alternative energy, but be-

FIGURE 16 Danish Oil Production, 1981 to 2008

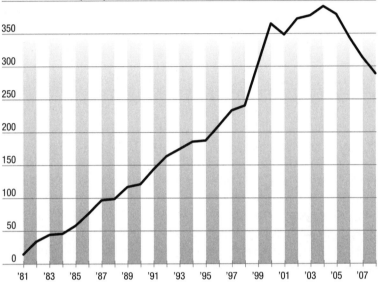

Source: Energy Information Administration, "Denmark Energy Data," http://tonto.eia.doe.gov/country/excel.cfm?fips=DA.

cause it has fully committed to drilling in the North Sea. Between 1981 and 2007, the country's oil production jumped from less than 15,000 barrels per day to nearly 314,000 barrels per day—an increase of nearly 2,000 percent.[53] The focus on sustained oil and gas exploration and production led to a corresponding increase in oil reserves, which jumped from about 500 million barrels to nearly 1.3 billion barrels.[54] Denmark has had similar success with its natural gas production. In 1981, the country was producing no natural gas. By 2007, natural gas production was nearly 900 million cubic feet per day—enough to supply all of the country's own consumption needs and to allow for substantial exports.[55]

And the Danes are continuing their development of the North Sea. In 2008, Denmark signed a new exploration license for more offshore drilling, and two applications for additional licenses were submitted to the Danish Energy Agency. In early 2009, the agency accepted another application for drilling. That focus on oil and gas is paying dividends. In 2008, the

Danish government took in about $7.1 billion in taxes and fees from oil and gas companies operating in its offshore waters in the North Sea.[56]

So how does Denmark's oil and gas production compare with its wind production? As stated above, in 2007, Denmark's wind-power sector produced 7.1 billion kilowatt-hours of electricity, which works out to about 19,400 megawatt-hours per day.[57] Remember that 1 barrel of oil is approximately equal to 1.64 megawatt-hours of electricity. Thus, in 2007, the total primary energy production from Danish wind was about 11,800 barrels of oil equivalent per day.

Now let's compare that to Denmark's oil and gas production. In 2007, the country's oil production was about 314,000 barrels per day. Natural gas production was almost 900 million cubic feet per day, or about 164,000 barrels of oil equivalent per day.[58] Add in the 94,400 barrels of oil equivalent that the Danes use in the form of coal, and the math is clear: Denmark's hydrocarbon diet consists of about 572,400 barrels of oil equivalent per day. Although some of the country's oil and gas production is exported, here's the punch line: Hydrocarbons provide Denmark with about forty-eight times as much energy as the country gets from wind power.

Atlantic Monthly's Joshua Green trumpeted Denmark's freedom from imported oil, but he didn't mention the country's coal imports. Nor did he mention the country's ongoing drilling programs in the North Sea. Thomas Friedman said the United States should be "as energy smart as Denmark!" but he didn't bother to compare the country's wind power production to its hydrocarbon use.[59]

Nevertheless, the facts—not the hype—show that the Danes are about as reliant on hydrocarbons as the United States and other countries are. In some ways, they are worse off than the United States when it comes to energy. The Danes themselves have begun to take note of the high cost of wind power and its inability to make significant reductions in carbon dioxide emissions. The September 2009 study by CEPOS said that Denmark's wind industry "saves neither fossil fuel consumption nor carbon dioxide emissions."[60] The final page of the report even offers a warning for the United States: "The Danish experience also suggests that a strong US wind expansion would not benefit the overall economy. It would entail substantial costs to the consumer and industry, and only to

FIGURE 17 Wind Power Versus Hydrocarbons: Denmark's Consumption in 2007

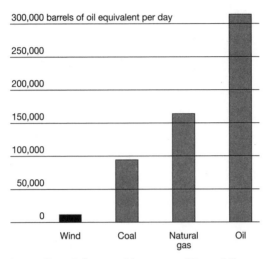

Sources: Energy Information Administration, "Denmark Energy Data," http://tonto.eia.doe.gov/country/excel.cfm?fips=DA; Danish Energy Agency, "Energy Statistics 2007," http://www .ens.dk/en-US/Info/FactsAndFigures/Energy_statistics_and _indicators/Annual%20Statistics/Sider/Forside.aspx.

a lesser degree benefit a small part of the economy, namely wind turbine owners, wind shareholders and those employed in the sector."[61]

The conclusion to be drawn is clear: Although Denmark has repeatedly been held up as a model to be copied, the numbers tell a markedly different story. But then, few policymakers have bothered to look closely at what has actually happened in Denmark. Instead, they've been seduced by the easy charms of Dallas billionaire T. Boone Pickens, who has launched a multimillion-dollar media campaign aimed at promoting, well, T. Boone Pickens.

The incredible media success of Pickens brings to mind a saying coined by another promoter, P. T. Barnum: "There's a sucker born every minute." Given the credulity with which politicians and media types gobbled up the Pickens Plan, it's obvious that old P.T.B. would be proud of T.B.P.

CHAPTER 11

T. Boone Pickens
Has a Plan (or a Clue)

FOR A FEW MONTHS, T. Boone Pickens was, once again, a rock star.

On July 4, 2008, Pickens unveiled his blueprint to rescue America from the evil clutches of foreign oil. Within days he was everywhere. He posed for the cover of *Texas Monthly*, sitting in the back of a vintage Rolls Royce kitted out like James Dean in the movie *Giant*.[1] The cover text boomed in huge black capital letters: BOONE. Below that, in smaller type: "He saved himself. Can he save America?" Inside the magazine, writer Skip Hollandsworth lauded Pickens as "the most famous wildcatter in Texas history."

Hollandsworth reported that when he caught up with Pickens in New York City on July 8, just four days after the launch of the Pickens Plan, "Boone had spent more than an hour that morning chatting about the plan on CNBC, and he had also appeared on ABC's *Good Morning America* and NPR's *Morning Edition*. Later that day, he was scheduled to meet with CNN, Fox, and the BBC, followed by visits to the offices of the Associated Press, the *Wall Street Journal*," and other news outlets.

About that same time, Slate named Pickens as number three in its ranking of the eighty most powerful octogenarians in America, behind only U.S. Supreme Court Justice John Paul Stevens and billionaire corporate raider Kirk Kerkorian.[2] Pickens testified on Capitol Hill, had private meetings with Al Gore and Barack Obama (then a senator), and was a near-constant presence on television—appearing on *The Jay Leno Show*, and *CBS Evening News with Katie Couric*, to name just a few.[3] Reporters swarmed the Dallas-based billionaire. They gushed over his money, his private jet—the leather seats are monogrammed!—and his ambitious plan to build the world's largest wind farm, a 4,000-megawatt project in the Texas Panhandle.[4]

Pickens' media team spent lavishly to court the public and the media. They covered all the new media bases, with presences on Twitter, My-Space, LinkedIn, and YouTube.[5] By November 2009, the Pickens Plan had close to 31,000 fans on Facebook.[6]

Politicians flocked to Pickens, straining themselves to amp up the hyperbole. In August 2009, at a "National Clean Energy Summit" in Las Vegas, Senate Majority Leader Harry Reid, the Democrat from Nevada, declared, "I now belong to the Pickens church. We're very thankful to have him here, he's been a good friend and a real visionary." At the same meeting, former president Bill Clinton extended a "special word of thanks to Boone Pickens," conveniently ignoring Pickens' lifelong support for Republicans and his financial support for the Swift Boat campaign against John Kerry in 2004. Former vice president Al Gore also chimed in, saying "I honest to God wish that more business leaders of your experience would care as much as you do, and throw themselves into this fight for the future of our country the way you are doing."[7]

Of course, seeing the rangy native of Holdenville, Oklahoma, in the limelight was nothing new. He's the energy industry's only rock star. Not even during Enron's heyday (and downfall) was Ken Lay as visible as Pickens has been over the past few years. Pickens loves the attention. He's friendly and cordial with reporters. And make no mistake, he's a savvy operator. Pickens is among a rare breed of entrepreneurs who just knows how to make money. And over the past few decades, he has applied that ability, as an oil man, corporate raider, dealmaker, and money manager, to amass a fortune that *Forbes* has estimated at $3.1 billion.[8] And those billions further burnish his rock-star credentials.

Unfortunately for Pickens, though, rock stars don't have a very long shelf life. Just a few months after Pickens rolled out his grand plan, the *Wall Street Journal* reported that two investment funds managed by Pickens had lost about $1 billion of his investors' money.[9] And exactly one year after Pickens launched his much-ballyhooed plan, he was forced to admit that his vision for the wind business was in tatters. In July 2009, Pickens told Elizabeth Souder of the *Dallas Morning News* that his wind-power project turned out to be "a little more complicated than we thought."[10]

Now there's an understatement. But it shouldn't be surprising. The Pickens Plan doesn't pencil. It never did. The entire concept was based on a set of faulty assumptions, the crux of which was this: Using more wind power would mean less natural gas consumption, and the extra natural gas could be used to fuel vehicles, meaning the United States could drastically cut its oil imports.

The reality is that Pickens launched his multimillion-dollar media campaign as part of an effort to backstop his own money-making ventures. In May 2008, he ordered $2 billion worth of wind turbines from General Electric.[11] Two months later, he began selling the Pickens Plan, the first tenet of which was that wind power was the answer to America's concerns about foreign oil—and by swaddling himself in Old Glory, many Americans were duped into believing him.

The main fallacy in Pickens' proposal: Using more wind power would mean more natural gas availability for the automotive sector. The problem with that logic is obvious: Utilities and power-plant owners now burning natural gas are not going to shut down their plants in order to save gas unless it makes economic sense for them to do so. In addition, and perhaps more important, is this: Over the past few years, state and federal regulators have blocked a large number of proposed coal-fired power plants. As a result, natural gas has become the preferred fuel for new power generation projects. Between 1997 and 2008, the volume of gas used for electricity production in the United States increased by 64 percent.[12] The same trend is evident around the world. Between 2000 and 2008, more than 75 percent of new global electricity demand was met with gas-fired power plants.[13]

Making fun of Pickens is easy, but to give him his due, he's right about wanting to increase the use of natural gas in the transportation sector. That concept makes economic sense for many fleet operators. But—and it's a big but—Pickens has grossly exaggerated the ability of the United States to make a quick transition to natural-gas-fueled vehicles. On the Pickens Plan website, the billionaire claims that using more wind power and "increasing the use of our natural gas resources can replace more than one-third of our foreign oil imports in 10 years."[14]

That's an easy claim to make, but Pickens can't do it. And he couldn't do it even if he were somehow able to manage a one-hundred-fold increase in the number of natural-gas-fueled vehicles in the United States and do so in just ten years. Building a large fleet of natural-gas-fueled vehicles—and more importantly, the refueling infrastructure to support them—will take decades, not years.

The numbers simply don't work. Let's look at oil imports: In 2008, the United States imported an average of 12.9 million barrels of oil and oil products per day.[15] One-third of that volume—the amount Pickens claims he can save—is about 4.25 million barrels of oil per day. Fine. Let's run the numbers.

According to Natural Gas Vehicles for America, a Washington, D.C.–based trade association, there are about 120,000 natural gas vehicles (NGVs) now in use in the United States.[16] Each of those vehicles consumes about 1,500 gasoline-gallon-equivalents per year.[17] Using that 1,500-gallon-per vehicle figure, those 120,000 NGVs conserve the equivalent of about 180 million gallons of oil per year.

Now let's multiply that number by 100. Doing so increases the U.S. fleet to 12 million NGVs, which could save 18 billion gallons of fuel per year, the equivalent of 1.17 million barrels of oil per day.[18] That type of reduction is significant. But remember, Pickens promised to cut oil use by 4 million barrels a day. Furthermore, creating a NGV fleet that size would require a Herculean effort. If the United States had 12 million NGVs, that fleet would be larger than the current *global* fleet of NGVs, which numbers about 9.6 million vehicles.[19]

Pickens led a gullible media and an even more gullible public to believe that the evils of foreign oil could be overcome if only the public provided him with a few more subsidies for his pet projects. And he put forward his plan without discussing any fuels other than wind and natural gas. The fact that his unrealistic plan was so readily accepted by so many journalists and politicos provides additional evidence of the lack of skepticism about green energy in general and wind power in particular.

In particular, politicians and the media willingly accepted Pickens' claim that adding wind power would reduce the need for natural gas. Nothing could be further from the truth.

Bird Kills? What Bird Kills?

> The Migratory Bird Treaty Act makes it illegal for anyone to kill a protected bird (including eagles and other raptors) by any means without first obtaining a permit.
>
> U.S. Fish and Wildlife Service press release, 2009[20]

On August 13, 2009, Exxon Mobil pled guilty in federal court to charges that it killed 85 birds—all of which were protected under the Migratory Bird Treaty Act (MBTA). The company agreed to pay $600,000 in fines and fees for the bird kills, which occurred after the animals came in contact with hydrocarbons in uncovered tanks and wastewater facilities on company properties located in five western states.[21]

The Exxon Mobil prosecution is the latest of hundreds of cases that federal officials have brought against oil and gas companies over the past two decades for violations of the MBTA, a statute that has been on the books since 1918.[22] But the oil and gas companies are not alone in having run afoul of this law. The U.S. Fish and Wildlife Service has also brought MBTA cases against electric utilities. On July 10, 2009, for example, Oregon-based PacifiCorp agreed to pay $1.4 million in fines and restitution for killing 232 eagles in Wyoming over a two-year period. The birds were electrocuted by the company's poorly designed power lines.[23]

Those cases are clearly justified. But they also underscore a pernicious double standard in the enforcement of federal wildlife laws: At the very same time that federal law enforcement officials are bringing cases against oil and gas companies and electric utilities under the MBTA, they have exempted the wind industry from any enforcement action under that statute and a similar one, the Bald and Golden Eagle Protection Act, enacted in 1940.

Once again, the numbers behind the story are essential. As it turns out, the number of birds being killed by wind turbines dwarfs the numbers involved in the prosecution of Exxon Mobil. A July 2008 study of bird kills by wind turbines at Altamont Pass, California, estimated that the massive wind farm was killing 80 golden eagles *per year*.[24] In addition, the study, funded by the Alameda County Community Development Agency, estimated that about 2,400 other raptors, including burrowing owls, American kestrels, and red-tailed hawks—as well as about 7,500 other birds, nearly all of which are protected under the MBTA—were being whacked every year at Altamont.[25]

To recap: Exxon Mobil was prosecuted for killing 85 birds over a five-year period. The wind turbines at Altamont, located about 30 miles east of Oakland, are killing more than one hundred times as many birds as were involved in the Exxon case, and they are doing it *every year*. Furthermore, the bird-kill problems at Altamont have been documented repeatedly. A 1994 study documented numerous raptor kills, a finding that has been corroborated by essentially all of the subsequent studies.[26]

To be sure, the number of birds killed by wind turbines is highly variable—and biologists believe the situation at Altamont, which uses older turbine technology, may be the worst example of bird kills by wind turbines.[27] That said, the carnage at Altamont likely represents only a fraction of the number of birds being killed by wind power every year. Michael Fry of the American Bird Conservancy estimates that between 75,000 and 275,000 birds per year are being killed by U.S. wind turbines. And yet, the Department of Justice won't press charges. "Somebody has given the wind industry a get-out-of-jail-free card," Fry told me.[28]

According to the American Wind Energy Association, each megawatt of installed wind power capacity results in the death of between one and six birds per year.[29] At the end of 2008, the United States had about 25,000 megawatts of wind turbines, and environmental and lobby groups are pushing for the country to be producing 20 percent of its electricity from wind by 2030. Meeting that goal, according to the Department of Energy, will require the United States to have about 300,000 megawatts of wind capacity, a twelve-fold increase over 2008 levels.[30] If that target is achieved, it will likely mean that at least 300,000—and perhaps as many as 1.8 million—birds will die each year from collisions with wind turbines.

The American Wind Energy Association dismisses the problem as insignificant, saying that the bird kills are a "very small fraction of those caused by other commonly accepted human activities and structures—

house cats kill an estimated 1 billion birds annually."[31] That may be true. But cats rarely get frog-marched to the courthouse in handcuffs. Nor are they killing many golden eagles.

Wind turbines are also killing other flying animals. A study of a 44-turbine wind farm in West Virginia found that up to 4,000 bats had been killed by the turbines in 2004 alone.[32] A 2008 study of dead bats found on the ground near a Canadian wind farm found that many of the bats had been killed by a change in air pressure near the turbine blades that caused fatal damage to their lungs, a condition known as "barotrauma."[33]

Bat Conservation International, an Austin-based group dedicated to preserving the flying mammals and their habitats, has called the proliferation of wind turbines "a lethal crisis." In September 2009, I interviewed Ed Arnett, who heads the group's research efforts on wind power. He said that the headlong rush to develop wind power is having major detrimental effects on bat populations, but few environmental groups are willing to discuss the problem because they are so focused on the issue of carbon dioxide emissions and the possibility of global warming. "To compromise today's wildlife values and environmental impacts for tomorrow's speculated hopes is irresponsible," Arnett said. He added that only a handful of bat species are protected by federal law, and therefore, the killing of bats by wind turbines gets little attention from the media.

But the bat-kill problem has begun getting some attention. In December 2009, a federal court halted the construction of a wind project in West Virginia over concerns that the 122-turbine facility would harm the Indiana bat, which is protected by the Endangered Species Act. In his ruling on the case, a federal judge wrote that "there is a virtual certainty that Indiana bats will be harmed, wounded, or killed" by the proposed wind farm. The developers of the wind farm must now apply for a special permit from the Fish and Wildlife Service that will allow the project to proceed.[34]

But the ruling in the West Virginia case brings up the obvious question: Why aren't federal wildlife officials doing more to protect birds from wind turbines? During the late 1980s and early 1990s, Rob Lee was one of the Fish and Wildlife Service's lead law-enforcement investigators on the problem of bird kills in western oil fields. Now retired and living in Lubbock, Texas, Lee told me that solving the problem of bird kills in the oil fields was relatively easy and not very expensive. Lee said the oil companies only had to put netting over their tanks and waste facilities, or close them. Asked why wind power companies aren't being prosecuted for killing eagles and other birds, Lee told me that "the fix here is not easy or cheap." Nor does Lee expect to see any prosecutions. "It's economics. The wind industry has a lot of economic muscle behind it."

In other words, what's good for the goose is not good for the gander. When it comes to energy production and the protection of America's wildlife, federal law-enforcement officials are favoring the wind industry over other industries. And that favoritism occurs because of the myth that wind power is "green."

CHAPTER 12

MYTH

Wind Power Reduces the Need for Natural Gas

THE PICKENS PLAN relies on the theory that increasing the use of wind power will allow the United States to reduce its oil imports, because lots of wind turbines will save natural gas that can then be redirected to the transportation sector.

As I've mentioned, there are two problems with that claim: U.S. natural-gas-fired electricity generation is soaring, and the United States doesn't have nearly enough natural-gas-fueled vehicles to make a significant dent in its oil imports. But there's another problem, one that Pickens and other wind promoters don't talk about: Wind power increases the need for natural gas. Pickens himself admits as much. In January 2010, when I challenged him about the merits of wind power, he replied, "I'm not saying wind replaces natural gas."

Wind power production is highly variable. For instance, on February 7, 2008, Colorado electric utilities had to scramble to keep the lights on after the state's wind power output declined by about 485 megawatts in an hour. And that sudden collapse in wind power availability occurred just as demand for electricity was reaching its morning peak. To keep the power grid stable, the state's utilities had to quickly start up a bank of gas-fired generators and buy electricity from neighboring utilities.[1]

Or consider the problems that hit the Bonneville Power Administration in January 2009. The federal agency, which supplies electricity to much of the Pacific Northwest, reported that between January 14 and 25, it could not use any of the 1,700 megawatts of installed wind power capacity in its service area. The agency said that wind generation during that time period was "essentially zero," a condition that was "due to extreme temperature inversion conditions" in eastern Washington and Oregon.[2] The agency was able to offset the lack of wind by using some of its hydropower capacity. But the Bonneville Power Administration is an exception. Most of the regions in the United States that have significant wind resources do not have hydropower assets that can be used to provide power when the wind stops blowing. And the United States isn't building dams, it's taking them down. Over the past decade, more than two hundred dams in the United States have been dismantled.[3]

In areas where wind turbines have been added to the electric grid, utilities and electric grid operators must have plenty of natural-gas-fired power generation available that can be switched on quickly when the wind stops blowing. Assuring that generation capacity is always available requires major investments in gas wells, gas pipelines, and gas storage fields to be certain that there's enough gas available for generators. And all of that infrastructure costs money, which means that consumers will ultimately have higher electric bills due to wind power.

Gas analysts and utility companies are still grappling with all of the ramifications of the variability of wind power. In the United States, gas-fired power plants operate about 36 percent of the time.[4] (The utilization rate, or capacity factor, for coal plants, which provide baseload, or always-on, power, is usually 70 percent or more. Many nuclear plants operate at about 90 percent.) But the addition of new wind-power resources may mean that those gas-fired generators are used even less, meaning their capacity factor may decline to just 25 percent or maybe 30 percent. And by reducing the amount of time those generators and other parts of the infrastructure get utilized, wind power reduces the capital efficiency of the equipment, which raises the effective cost of that equipment. Put another way, the relative inefficiency of wind as a power source requires traditional electricity producers to invest more capital in expensive hardware—such as gas-fired generators, gas storage equipment, and other items—that may be used less often than they would be in the absence of wind power.

The concept is fairly simple to understand. Let's assume that ABC Utility is required to install 100 megawatts of wind turbines. To assure that those turbines are able to provide reliable quantities of electricity, ABC must also install 100 megawatts of gas turbines. (As stated in earlier chapters, every megawatt of wind power that is added to a given electricity system must be backed up with a megawatt of gas-fired generation.) Let's further assume that the cost of those 100 megawatts of gas generation is $100 million.

Without the wind mandates, ABC could operate those gas turbines at a capacity factor of 36 percent or more. In doing so, it could calculate a certain return on its $100 million investment. But as the wind turbines are added to ABC's grid, they reduce the gas turbines' utilization rate and make them less efficient from a capital standpoint. And somehow, somewhere, consumers will have to pay for that inefficiency.

This problem was discussed in a July 2009 report by Pöyry, a Helsinki-based consulting firm. In an analysis of how the variability of wind power will affect the British and Irish electricity markets, Pöyry concluded that new power plants designated to back up wind power "will have to operate at low, and highly uncertain loads, and under the current market arrangements the likely returns [on capital] do not appear good." The report goes on to say that revenues from any backup generation "will be volatile and uncertain to the point where a plant may only operate for a few hours one year, and then some hundreds the next. Generating companies will need to factor this possibility into their investment strategies."[5]

The Pöyry conclusions are nearly identical to those made by the International Energy Agency. In its "Natural Gas Market Review 2009," the agency said that as renewable capacity is added, "gas-fired capacity will increase while its overall load factor may be reduced. . . . This switching will have an impact on the profitability of new investments."[6] The agency used Spain as an example of the need for gas-fired generation and the variability of wind. "Peak summer loads coincide with periods when there usually isn't much wind, while the opposite happens during winter, meaning large differences in terms of gas-fired plant load: on 20

June 2008, Spain broke a record in terms of peak daily gas demand from the power sector."

On that day in June 2008, Spain's gas-fired generators were operating at a capacity factor of 75 percent. But six months later, on December 20, when the wind was blowing strongly, the capacity factor for the country's gas-fired generators was just 18 percent. "Such fluctuations require heavy investments in gas storage, particularly of the type needed to respond quickly to large gas demand movements from power generators," said the IEA.[7]

Though it is true that gas consumption declines during periods when the wind is providing lots of electricity, it's not yet clear how large those savings will be. Nor is it clear that the savings in fuel costs will be enough to offset the capital costs incurred to install the needed gas storage capacity, pipelines, and generators. Furthermore, all of that gas- and power-delivery infrastructure—and the generators, in particular—must be staffed continually. The utilities cannot send workers home only when the wind is blowing. The generators must be available and staffed to meet demand 24/7.

The additional costs imposed by wind power can be seen by looking again at Colorado, one of the states in the vanguard of the push for renewable energy. In 2004, Colorado voters approved a ballot measure requiring the utilities in the state to generate or purchase enough renewable energy to supply at least 10 percent of their retail electricity sales. Since that time, the state's legislature has raised that target to 20 percent of retail electricity sales, and that target must be met no later than 2020.[8] In December 2008, Xcel Energy, a natural gas and electric utility that serves customers in eight states, issued a report on the costs associated with integrating wind power into its mix of generation assets in Colorado.[9] The study says that the utility expects "the costs of integration to be predominantly fuel costs resulting from 1) the inefficiency of generation due to wind generation uncertainty," and "2) the cost of additional gas storage."[10]

To be clear, the integration of wind power into a specific electricity grid will vary widely depending on the size of the grid, what types of fuels it uses, and the amount of wind being added. But Xcel is correct about the need for gas storage. The gas-fired generators needed to back up

TABLE 3 States That Get More Than 80 Percent
of Their Electricity from Coal

State	Electricity from Coal
West Virginia	97%
Indiana	95%
Wyoming	94%
North Dakota	93%
Kentucky	93%
Utah	89%
Ohio	86%
Missouri	84%
New Mexico	80%

Sources: EIA, American Coalition for Clean Coal Electricity,
"Your State" (clickable map), http://www.cleancoalusa.org/
docs/state/. Data current as of July 2009.

wind power must be switched on or off several times per day, and underground gas storage facilities must be located relatively close to the generators so that there is enough pressure in the gas pipeline to keep the generators going. For parts of the United States that have favorable geology—consider western states such as Colorado, Texas, Oklahoma, and Kansas, where gas production is common—the gas storage issue may not pose a problem. But in other states, and particularly in other countries with less-favorable geology, the lack of gas storage may pose a real barrier to wind power.

Despite the situation in Colorado, there have been no comprehensive studies, on either the federal level or the state level, exploring how increasing deployment of wind power will affect natural gas infrastructure and demand. Furthermore, the dynamics of the wind–natural gas relationship will vary widely. It will likely be particularly problematic in the states that are heavily dependent on coal-fired power plants for their electricity. If states such as Wyoming and West Virginia—which are 94 percent and 97 percent reliant on coal, respectively—are required to add renewable electricity to their grids, those states will likely have to make large parallel investments in new gas generation and gas-delivery infrastructure.

The reason those states will need to add gas infrastructure is that coal plants are not designed to be turned on and off. Instead, they are designed to run at a constant rate. When they run below their optimum design rates, they become less efficient, and they may emit more carbon dioxide and air pollutants at lower power output levels than when they operate at maximum output.

All of this matters because Americans have been repeatedly told that electricity generated from wind costs less than electricity produced by other forms of power generation. That's only true if you don't count the investments that must be made in other power-delivery infrastructure that assures that the lights don't go out.

The extreme variability of wind generation means that wind turbines are simply a supernumerary—an extra—element of our electricity generation system. They don't displace existing power plants at all. Instead, when they are added they must be carefully integrated into the electricity grid—and backed up with gas-fired generators—so that they won't cause too much disruption. And therein lies the punch line: The costs of all the new gas-related infrastructure that must be installed in order to accommodate increased use of wind power should be included in calculations about the costs of adding renewable sources of energy to the U.S. electricity grid. Those calculations should be done on a state-by-state basis. But so far, little, if any, of that type of work has been done.

So why hasn't it been done? Part of the blame should be aimed at promoters like Pickens, who have continually understated the costs of moving the U.S. toward wind power. In addition, Pickens confuses the issues of wind-generated electricity and oil by claiming that more wind power would mean less oil use and therefore less need for imports.

That claim leads to the next myth-busting opportunity. Americans have repeatedly been told that increasing the production of "green" energy in the United States will result in a reduction in U.S. imports. That's just not true. Nearly all of the wind turbines now being produced depend on a rare element called neodymium. One of the most confounding aspects of the push for "green" energy in America is that, by rushing headlong into efforts to reduce the use of hydrocarbons, the United States is making itself even more beholden to China, which has a stranglehold on the world's supply of neodymium and other "green" elements.

CHAPTER 13

Going "Green" Will Reduce Imports of Strategic Commodities and Create "Green" Jobs

FOR DECADES, THE MOST important invention attributed to Baron Carl Auer von Welsbach was the flint used in cigarette lighters. In the world of inventions, that bit of acclaim is notable, though it doesn't necessarily rise to the level of world-changing. Nevertheless, the work that von Welsbach—a Viennese chemist born in 1858—did on the flint (an alloy made of 70 percent cerium and 30 percent iron) was part of his ongoing interest in a group of elements known as the lanthanides.[1] And his predilection for the lanthanides led him to discover two cornerstones of the "green" economy: neodymium and praseodymium.

Those two materials, along with the other elements found in the lanthanides' row of the periodic table, are essential commodities in nearly all of the technologies that are seen as solutions to our energy challenges, from wind turbines and hybrid cars to solar panels, computers, and batteries. Why are they so important? The lanthanides—which are also called "rare earths"—have special features at the quantum mechanics level. The configuration of their electrons allows them to have unique magnetic interactions with other elements.

These characteristics make the lanthanides a key chokepoint in the development of the "green" economy. And that leads to one of the biggest myths about going green: the notion that if only we would use more hybrid cars, wind turbines, solar panels, and other such devices, we would be free of messy international entanglements and the need to import oil and other strategic commodities. But here's the reality: China has a de facto monopoly on the global trade in the lanthanides, and about 90 percent of the world's lithium, an essential element in high-capacity batteries, comes from just three countries: Argentina, Chile, and China.[2] Huge lithium deposits exist in Bolivia, but that country's populist leader, Evo Morales, has made it clear that he won't be selling it on the cheap.[3]

In its headlong rush to go "green," the United States may simply be trading reliance on one type of import for reliance on another. Instead of requiring oil supplied by dozens of producers located in the Persian Gulf and elsewhere, it will need rare earth commodities produced by the Chinese as well as lithium mined by a handful of foreign countries.

Of course, we live in a global economy, particularly when it comes to energy. The petrostates of the Persian Gulf and elsewhere must sell their oil. They can't drink it or use it to water their geraniums. The same holds true for the countries that produce lithium and the lanthanides. And given the ongoing globalization of the world economy, it stands to reason that the marketplace will help to assure that buyers and sellers will reach agreed-upon prices for whatever goods or services are on offer. That said, the difference between the hyper-global oil sector and the rare earths business is akin to the difference between aluminum and dysprosium.[4]

The global oil market allows price discovery for crude oil and a myriad of oil commodities on a near-instantaneous basis, and that market is fully integrated with a large network of buyers and sellers. For instance, in 2007, the United States imported oil or oil products from ninety different countries.[5] In 2008, twenty-one nations were each producing more than 1 million barrels of oil per day.[6] If one producer's price is too high, buyers can almost certainly find another seller.

Although there's plenty of rhetoric about the evils of the biggest oil producers—countries such as Venezuela, Saudi Arabia, and Iran—the simple truth is that those nations are exporting much of their oil production because their domestic economies are not big enough or sophisticated enough to utilize all of the oil they produce. Over the past several years, some vocal American neoconservatives have repeatedly demonized Saudi Arabia for what they view as the Saudis' excessive control over the world oil market and their influence on OPEC. Though it's true that the Saudis are influential, they only control about 10 percent of daily world oil production. These same neoconservatives hate OPEC—but OPEC only controls about one-third of world oil production.[7]

Now compare the diffused global oil market with the constricted market for the lanthanides—lanthanum, cerium, praseodymium, dysprosium, neodymium, and the others. China controls—depending on whose numbers you believe—between 95 and 100 percent of the global market in those elements. The fact that China sits atop such favorable geology for mining the lanthanides is pure luck. But the Chinese are aiming to make the most of that luck. Deng Xiaoping, the former Communist Party boss, once said, "There is oil in the Middle East. There are rare earths in China. We must take full advantage of this resource."[8]

China is doing just that. It has about 1,000 Ph.D.-level scientists working on technologies related to the mining and separation of rare earth elements as well as on ways to turn those elements into salable products.[9] At the same time that China is increasing its knowledge base on rare earths, it is cutting its exports of those same materials. In addition, China has raised export taxes on some of the rare earths and has prohibited foreign companies from investing in China's rare earth mining and processing sector. Meanwhile, it is encouraging the companies that need steady supplies of lanthanides to move their production facilities to China.[10]

China's control over the green elements has begun to attract attention, and few countries are more worried than Japan, which is heavily dependent on the lanthanides for its high-technology manufacturing. In May 2009, an official at Japan's Ministry of Economy, Trade, and Industry told the *Times* of London that "all green technology depends on rare-earth metals and all global trade in rare earth depends on China."[11]

While all of the lanthanides are important, neodymium has particular significance. Neodymium-iron-boron magnets are powerful, lightweight, and relatively cheap—at least they are when compared to the magnets they replaced, which were made with samarium (another lanthanide) and cobalt. The Toyota Prius uses neodymium-iron-boron magnets in its motor-generator and its batteries. Analysts have called the Prius one of the most rare-earth-intensive consumer products ever made, with each Prius containing about 1 kilogram (2.2 pounds) of neodymium and about 10 kilograms (22 pounds) of lanthanum.[12] And it's not just the Prius. Other hybrids, such as the Honda Insight and the Ford Fusion, also require significant quantities of those elements.[13]

The American wind sector is almost wholly dependent on neodymium-iron-boron magnets, which are used inside wind turbines. General Electric reportedly buys all of the neodymium-iron-boron magnets used in its wind turbines from China. Other metals, including europium (a lanthanide) and yttrium (a non-lanthanide), are essential elements in video displays. As David Trueman, a Canadian geologist who has spent more than three decades finding, mining, and exploiting rare metal deposits, explained to me, "without rare earths, you wouldn't have color TV." Trueman said that China is not only cutting its exports of rare earths but also reducing exports of other key elements, including tungsten, antimony, and indium. The Chinese, said Trueman, "are the world's oldest capitalists. They'd rather build the TV for you than sell you the metals" needed to make the unit.[14]

China's increasing grip on the rare earths is coming at the same time that the Chinese are capturing a bigger and bigger share of the global high-tech manufacturing business. Between 1985 and 2005, China's export volume of high-technology goods went from near zero to about $450 billion.[15] During that same period, China's share of global high-tech manufacturing grew from about 1 percent to about 16 percent.[16]

China's low-cost labor, lax environmental policies, and abundance of rare earths have allowed it to surpass both Japan and the United States in terms of the export value of high-tech manufactured goods. According to data from the National Science Foundation, China surpassed Japan as a leading high-tech manufacturer in about 2001. And by 2005, China's high-tech exports were twice as valuable as Japan's.

Non-lanthanide rare elements are also essential in the solar power sector. For instance, Arizona-based First Solar (2008 revenues: $1.2 billion), one of the biggest producers of photovoltaic cells in the United States, relies on the compound cadmium telluride. First Solar's business hinges on the availability of tellurium (atomic number 52) which is usually

FIGURE 18 World Share of High-Tech Manufacturing Exports, by Region/Country, 1985 to 2005

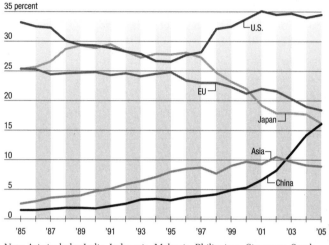

Note: Asia includes India, Indonesia, Malaysia, Philippines, Singapore, South Korea, Taiwan, and Thailand. China includes Hong Kong.
Source: National Science Foundation, "Science and Engineering Indicators 2008: Presentation Slides," January 2008, http://www.nsf.gov/statistics/seind08/slides.htm.

FIGURE 19 Export Volume of High-Tech Manufactured Goods, by Region/Country, 1985 to 2005

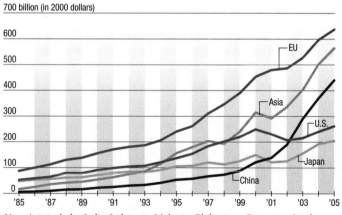

Note: Asia includes India, Indonesia, Malaysia, Philippines, Singapore, South Korea, Taiwan, and Thailand. China includes Hong Kong.
Source: National Science Foundation, "Science and Engineering Indicators 2008: Presentation Slides," January 2008, http://www.nsf.gov/statistics/seind08/slides.htm.

produced as a by-product of the copper-refining process. In its 2009 annual report, First Solar says that if its suppliers could not obtain adequate supplies of tellurium, those suppliers "could substantially increase prices or be unable to perform under their contracts." If such a shortage happened, First Solar admits, it could get squeezed, "because our customer contracts do not adjust for raw material price increases and are generally for a longer term than our raw material supply contracts. A reduction in our production could result."[17]

Though copper refining may be able to produce enough tellurium for First Solar and other producers of photovoltaic panels, it's apparent that China, once again, has an advantage. According to experts on rare metals, the Chinese have the world's only tellurium mine.[18] And the Chinese are using their access to rare metals to pump out large quantities of solar panels. Several Chinese solar-panel factories have recently boosted their output of panels. The result: Between mid-2008 and mid-2009, the price of solar panels in the United States fell by about 40 percent.[19]

This is a critical issue for the United States when it comes to competitiveness: If U.S. policymakers decide to support an indigenous rare earths business as a way to hedge against Chinese supplies, China could simply drop the price of their rare earths, and in doing so make any upstart mining operation unprofitable.

At the moment, the only hope for the United States when it comes to domestic lanthanide production appears to be Molycorp Minerals, which owns America's only operable rare earths mine, located near Mountain Pass, California. In 2008, Molycorp was sold to a group of private investors, including Goldman Sachs. The mining outfit was purchased from Chevron, which took control of Molycorp when it bought Unocal. (By the way, the other suitor for Unocal was the Chinese National Offshore Oil Company.) Molycorp has begun processing some of the ore that was stockpiled at the mine, but the company says it won't be able to resume mining operations until 2011 or 2012. Even if the Molycorp mine can return to full output, company officials believe it will only be able to capture perhaps 15 percent of the world market.[20]

The availability of rare earths is not just about balance of trade, it's also about national security. The U.S. military is heavily reliant on high-tech weaponry, which means navigation systems, guidance systems, radios, and computers—all of which require rare earths. Now, suppose the United States decided to impose trade sanctions on China, perhaps due to some type of dispute over carbon emissions. What might China do in retaliation? Well, one obvious way to retaliate would be to cut off the flow of rare earths to the United States and other countries, thus pinching the U.S. Defense Department's ability to obtain the high-tech equipment it needed.

China's near-monopoly control of the green elements likely means that most of the new manufacturing jobs related to "green" energy products will be created in China, not the United States. Chinese companies have made it clear that—thanks to huge subsidies provided by the Chinese government—they are willing to lose money on their solar panels in order to gain market share.[21] And the government is subsidizing more solar-panel manufacturing capacity, which will likely allow the Chinese to undercut the prices of panel makers all over the world. Consider a deal that the city of Austin's municipal utility, Austin Energy, made in early 2009: The utility agreed to build a solar farm that will use 220,000 solar panels. All of them will be made in China.[22] Or consider wind turbines: In late 2009, the backers of a $1.5 billion wind project in West Texas announced that they were planning to install 240 wind turbines on the 36,000-acre site. The project backers were seeking $450 million in federal stimulus money to make the deal happen. All of the wind turbines for the project are to be built in China.[23]

Environmental activists in the United States and other countries may lust mightily for a high-tech, hybrid-electric, no-carbon, super-hyphenated energy future. But the reality is that that vision depends mightily on lanthanides and lithium. That means mining. And China controls nearly all of the world's existing mines that produce lanthanides. These facts demonstrate, once again, the need to accept the interconnectedness of the global economy. Simply trading one type of strategic commodity import (such as oil) for another (lanthanides and lithium) makes little sense. The reality is that the United States, like every other country, will continue to depend on the global marketplace to obtain the commodities it needs.

Of course, the United States will need more of the green elements. But the good news is that even without huge increases in imports of the

FIGURE 20 The "Green Elements" and the Periodic Table

The Periodic Table

H																	He
Li	Be											B	C	N	O	F	Ne
Na	Mg											Al	Si	P	S	Cl	Ar
K	Ca	Sc	Ti	V	Cr	Mn	Fe	Co	Ni	Cu	Zn	Ga	Ge	As	Se	Br	Kr
Rb	Sr	Y	Zr	Nb	Mo	Tc	Ru	Rh	Pd	Ag	Cd	In	Sn	Sb	Te	I	Xe
Cs	Ba		Hf	Ta	W	Re	Os	Ir	Pt	Au	Hg	Tl	Pb	Bi	Po	At	Rn
Fr	Ra		Rf	Db	Sg	Bh	Hs	Mt	Ds	Rg	Cp	Uut	Uuq	Uup	Uuh	Uus	Uuo

lanthanides →

La	Ce	Pr	Nd	Pm	Sm	Eu	Gd	Tb	Dy	Ho	Er	Tm	Yb	Lu
Ac	Th	Pa	U	Np	Pu	Am	Cm	Bk	Cf	Es	Fm	Md	No	Lr

Lanthanides description

Name	Symbol	Uses
Lanthanum	La	Batteries, catalyst in oil refining
Cerium	Ce	Glass and lens production, catalytic converters
Praseodymium	Pr	High-strength magnets, electronics, pigments
Neodymium	Nd	High-strength magnets, electronics, lighting
Promethium	Pm	X-ray units
Samarium	Sm	High-strength magnets, glass
Europium	Eu	Lighting, video screens
Gadolinium	Gd	Magnetic resonance imaging, video screens
Terbium	Tb	High-strength magnets, lighting, video screens
Dysprosium	Dy	High-strength magnets, video screens
Holmium	Ho	Glass tint
Erbium	Er	Metal alloys
Thulium	Tm	Lasers
Ytterbium	Yb	Metal alloys
Lutetium	Lu	Catalyst, metal alloys, nuclear technology

lanthanides, the U.S. economy is steadily becoming more efficient in its energy use. In fact, over the past three decades, the United States has been among the best nations in the world when it comes to improving the efficiency of its economy.

CHAPTER 14

The United States Lags in Energy Efficiency

YOU'VE HEARD IT a thousand times: The United States wastes energy. Americans are energy hogs. Our cars are too big. Our houses are too big. And, of course, our collective butts are too big, too.

Aside from the last one, which can likely be verified with a tape measure, those claims are largely wrong. Over the past three decades or so, the United States has been as good as—or better than—nearly every other developed country on Earth at improving its energy efficiency. It has been among the best at reducing its carbon intensity, its energy intensity, and its per-capita energy use. And here's the most important thing to remember when considering those facts: The United States has achieved these reductions without participating in the Kyoto Protocol (which would have set targets for reductions in carbon emissions) or creating an emissions trading system like the one employed in Europe. In fact, the United States has been better than nearly every other country on the planet at reducing its carbon intensity and its energy use without doing any of the things that environmental groups and renewable energy lobbyists contend are essential.

Nevertheless, Americans are being steamrolled with claims that the United States is a laggard—and Congress has latched on to the idea. In 2007, it passed the Energy Independence and Security Act, a 310-page bill in which the word "efficiency" appears 331 times and "efficient" appears

111 times.[1] In 2009, the original version of the cap-and-trade bill (also known as Waxman-Markey or the American Clean Energy and Security Act of 2009) contained the word "efficiency" 218 times and the word "efficient" 44 times.[2] Environmental groups such as the Sierra Club, Greenpeace, and the Environmental Defense Fund, repeatedly say that America's first priority on energy policy should be greater efficiency. For decades, Amory Lovins, the Colorado-based energy pundit, has been claiming that efficiency is the essential solution to all of America's energy challenges.

Though nearly everyone is convinced that the United States is lagging on energy efficiency, the numbers, as usual, tell a different story.

Let's look first at the carbon intensity of the U.S. economy—that is, the amount of carbon dioxide per unit of economic output, measured in metric tons of carbon dioxide per $1,000 of GDP. Between 1980 and 2006, U.S. carbon intensity fell by 43.6 percent. That's far better than the performance of the EU-15, which managed a reduction of 30.1 percent over that same time period. The U.S. reduction in carbon intensity over that period nearly matched that of Western European countries that are often held up as models of aggressive national energy policy implementation, namely, Denmark and France, which managed to reduce their carbon intensity by 47 percent and 50.2 percent, respectively. Somewhat surprising is the performance of China, which, thanks to improvements in the efficiency of its machinery, reduced its carbon intensity by 64 percent.[3]

The United States is reducing its carbon intensity because it continues to become more efficient in how it uses energy. Despite bingeing on SUVs and ever-faster cars with ever-increasing numbers of cupholders and seat heaters, the United States has become far more efficient in how it consumes oil and natural gas. And that efficiency is occurring in myriad ways, from more efficient cars to better home furnaces and more efficient power plants that are able to wring more electricity out of a molecule of natural gas.

Now let's consider energy intensity. (Note that carbon intensity and energy intensity, while closely related, are not the same thing. A given country may have greater carbon intensity than its neighbors if it relies more heavily on coal. Conversely, it may have lower carbon intensity if it

FIGURE 21 Change in Carbon Intensity of Major World Economies, 1980 to 2006

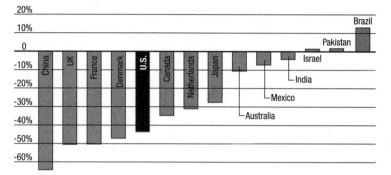

Source: Energy Information Administration, "World Carbon Intensity—World Carbon Dioxide Emissions from the Consumption and Flaring of Fossil Fuels Per Thousand Dollars of Gross Domestic Product Using Market Exchange Rates, 1980–2006," http://www.eia.doe.gov/pub/international/iealf/tableh1gco2.xls.

relies more heavily on nuclear power or hydropower.) Between 1980 and 2006, America's energy intensity—the amount of energy needed to produce $1 of gross domestic product (GDP)—fell by about 42 percent. Among the major countries of the world only one did better: China, where energy intensity fell by 63 percent. Over that same time period, Britain matched the United States, seeing its energy intensity fall 42 percent. The reduction in U.S. energy intensity is particularly notable given that it bested countries such as France and Japan, which saw their energy intensities decline by 20.4 percent and 17.4 percent, respectively. Meanwhile, in some developing countries, energy intensity actually increased. For instance, in Indonesia, the amount of energy needed to produce $1 of GDP increased by 5.4 percent, and in Brazil, energy intensity increased by 33.3 percent.[4]

The decline in U.S. energy intensity becomes even more interesting when you consider that between 1980 and 2006, the U.S. GDP more than doubled, going from $5.8 trillion to about $12.9 trillion.[5] (Those figures are in constant year 2005 dollars.)[6] Furthermore during that same period the U.S. population increased by about 31.5 percent, going from about 228 million people to about 300 million.[7] Put another way, between 1980 and 2006, the U.S. economy grew by 122 percent, its population

FIGURE 22 Change in Energy Intensity of Major World Economies, 1980 to 2006

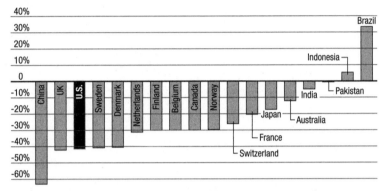

Source: Energy Information Administration, "World Energy Intensity—Total Primary Energy Consumption Per Dollar of Gross Domestic Product Using Purchasing Power Parities, 1980–2006," http://www.eia.doe.gov/pub/international/iealf/tablee1p.xls.

grew by 31.5 percent, and yet the total amount of energy needed to produce $1 of GDP fell by about 42 percent. Why did that happen? Much of it can be explained simply by the fact that consumers, engineers, and entrepreneurs are always working to do things more efficiently, not because it is better for the environment, necessarily, but because it saves them money and increases profits. In other words, it's just good business.

Now let's look at one of the most important energy metrics: per-capita energy consumption. From 1980 through 2006, the average per-capita energy consumption in the United States fell by 2.5 percent. That decline was greater than in any other developed country in the world except for Switzerland and Denmark, which saw their per-capita energy use fall by 4.3 percent and 4.2 percent, respectively.[8] During that same time period, per-capita energy use in major European countries rose significantly. For instance, per-capita energy use in France rose by 18.8 percent. Spain's per-capita energy use jumped by 93.4 percent, and Norway's increased by 25.1 percent. (Britain's per-capita energy use increased by 3 percent.) During that same time period, the world average for per-capita energy use rose by 13.7 percent.[9]

While it's true that the United States uses significantly more energy on a per-capita basis than the rest of the world (334.6 million Btu in

FIGURE 23 Change in Per-Capita Energy Use, Major World Economies, 1980 to 2006

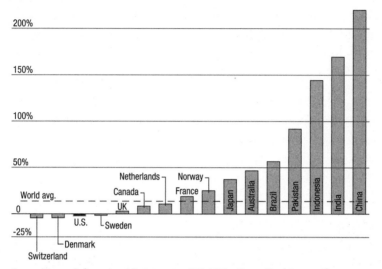

Source: Energy Information Administration, "World Per Capita Total Primary Energy Consumption, 1980–2006," revised December 19, 2008, http://www.eia.doe.gov/pub/international/iealf/tablee1c.xls.

2006 versus a world average of 72.4 million Btu), that difference is largely a function of America's higher standard of living as well as the much greater distances that Americans have to travel. For instance, the state of Texas sprawls over an area covering some 268,000 square miles.[10] That means it's bigger than France, the largest country in Western Europe, which encompasses 220,000 square miles.[11] Or consider the area displaced by everyone's favorite banker and chocolatier: Switzerland, which covers less than 16,000 square miles.[12] Though few people would swap Zurich for Enid, you could fit four Switzerlands inside Oklahoma's friendly confines and still have enough space left over to squeeze in Puerto Rico.[13]

So what's going on? Why is America doing so well on these measures of carbon intensity, energy intensity, and per-capita energy use?

There are several answers to those questions. Among the most important is that the U.S. economy has moved toward more service-based production. It may be lamentable for some people, but significant segments of America's heavy industry have moved overseas where labor and raw materials are cheaper. Energy-intensive businesses such as steel making, aluminum smelting, and auto manufacturing have become far more globalized than before, and that has resulted in the loss of factory jobs to China, Mexico, the Middle East, and other countries that have certain advantages, such as lower labor costs or cheaper energy.

Another key factor: Engineers keep doing what they do best, making products that are faster, cheaper, and produce more horsepower, more efficiently, than the ones they made the year before. Modern refrigerators use far less energy than the ones that were made back in the 1970s and 1980s. Modern personal computers use processors that are far more powerful and efficient than the ones made just three or four years ago. The proliferation of programmable thermostats has allowed consumers to better manage the cooling and heating of their homes. The integration of microchips into automobiles has helped to make modern cars faster, more luxurious, and more powerful while allowing them to achieve better fuel economy than the ones produced thirty years ago. Modern long-haul trucks are more efficient thanks to better aerodynamics and engines. The same is true of modern jet airplanes.

Similar trends are discernible in America's oil consumption. In 1973, there were about 130 million registered vehicles in the United States and Americans were driving about 1.3 trillion miles per year.[14] That year, the United States was using 17.3 million barrels of oil per day.[15] By 2007, the United States had 254 million motor vehicles on the road and they were being driven 2.9 trillion miles per year.[16] Thus, between 1973 and 2007, the number of miles driven in the United States increased by 123 percent, but the amount of oil needed to do that driving had increased by just 20 percent, to 20.7 million barrels per day.[17]

The relatively small increase in oil consumption since 1973 is a product of several factors. Cars, trucks, and airplanes are becoming more efficient, and the United States has become smarter about its oil consumption. Back in 1973, nearly 17 percent of the electricity produced in the United States was generated with oil. By 2008, that number had fallen to 1.1 percent.[18]

None of this means the United States should quit pursuing gains in energy efficiency. Asking engineers and entrepreneurs to stop seeking ef-

FIGURE 24 U.S. Energy Use Per Capita and Per Dollar of Gross
Domestic Product, 1980 to 2030

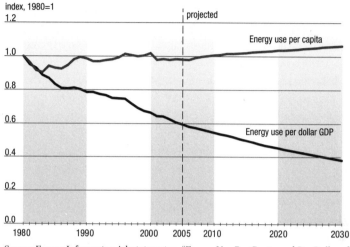

Source: Energy Information Administration, "Energy Use Per Capita and Per Dollar of
Gross Domestic Product, 1980–2030," n.d., http://www.eia.doe.gov/oiaf/aeo/ppt/
fig004.ppt.

ficiency would be akin to telling them to go out of business. Increasing
efficiency is a natural product of the competitive economic system, and
it will continue regardless of any mandates put forward by government.
So the next time you hear someone claim that the United States wastes
energy, understand that the opposite is true: The United States is among
the world's best at energy efficiency, and it's getting better, a lot better.

Furthermore, U.S. efficiency gains are accelerating. The increasing use
of hybrid-electric cars, as well as more efficient lightbulbs, air-conditioning
systems, refrigerators, and other equipment, is adding velocity to America's efficiency gains. In the wake of the oil price spike in mid-2008,
Americans appear to be less enamored with SUVs and big pickup trucks.
Although those vehicles are still selling in large numbers, U.S. consumers
are buying more fuel-efficient cars.[19] And in May 2009, President Obama
announced new federal regulations that will require automakers' fleets to
average 35.5 miles per gallon by 2016.[20] That's a significant increase over
the 2009 fleet average of about 28 miles per gallon.[21]

Numerous high-profile efficiency upgrades are under way, including the retrofit of the Empire State Building. In April 2009, Johnson Controls announced that it had teamed up with the Clinton Climate Initiative and the Rocky Mountain Institute on a $20 million program to modernize the 103-story building's insulation, windows, lighting, ventilation, chiller, and energy management systems. The project is expected to reduce the eight-decade-old skyscraper's energy use by nearly 40 percent and cut energy costs by about $4.4 million per year.[22]

Energy efficiency is the one energy policy issue upon which both Republicans and Democrats can agree. That bipartisan support is providing momentum to upgrades of the U.S. electric grid. Although there's been a lot of hype around the phrase "smart grid," there are significant gains to be had by providing consumers with more information about their usage and by giving utilities better information about the amount of voltage they are pumping into their wires. For instance, if a utility has reliable data showing that it is providing enough voltage to its most-distant customers on a given section of the grid, it can reduce the voltage on its generators and thereby reap energy savings of as much as 4 percent.[23]

In July 2009, the consulting firm McKinsey & Company released a report that predicted that if the United States adopted aggressive efficiency policies it could reduce primary energy consumption by about 20 percent when compared to a "business as usual" scenario.[24] The consulting firm determined that there are big gains to be had from efficiency upgrades in the residential, commercial, and industrial sectors. The report concludes that, "in the nation's pursuit of energy affordability, climate change mitigation, and energy security, energy efficiency stands out as perhaps the single most promising resource."[25]

While that may be true, the McKinsey report contains an enormous caveat, a warning that efficiency is not a panacea, not easily funded, and not easily measured or verified:

By their nature, energy efficiency measures typically require a substantial upfront investment in exchange for savings that accrue over the lifetime of the deployed measures. Additionally, efficiency potential is highly fragmented, spread across more than 100 million locations and billions of devices used in residential, commercial, and industrial settings. This dispersion ensures that efficiency is the highest priority for virtually no one. Finally, measuring and verifying energy not consumed is by its nature difficult.[26]

That last sentence should be read again. "Measuring and verifying energy not consumed is by its nature difficult." In other words, knowing whether a given efficiency project has made a difference depends to a great degree on how the measuring is done, what assumptions were used, and what was considered "business as usual."

Though the United States should, and no doubt will, continue pursuing efficiency, efficiency is not a cure-all. We cannot fuel our cars, homes, and airplanes with efficiency; we must supply them with real fuel. And that means digging something out of the ground, manufacturing electricity from hydrocarbons or fission, or wringing juice from the kinetic power of the wind or the radiation from the sun.

Increasing efficiency merely paves the way for greater energy consumption, a situation known as the Jevons Paradox. Named for the world's first energy economist, a Brit named William Stanley Jevons, the paradox has only gained credibility in recent years as increasing numbers of researchers have corroborated Jevons's work, confirming what Jevons said in 1865: "It is wholly a confusion of ideas to suppose that the economical use of fuels is equivalent to a diminished consumption. The very contrary is the truth."[27]

Perhaps the most exhaustive analysis of energy-efficiency efforts and their effects on consumption was done in 2008 in a book called *The Jevons Paradox and the Myth of Resource Efficiency Improvements*. The authors, led by economist John Polimeni, looked at dozens of studies on how energy efficiency affects consumption. Their conclusion: Jevons was right. In 2008, Polimeni told me that understanding the paradox is not difficult. "As you become more efficient you do not have to spend as much to consume the same amount of resources (energy). Thus, you can consume more with the same budget constraint."[28] Put another way, any time you reduce the cost of consuming something (in this case, by increasing the efficiency of a machine, home, or vehicle), then people will respond by consuming more of it. Over time, the gains in efficiency get swamped by the increased consumption that follows each gain. Numerous other analysts have come to the same conclusion as Polimeni.[29]

We can also look at historical trends for evidence of the Jevons Paradox. James Watt's improvements to the steam engine led to huge improvements in energy efficiency, with the immediate result being a sharp drop in coal consumption. Watt continued making improvements in the steam engine until he died in 1819, before he was fully able to appreciate the revolution he helped to ignite.[30] And the dimensions of that change can be seen in the amount of energy that was consumed: Between 1830 and 1863, British coal use increased by about 1,000 percent.[31]

Given that energy efficiency results in increased energy use, it's obvious that, although energy efficiency should be pursued, it cannot be expected to solve the dilemmas posed by the world's ever-growing need for energy. And as the people of the world continue consuming more energy, they will continue releasing lots of carbon dioxide.

That harsh reality leads to my next heresy: The United States should forget about trying to cut carbon dioxide emissions, forget about carbon capture and sequestration, and focus on adapting to the ever-changing global climate.

CHAPTER 15

The United States Can Cut CO$_2$ Emissions by 80 Percent by 2050, and Carbon Capture and Sequestration Can Help Achieve That Goal

WHEN IT COMES to global warming, there are two camps: the believers and the skeptics. Both sides put forth complex studies that discuss topics such as "albedo," "forcings," and "flux" in the effort to prove their claims—for or against—the idea that unrestrained manmade carbon dioxide emissions will lead to global cataclysm.

I've seen Al Gore's movie, *An Inconvenient Truth*, and I've read some of the reports from the Intergovernmental Panel on Climate Change. I've also listened to and read some of the things published by the climate "skeptics," including Richard Lindzen, the climate scientist from Massachusetts Institute of Technology. I have interviewed climate scientists.[1] I regularly read blogs on climate from both camps.

My position on the science of global warming and climate change: I don't know who's right. And I don't really care. What can be demonstrated without any caveats is this: The carbon dioxide reduction targets being advocated by the U.S. Congress and the Democratic leadership in Washington are pure fantasy. In April 2009, President Barack Obama gave a speech to the National Academy of Sciences during which he said that he has "set a goal for our nation that we will reduce our carbon pollution by more than 80% by 2050."[2]

Before showing why that target won't be possible, let me explain my stance on climate change in more detail. As a journalist, I'm skeptical of nearly everything. When my mother told me she loved me, I double-checked it with my dad. And now, with legions of greens, politicos, and pundits all parroting the same message about the dangers of global warming, my reflexive skepticism only increases.

My skepticism about the conventional wisdom on global warming arises from two main points. First, I adhere to one of the oldest maxims in science: Correlation does not prove causation. Carbon dioxide levels in the atmosphere may be increasing, but that does not necessarily prove that the carbon dioxide is causing any warming that may be occurring. Second, models are only as good as the data going into them. All of the alarm bells now being sounded are based on atmospheric and climatic models about how temperatures in the future are expected to react, given the data fed into the models. But as Vaclav Smil put it in his 2008 book, *Global Catastrophes and Trends*, "even our most complex models are only elaborate speculations. We may get some particulars right, but it is beyond us to have any realistic appreciation of what a world with average temperature 2.5° C warmer than today's would be like."[3]

There's no question that carbon dioxide plays a significant role in the atmosphere. Just how significant, we don't know. And it's just as obvious that there's a huge amount of political pressure to pass regulations that limit carbon dioxide emissions. In June 2009, the U.S. House of Representatives narrowly passed a 1,400-page bill on a cap-and-trade plan designed to do just that.[4]

Whatever. For me, in many ways, the science no longer matters because discussions about the science have become so vituperative and politicized.[5] Thus, my position about the science of global climate change is one of resolute agnosticism. When it comes to climate change, the key issues are no longer about forcings, albedo, or the ideal concentration of atmospheric carbon dioxide. Instead, the key question is about policy, namely: If we are going to agree that carbon dioxide is bad, what are we supposed to do? And that question—as the Duke of Bilgewater memorably put it in *Huckleberry Finn*—"is the bare bodkin."[6]

Al Gore, James Hansen, and dozens of other people who are sounding the alarm about global climate change are not offering any viable alternatives to our existing energy sources. Gore says we should transition to "carbon-free" electricity within a decade. Hansen says we should shut down all coal-fired power plants. But those are glib responses to a set of enormously complex problems. It's far easier to say "stop using coal" than it is to actually follow that course of action.

The key question, and the one that precious few are willing to discuss in depth—If we are going to agree that carbon dioxide is bad, then what?—leads directly to two more:

- Where are the substitutes for hydrocarbons? Hydrocarbons now provide about 88 percent of the world's total energy needs.[7] Replacing them means coming up with an energy form that can supply 200 million barrels of oil equivalent per day.
- Increasing energy consumption equals higher living standards. Always. Everywhere. Given that last fact, how can we expect the people of the world—all 6.7 billion of them—to use less energy?[8] The answer to that question is obvious: We can't.

The developed countries of the world can talk all they like about solar panels and wind turbines, but what the world's poor desperately need—and quickly—are common fuels such as kerosene, propane, and gasoline. And, of course, they want reliable electricity. The people in the industrialized countries cannot and should not hinder the efforts of the world's poor to gain access to cheap, reliable sources of energy. Sure, solar panels and windmills are appropriate for some locations, and in many cases, they may be the best choices. But it's also true that the cheapest and most reliable forms of energy, in most cases, are hydrocarbons.

That's why the world's most authoritative forecasters expect the developing world to surpass the United States and the European Union in carbon dioxide emissions. When you consider the huge numbers of people

FIGURE 25 Total Carbon Dioxide Emissions, OECD Countries
Versus Non-OECD Countries, 1990 to 2030

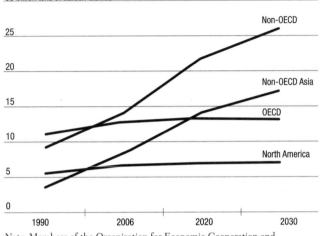

Note: Members of the Organization for Economic Cooperation and
Development include Australia, Austria, Belgium, Canada, Czech Republic,
Denmark, Finland, France, Germany, Greece, Hungary, Iceland, Ireland, Italy,
Japan, South Korea, Luxembourg, Mexico, Netherlands, New Zealand,
Norway, Poland, Portugal, Slovak Republic, Spain, Sweden, Switzerland,
Turkey, United Kingdom, United States (OECD.org , "OECD Member
Countries," http://www.oecd.org/document/58/0,3343,en_2649_201185
_1889402_1_1_1_1,00.html).
Source: International Energy Agency, World Energy Outlook 2008, Table 16.2,
"Energy-Related CO_2 Emissions by Region in the Reference Scenario."

who live in the developing world, it's apparent that they *should* surpass
us. In 2008, the International Energy Agency estimated that the carbon
dioxide emissions of the developed countries of the world—that is, the
members of the Organization for Economic Cooperation and Development (OECD)—were surpassed by those of the non-OECD countries
in 2006. And by 2030, the agency expects that carbon dioxide emissions
from the non-OECD countries will be nearly double those of the
OECD countries.[9]

Those numbers should not be surprising. A bit of arithmetic can cor-
roborate the estimates by simply looking at the world's six most populous
countries. They are, in descending order of population, China, India, the
United States, Indonesia, Brazil, and Pakistan.

The energy disparity among the residents of the Big Six is stark. The United States, with about 300 million residents, consumes almost as much energy as the other five most populous countries, the Big Five, combined. The total population of the Big Five—Brazil, China, India, Indonesia, and Pakistan—is about 3 billion, and between 2000 and 2006, that population increased its energy use by more than 50 percent. Despite that huge increase, the average resident of the Big Five uses about one-tenth as much energy as the average American.[10] Of course, many Americans believe the United States uses too much energy. And various environmental groups suggest that the United States should simply "use less."

But the citizens of the United States could not stop the ongoing rise in carbon dioxide emissions even if they all reached a consensus, because so many people are still living in energy poverty. Furthermore, the industrialized countries in general, and the United States in particular, have no moral standing to tell the developing countries that they should slow the growth of their own energy consumption.

Bringing hundreds of millions of people out of poverty—and thus, into higher standards of living—means providing them with access to cheap, plentiful energy. And like it or not, that largely means hydrocarbons. Of course, increasing the use of hydrocarbons will mean further increases in carbon dioxide emissions. The leaders of the developing world understand this, and they have very clearly stated that they are not going to quit using hydrocarbons.

In June 2009, just a few days after the House of Representatives passed its cap-and-trade bill, Indian Environment Minister Jairam Ramesh said that India "will not accept any emission-reduction target— period. This is a non-negotiable stand." The Indian leader was very clear, saying that "there is no way India is going to accept any emission reduction target, period, between now and the Copenhagen meeting [held from December 7–18, 2009] and thereafter."[11] Chinese officials have made similar statements.

One of the best arguments against any effort to cut carbon dioxide emissions from the use of hydrocarbons comes from Freeman Dyson, a renowned professor of physics at the Institute for Advanced Study at Princeton University. In August 2007, Dyson wrote an essay for Edge.org that made me reconsider my own thinking on energy use and climate issues. Dyson quickly conveyed his skepticism about the models being used by climate scientists:

> My first heresy says that all the fuss about global warming is grossly exaggerated. Here I am opposing the holy brotherhood of climate model experts and the crowd of deluded citizens who believe the numbers predicted by the computer models. Of course, they say, I have no degree in meteorology and I am therefore not qualified to speak. But I have studied the climate models and I know what they can do. The models solve the equations of fluid dynamics, and they do a very good job of describing the fluid motions of the atmosphere and the oceans. They do a very poor job of describing the clouds, the dust, the chemistry and the biology of fields and farms and forests. They do not begin to describe the real world that we live in. The real world is muddy and messy and full of things that we do not yet understand.[12]

But the essence of Dyson's essay isn't about the computer models. Instead, it's about equity and human development. And that's where Dyson points to the need for clean, cheap, abundant energy. "The greatest evils are poverty, underdevelopment, unemployment, disease and hunger, all the conditions that deprive people of opportunities and limit their freedoms," he wrote. "The humanist ethic accepts an increase of carbon dioxide in the atmosphere as a small price to pay, if world-wide industrial development can alleviate the miseries of the poorer half of humanity."[13]

To that, I say amen.

The hard truth is that the people of the world are going to have to adapt to a changing planet—regardless of the causes of those climatic changes. If the climate gets colder, hotter, wetter, or drier, we're simply going to have to figure out how to cope, because the billions of people now living in poverty desperately want to improve their standards of living. And one of the cheapest, fastest ways for them to do that is by burning hydrocarbons. What will adaptation to the changing climate mean? Well, for one thing, it may mean relocating large swaths of the population away from areas most affected by the symptoms of global warming. For example, those living in coastal cities may have to move further inland, while those in desert cities may have to go to wetter regions.

In short, I don't side with the "climate alarmists" like Al Gore, who famously—and nonsensically—declared in his movie *An Inconvenient Truth* that "you can even reduce your carbon emissions to zero."[14] Nor do I side with the "climate skeptics" who maintain that nothing is happening. Instead, I consider myself a realist, a pragmatist who has done the math on carbon dioxide emissions and understands that no matter what course of action the United States takes—short of completely shutting down its economy and consigning the vast majority of its citizens to the drudgery of scratching out an existence with something approximating 40 acres and a mule—it cannot, and will not, make a significant difference to concentrations of atmospheric carbon dioxide.

And that's the key issue. Over the past few decades, nearly the entire discussion about global warming and climate change has focused on the desire to reduce carbon dioxide emissions. But for all the talk about reducing carbon emissions, the reality is starkly obvious: Those efforts have failed miserably.

The Kyoto Protocol was supposed to address global climate change. Adopted in 1997, the agreement took force in 2005. By 2009, some 188 countries had signed the agreement.[15] The United States was not among them. In the late 1990s, the U.S. Senate voted 94 to 2 against ratifying the agreement.[16] How well has Kyoto worked? By 2012, it is expected that just 6 of the signatories will have achieved their goal; emissions in the other 182 will likely still be well above target levels. Among the countries that have failed to achieve the targets, for example, is Japan. By late 2007, Japan's carbon dioxide emissions were 14 percent above the Kyoto target.[17] By early 2009, with the country still far above the target level, the Japanese had begun buying carbon offsets.[18] The inability of Japan—which has high population density, a homogeneous society, and an extremely energy-efficient economy—to achieve the carbon dioxide reductions outlined in Kyoto may be the single best indicator of the impracticality of the proposed carbon-reduction schemes. And Japan's failure to achieve its carbon reduction targets provides a stark warning about the ability of the United States to achieve an 80 percent reduction by 2050.

The problem, once again, can be shown by doing basic math. Here are the numbers: In 2006, the United States emitted about 5.9 billion tons of carbon dioxide.[19] That means that the average American is responsible for the production of about 20 tons of carbon dioxide annually. (The U.S. population is about 307 million.)[20] An 80 percent reduction in U.S. emissions would mean that the United States would only be allowed to emit about 1.2 billion tons of carbon dioxide by 2050. That level of emissions would take the United States back to the levels achieved in 1910, when the country's factories and households emitted about 1 billion tons of carbon dioxide per year.[21] But in 1910, the United States only had about 92 million people, and the per-capita income (in 2009 dollars) was about $6,000. Today, the population is more than three times greater than it was a century ago, and thankfully, per-capita incomes have jumped to more than $39,000.[22] By 2050, the United States will likely have about 439 million people.[23] At that population level—and remember, total emissions are to be no more than 1.2 billion tons per year—per-capita carbon dioxide emissions would have to be about 2.7 tons per year, or about one-seventh of current per-capita emission levels of about 20 tons per year.

Which countries are close to achieving Obama's implied per-capita target of 2.7 tons of carbon dioxide per year? In 2006, countries that had emissions in that range included Cuba (2.36 tons), North Korea (3.18), and Syria (2.65). In 2006, global average per-capita carbon dioxide emissions amounted to 4.28 tons per year—or about 50 percent higher than what the United States is aiming to achieve by 2050.[24]

Under the target identified by Obama and congressional Democrats, by 2050, U.S. per-capita emissions would have to be far lower than those of current-day China, where each citizen now emits the world average of about 4.27 tons of carbon dioxide per year.[25] As Steven Hayward, a fellow at the American Enterprise Institute, pointed out in a 2008 op-ed piece in the *Wall Street Journal*, the "enthusiasm for an 80% reduction target is often justified on grounds that national policy should set an ambitious goal."[26] But as Hayward noted, there's a difference between an ambitious goal and an absurd one. The chances of the United States actually achieving an 80 percent cut in carbon dioxide output by 2050 ranges somewhere between slim and none. And, as my father used to say, "Slim left town."

FIGURE 26 Per-Capita Carbon Dioxide Emissions in the United States and Other Countries, 2006, with Implied Projection for U.S. Emissions in 2050

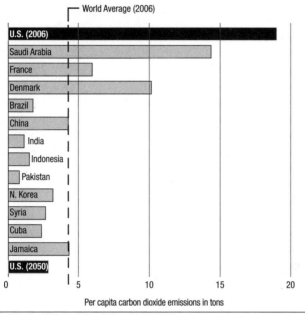

Per capita carbon dioxide emissions in tons

* Assumes U.S. population of 420 million by 2050.

Source: International Energy Agency, "Key World Energy Statistics 2008," http://www.iea.org/textbase/nppdf/free/2008/key_stats_2008.pdf, 49–57.

But just for the sake of discussion, let's run the numbers one more time, and let's do it in a way that is favorable to the 80 percent reduction target. Let's assume that the U.S. population doesn't grow at all over the next four decades, so that by 2050, there are still about 307 million Americans. Again, the math is straightforward: With 1.2 billion tons of total annual carbon dioxide emissions, divided among 307 million people, the United States would have per-capita emissions of about 4 tons. That's fine. But it's still less than the world average per-capita carbon dioxide emissions in 2006. Indeed, it's nearly 10 percent below the level in Jamaica, a country where per-capita GDP in 2006 was about $7,500.[27]

The reality is that, for all the talk about drastic cuts in carbon dioxide, the United States cannot—and it will not—make any radical reductions in its carbon output. And the reasons it won't are obvious: There are no cheap—and that is the most essential qualifier—viable technologies that will allow it to do so; Americans are not willing to change their lifestyles to make it happen; and any government-mandated restrictions on hydrocarbon use that would be severe enough to achieve the 80 percent reduction target would almost certainly ruin the economy.

The problem of how to make such drastic cuts in global carbon dioxide emissions has led some of the world's leading thinkers on climate and science to question the conventional wisdom. In 2006, Roger Pielke Jr., a professor in the environmental studies program at the University of Colorado, made a concise statement about climate change.[28] In testimony before the House Committee on Government Reform, Pielke said: "Even if society takes immediate and drastic action on emissions, there can be no scientifically valid argument that such actions will lead to a perceptibly better climate in the coming decades. For the foreseeable future the most effective policy responses to climate-related impacts (e.g., such as hurricanes and other disasters or diseases such as malaria) will necessarily be adaptive."

Pielke went on, making it clear that he wasn't arguing that the United States should ignore the problem: "The point of this analysis is not to throw up our hands and do nothing about mitigation," he said. But "if meaningful action is to occur on mitigation we must think about different strategies, and in particular policy options that have more symmetry between the timing of costs and benefits."[29]

Other analysts are coming to the same conclusion. In February 2009, Britain's Institution of Mechanical Engineers issued a report, "Climate Change: Adapting to the Inevitable?" which pointed out that the Kyoto Protocol has been "a near total failure with emissions levels continuing to rise substantially."[30]

Given the continuing use of hydrocarbons, and carbon dioxide levels, the British engineering group, which has some 75,000 members in 120 countries, determined that adaptation to changing weather patterns will be an essential strategy for the countries of the world.[31] The report said, "We are unlikely to be far more successful at curbing our carbon dioxide emissions in the near future than we have been over the past decade or so. And even with vigorous mitigation effort, we will continue to use fossil fuel reserves until they are exhausted."[32] Instead of focusing on emissions, governments should "invest significantly more effort in adaptation for the long term, which enables each nation to undertake the necessary steps to ensure its future prosperity and survival." It went on to say that although efforts to curb emissions are "vital," those efforts have "questionable" efficacy. Therefore, "It is a duty of government to embrace adaptation and protect the nation against the potential risks of a 'business as usual' outcome."[33]

When the report was released, the institution's point person on climate change issues, Tim Fox, said that too much time was being spent considering how to reduce carbon emissions and too little on how to cope with climatic changes. "But by researching and developing adaptation strategies we have a chance to cope with what is around the corner," he explained.[34]

About the same time that the Institution of Mechanical Engineers released its report, the National Academy of Sciences released a similar report that came to similar conclusions: "We must consider how climate change research should evolve in the United States. A federal science program is needed to comprehend the nature and extent of the climate change threat, to quantify the magnitude of the impacts, and to provide a data and knowledge foundation for identifying effective adaptation and mitigation options."[35]

My thinking about the science of climate change is not that we should do nothing with regard to carbon dioxide emissions. The United States, even in the absence of any Kyoto-style mandates to reduce carbon emissions, has made substantial reductions in its carbon dioxide emissions. And global carbon emissions are getting a big nudge downward thanks to the sorry state of the U.S. and world economies. In August 2009, the Energy Information Administration predicted that 2009 emissions would fall by about 6 percent from 2008. And that follows a drop of about 3 percent in 2008.[36]

Those are sizable reductions, but they are still light-years away from the 80 percent reduction by 2050 that Obama claims should be a U.S. goal. The sad but true state of carbon politics is that the only realistic way to achieve that goal would be for the United States to intentionally destroy its economy and the jobs that go with it. Of course, no savvy politician will say that the goal should be a weak economy. So politicos and environmentalists are advocating another approach to the carbon dioxide emissions issue, one that will allow the United States to continue using large amounts of coal while also making huge volumes of carbon dioxide simply disappear.

But as I mentioned in the introduction, we must remember Jim Collins's admonition that "facts are better than dreams." And when it comes to carbon capture and sequestration, Americans are hearing lots of dreams and precious few facts.

▢▢▢

Carbon capture and sequestration (CCS) is the Holy Grail of carbon strategies. Given the right technologies, CCS will be a key part of the solution to the world's ever-increasing quantities of carbon dioxide emissions, or at least that's what the promoters have been telling us. And there are plenty of promoters. In 2008, David Hawkins, the head of the climate change program at the Natural Resources Defense Council, said that "burying billions of tons of carbon dioxide is a huge job, but that is not necessarily an argument against CCS. You can't solve a big problem without a big effort."[37]

In February 2009, Fred Krupp, the head of the Environmental Defense Fund, one of America's biggest environmental groups (the group's annual budget is about $100 million), said that carbon capture provides "a future for coal" and that "it will be an important technology for reducing carbon dioxide emissions because it offers the possibility of retrofitting some of the existing power plants."[38] At about that same time, a research engineer at the Laboratory for Energy and the Environment at the Massachusetts Institute of Technology, Howard Herzog, said that, given the fact that coal-fired power plants will be around for a long time, CCS is "the only real alternative": "Therefore, we have to make CCS work."[39]

Herzog was repeating some of the findings of a 2007 report done at MIT called "The Future of Coal: Options for a Carbon-Constrained World," which concluded that coal was going to be part of the world's generation mix for a long time to come. "Coal use will increase under any foreseeable scenario because it is cheap and abundant," said the authors of the interdisciplinary study. "Coal can provide usable energy at a cost of between $1 and $2 per MMBtu [million Btu] compared to $6 to $12 per MMBtu for oil and natural gas."[40] The report also claimed that CCS was "the critical enabling technology that would reduce carbon dioxide emissions significantly while also allowing coal to meet the world's pressing energy needs.[41]

In March 2009, U.S. Secretary of the Interior Ken Salazar told the *New York Times* that his agency was investigating the potential for using public lands for carbon capture and storage. The U.S. Geological Survey, he said, had recently produced a report designed to help find the best areas for CCS.[42]

In May 2009, Nobuo Tanaka, the executive director of the International Energy Agency, called CCS a "vital" technology for greenhouse gas control "that will be needed to make power generation and heavy industry sustainable."[43] About that same time, U.S. Secretary of Energy Steven Chu gave a speech in Rome in which he declared that "we need to capture the carbon" and sequester the emissions "safely," adding, "and we have to do this in an economically viable way."[44]

In August 2009, David Sandalow, an assistant secretary of energy, testifying before the Senate, declared that "it is technically feasible, through retrofitting and new construction, to ensure that the entire US coal fleet employs CCS by 2035."[45]

Note that Sandalow said CCS is "technically feasible." He didn't say it made economic sense. And that's the rub: No one knows how to do CCS in an economically viable way. The idea of CCS is simple: Capture carbon dioxide from the flue of a power plant and then inject that gas (after it has been compressed and cooled) into a geologic formation. And while the idea is simple, making it into a reality on a large scale is extremely tricky. That point was made clear by the Congressional Research Service in December 2008: "Developing technology to capture CO_2 in an environmentally, economically, and operationally acceptable manner—especially from coal-fired power plants—has been an ongoing interest of the federal government for a decade. Nonetheless, the technology on the whole is still under development: no commercial device is currently available to capture carbon from coal plants."[46]

That's not to say there's no money available. Shortly before Chu gave his speech in Rome, the U.S. Department of Energy announced that it was going to provide $2.4 billion in funding for CCS projects. That money was made available from the massive stimulus package passed by Congress known as the American Recovery and Reinvestment Act.[47] The European Union is also throwing huge amounts of money at CCS. In mid-2009, the European Commission announced that it was providing about $1.4 billion to thirteen CCS projects across Europe. Europe believes it will be able to collect large amounts of money to support additional CCS projects through the auctioning of carbon credits.[48]

But here's the reality: No matter how much money the United States and the European Union throw at CCS, it won't work. The volumes of carbon dioxide are too large, and the technical problems—and therefore the costs—associated with sequestering that much gas are just too big.

Only a handful of energy projects are now using CCS. The most familiar of those is one being done by the Norwegian oil giant, StatoilHydro, in the North Sea at the company's Sleipner platform. Since 1996, StatoilHydro has been capturing, compressing, and reinjecting about 1 million tons of carbon dioxide per year into a formation under the ocean.[49] (The carbon dioxide is stripped out of the natural gas the company is producing at the site.) But Sleipner is tiny compared to overall global emissions. And as author and journalist Jeff Goodell pointed out in a 2008 article on CCS, it would take ten Sleipner-sized projects to offset the emissions from just one large coal-fired power plant.[50]

The environmental group Greenpeace raises a number of valid points about the problems of CCS, including the length of time needed to deploy it, the additional energy required to fuel the CCS process, the viability of underground storage, cost issues, and long-term liability questions.[51]

Greenpeace's criticisms are valid, and they are all essentially related to the problem of scale. The volume of global carbon dioxide emissions is staggering. In 2006, global carbon dioxide emissions totaled 29.1 billion tons.[52] Let's assume that policymakers mandate a program requiring the annual sequestration of 10 percent—about 3 billion tons—of global carbon dioxide emissions. That figure is a reasonable starting point, and it's equal to about one-half of U.S. carbon dioxide emissions, which totaled 5.9 billion tons in 2006.[53]

But how can we get our minds around that figure? Three billion tons is a difficult number to comprehend, especially when it represents something that is widely dispersed the way carbon emissions are in the atmosphere. According to calculations done by Vaclav Smil, if that amount of carbon dioxide (remember, it's just 10 percent of global annual carbon dioxide emissions) were compressed to about 1,000 pounds per square inch, it would have about the same volume as the total volume of global annual oil production.[54]

In other words, the volume of that highly pressurized gas would be approximately equal to the volume of *all* of the oil produced around the world in a year. Of course, since no one has ever seen all the world's annual oil production stacked up in one spot, that amount of material is still too large to be understandable. We can bring it into sharper focus by cutting it down to daily figures.

In 2008, global oil production was about 82 million barrels per day.[55] Thus, 10 percent of global carbon dioxide emissions in one day would be approximately equal to the daily volume of global oil production. So here's the punch line: Getting rid of just 10 percent of global carbon dioxide per day would mean filling the equivalent of forty-one VLCC supertankers every day. (Each VLCC, or very large crude carrier, holds about 2 million barrels.)[56] Given that huge volume of carbon dioxide, the immediate questions are obvious: Where will we put it? And how will we put it there?

Even if some locations were found that could swallow that much material, the cost of handling it would be enormous. Smil emphasized the tremendous difficulty of "putting in place an industry that would have to force underground every year the volume of compressed gas larger than or (with higher compression) equal to the volume of crude oil extracted globally by [the] petroleum industry whose infrastructures and capacities have been put in place over a century of development." "Such a technical feat," he said, "could not be accomplished within a single generation."[57] Even if we could somehow accomplish that technical feat, CCS presents another technical challenge that is equally daunting: How do you capture all those carbon dioxide molecules in the first place in a way that is cost-effective? Most CCS projects envision a system where the carbon dioxide gets captured from the flue gas of an electric power plant. But capturing that carbon dioxide is an energy-intensive process. Analysts estimate that capturing that gas adds "parasitic load" of up to 28

FIGURE 27 Carbon Capture and Sequestration: The Forty-One-Supertanker-Per-Day Challenge

Sequestering 10 percent of daily global carbon dioxide emissions would result in a volume of gas approximately equivalent to the volume of global oil production. In 2008, world oil production averaged about 82 million barrels per day. Thus, sequestering 10 percent of the world's daily carbon dioxide emissions (assuming it is fully compressed) would require handling enough gas to fill 41 VLCC supertankers per day.

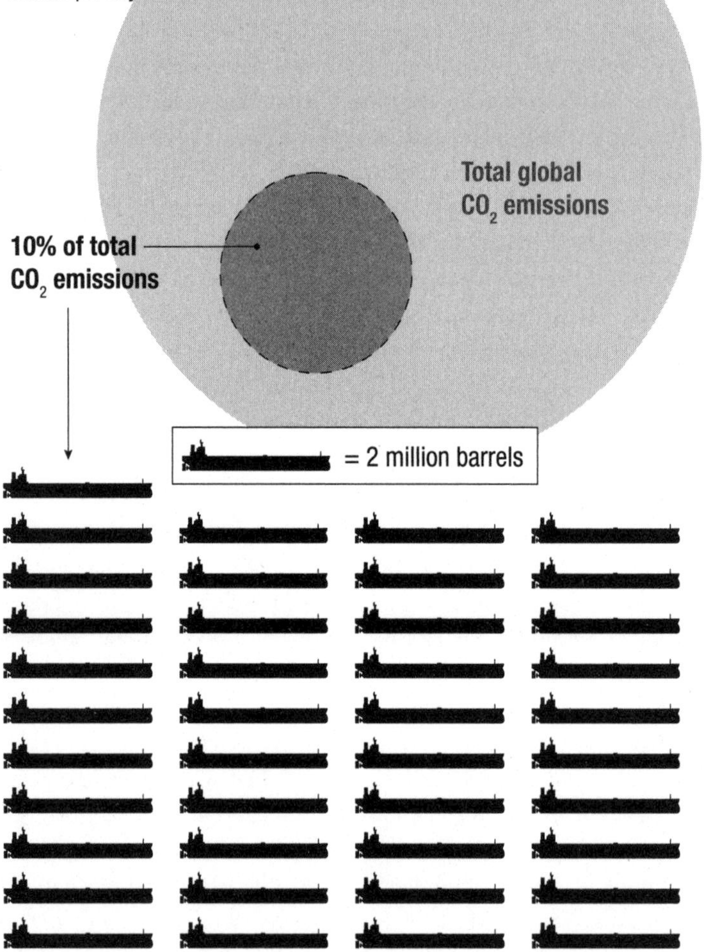

Total global CO_2 emissions

10% of total CO_2 emissions

= 2 million barrels

Sources: Vaclav Smil, "Energy at the Crossroads: Background Notes for a Presentation at the Global Science Forum Conference on Scientific Challenges for Energy Research, Paris," May 17–18, 2006, http://home.cc.umanitoba.ca/~vsmil/pdf_pubs/oecd.pdf, 21; author calculations.

percent to the power plant.[58] Parasitic load is additional power that must be generated at the site but never gets sold to consumers. Meeting such additional loads would be a Herculean effort, regardless of whether the fuel in question is natural gas or coal. Given that the global energy sector is already straining to meet growing demand, the idea of increasing electricity production—or reducing electricity output—by more than one-fourth, for the sole purpose of attempting CCS, is dubious at best.

Even if the technology were available and cost-effective, it faces significant public opposition. In Germany and Denmark, opposition to CCS projects has helped stop plans by two major electricity producers to bury captured carbon dioxide in saline aquifers deep below the surface of the Earth.[59]

Despite the problems, the coal industry continues to perpetuate the myth that CCS is viable. In September 2009, Vic Svec, a senior vice president at St. Louis–based Peabody Energy, the world's largest private-sector coal company, told the *New York Times* that "coal with carbon capture and storage is the low cost, low carbon solution and has fantastic implications for the nation's energy security."[60]

The myth that the United States can (or will) substantially cut its carbon dioxide emissions is closely related to the belief that placing a tax on carbon will ignite a revolution in alternative-energy technologies. And that belief leads to the next myth-busting opportunity: Any scheme to tax carbon is doomed to failure. The better strategy: taxing neurotoxins.

CHAPTER 16

Taxing Carbon Dioxide Will Work

FORGET IT. No matter how many times world leaders meet in places such as Rio, Kyoto, Copenhagen, Mexico City, or Tulsa, the countries of the world will never agree on a global scheme to tax carbon dioxide. The disparity between the wealthy countries and the developing countries—particularly when it comes to the availability of electricity—is simply too great to expect that the developing countries, such as China, India, and Indonesia, will agree to policies that will effectively restrain their economic growth.

Policymakers should forget about attempting to tax carbon dioxide or limit emissions of that gas. Instead, they should aggressively pursue taxes or caps on the emissions of neurotoxins, particularly those that come from burning coal. The rationale here is simple: Many scientists and policymakers can, and will, argue about the relative dangers of carbon dioxide emissions, but no one in their right mind is willing to stand up and say, "We need more mercury and lead in our ecosystems."

Mercury and lead—which are released in significant quantities during coal combustion—are among the most dangerous neurotoxins.[1] U.S. emissions from coal-fired power plants total some 48 tons of mercury and 88 tons of lead per year. Global emissions of those metals from coal-fired plants are several multiples of those quantities. And those global emissions of neurotoxins—from coal-fired power plants, cement kilns, and other industrial facilities—are leading to global contamination of the environment with heavy metals. No one knows just how damaging that contamination will be over the long term. But there are plenty of reasons to be concerned.

In August 2009, the U.S. Geological Survey released a report that analyzed hundreds of fish, from 34 species, caught in 291 streams around the United States. Every one of the fish sampled by the Geological Survey's scientists contained mercury.[2] The agency also found that concentrations of mercury in "about a quarter of the fish sampled exceeded the criterion for the protection of humans who consume average amounts of fish."[3] The seven-year-long study placed most of the blame for the elevated mercury levels on "atmospheric deposition"—that is, airborne mercury particles produced by industrial plants that have been deposited on the Earth's surface.[4] After the airborne mercury hits the surface, microorganisms turn it into its organic form, methylmercury, which then builds up in fish, shellfish, and animals that consume the fish. The highest levels of mercury were found in fish caught in remote coastal-plain streams in the eastern and southern United States.

Of course, coal-fired power plants are not the only sources of airborne mercury contamination. Gold mining and mercury mining also are contributors.[5] That said, coal-fired power generation has been repeatedly identified as perhaps the single most dangerous spreader of heavy metals. The Environmental Protection Agency (EPA) estimates that coal-fired power plants account "for over 40% of all domestic human-caused mercury emissions." Furthermore, the EPA says, "about one quarter of US emissions from coal-burning power plants are deposited within the contiguous US and the remainder enters the global cycle."[6] In addition, the ash that comes out of coal-fired power plants is usually contaminated with heavy metals.

It's the global nature of neurotoxins such as mercury and lead that make them so worrisome and so difficult to control. Scientists have estimated that about 30 percent of the mercury that settles onto the ground in the United States comes from other countries. And of those other countries, China is the most problematic. Every year, China spews some 600 tons of mercury into the air—and the majority of that volume comes from the country's 2,000 coal-fired power plants.[7] Scientists in

Oregon have estimated that about 20 percent of the mercury that enters the Willamette River comes from overseas, and some of that, no doubt, is from China.[8]

Mercury, lead, and other heavy metals are accumulating in the bones and tissues of humans, and we do not have a full understanding of how those accumulations may be affecting our long-term health. Slow poisoning from neurotoxins has a long, unfortunate history. The Romans used lead cooking vessels that contaminated their food. We make jokes about the "mad hatter," but few people understand that the hatters of yesteryear were often driven into madness by acute mercury poisoning, a condition caused by using mercury in the finishing of their hats.[9] Or consider the case of cadmium, another neurotoxin.[10] Cadmium concentrations in the bones of modern-day Americans are about fifty times as high as those found in bones of American Indians who roamed the deserts of the American Southwest some six hundred years ago. These increasing concentrations of heavy metals in the human body could help explain why certain populations of people are more violent than others.[11]

The electric utility industry has technologies that can reduce their emissions of neurotoxins. Of course, deploying those technologies on a global basis would undoubtedly cost tens of billions of dollars. And that deployment would raise the cost of electricity for all of the consumers who rely on coal-fired electric power. That said, the benefits to the environment— for both people and wildlife—would be measurable and almost certainly positive.

The need to reduce mercury exposure is becoming ever more obvious. In mid-2009, a study by Dan Laks, a neuroscience researcher at the University of California, Los Angeles, found that one-third of all U.S. women now have inorganic mercury in their blood and that those mercury levels have been rising over the past few years. After the study was released, Laks said that the "results suggest that chronic mercury exposure has reached a critical level where inorganic mercury deposition within the human body is accumulating over time. . . . It is logical to assume that the risks of associated neurodevelopmental and neurodegenerative diseases will rise as well."[12]

A truly motivated "green" movement would push for a global effort to cut the emissions of neurotoxins. Achieving those cuts will take enormous political will, and it will take time. But arguing against heavy metal contaminants with known neurotoxicity will be far easier than arguing against carbon dioxide emissions. Cutting the output of mercury and the other heavy metals may, in the long run, turn out to have far greater benefits for the environment and human health.

And that brings us to a critical point: Most, if not all, of energy policy should be aimed at improving human health, and, generally speaking, that means working toward improvements in the environment. Thus, rules that have forced heavy industry to clean up the emissions they send out of their smokestacks have resulted in cleaner air, a move that has benefited the entire population. Back in the 1970s, the U.S. government ordered refiners to remove lead from their gasoline. The result: dramatic reductions in the amount of lead that was released into the environment.

Obviously, cutting air pollution and reducing the volume of neurotoxins makes sense. But what other moves would result in dramatic improvements to both human health and the environment? Though it may be counterintuitive, one of the best moves would be to promote the increased use of oil and liquified petroleum gas, particularly among the rural poor.

CHAPTER 17

Oil Is Dirty

YOU'VE HEARD IT dozens of times: Oil is a dirty fuel. Oil is polluting the air and the water, and it is despoiling the planet.

Here's the reality: The world isn't using too much oil. It's not using enough.

Whether the issue is gorillas in the Democratic Republic of the Congo, tropical rainforests in Sumatra, or the health and welfare of women and young children in the developing world, the solution is abundantly clear: The world needs more oil consumption, not less. While that may sound like a statement designed to cheer the hearts of Houston-based energy tycoons, the easily provable truth is that hundreds of millions of people desperately need better fuels and stoves to help them cook their food. Simply put, they need more power in their kitchens.

Oil should be seen as an essential ingredient in the effort to save the world's most endangered animals as well as huge swaths of tropical forests. More oil consumption among the world's energy poor would help save the lives of hundreds of thousands of impoverished people every year who die premature deaths because of indoor air pollution caused by burning biomass.[1]

The issue, once again, is one of density. The world's most impoverished people have no choice but to cook their food and heat their homes with fuels that have low energy density, such as straw, dung, twigs, wood, and leaves. They are denuding the landscape of biomass in their struggle to survive. But in doing so, they are also contributing to deforestation and to the problem of airborne soot, which is a major contributor to global climate change. Furthermore, those same low-density fuels, which are used in low-power cooking stoves, are sickening—and killing—thousands of people per day.

Clean-burning, high-energy-density liquid petroleum gas, such as propane and butane, as well as kerosene and gasoline, are the best solution to these problems. I do not make this heretical claim—that we should be using more oil—lightly. Nor am I denying oil's many deleterious effects on the environment. Over the past century, the oil industry has had countless spills, both offshore and onshore, that have caused serious damage. Burning refined oil products pollutes the air. Oil refining is a gritty, dangerous, capital-intensive business. Accidents at refineries, pipelines, and drilling rigs have left many people hurt, and many others dead. Sabotage and carelessness have led to messy land contamination problems, such as the ones related to Texaco's shoddy production practices in Ecuador when it operated in that country from the 1960s to the early 1990s.[2] I've personally seen—and written about—dozens of examples of lousy environmental practices in the oil fields of Texas and Oklahoma.[3]

Nor am I denying the corruption that comes with oil. Petroleum has spawned kleptocracies all over the world. Most of the Middle Eastern petrostates are run by corrupt royal families who have effectively been stealing the mineral wealth of their countrymen for decades. Or consider the corruption in Equatorial Guinea, a tiny, oil-rich country in West Africa that has been ruthlessly ruled by Teodoro Obiang since the 1970s. In 2009, Transparency International named Obiang's country the twelfth most corrupt locale in the world, ranking it 168th in the Corruption Perceptions Index, a rank that puts it in a tie with another petrostate, Iran. Two other petrostates, Chad (175th) and Iraq (176th), also made the list of the most corrupt places.[4]

The current poster boy among the petro-kleptocrats: Teodoro Nguema Obiang, the son of Equatorial Guinea's president. In November 2009, the *New York Times* reported that the younger Obiang, who serves as the forest and agriculture minister in his native country, owns a $35 million estate in Malibu as well as a fleet of luxury cars and a private jet. A U.S. Justice Department memorandum says that much of Obiang's wealth has come "from extortion, theft of public funds, or other corrupt conduct." The *Times* reported that during one twelve-month period ending in April 2006, Obiang "funneled at least $73 million into the United States, using shell corporations and offshore bank accounts to launder the money and ultimately buy his Malibu estate and a luxury jet."[5] Obviously, the environmental and societal ills caused by petroleum cannot be denied. Oil is not a perfect fuel. There is no such thing. But oil is—in nearly every case—greener than any of the alternative energy forms that might replace it. No matter whether the replacement is ethanol from corn, biomass—such as wood, straw, or dung—or biofuels made from palm oil or other feedstocks, the conclusion is apparent: Oil (and if you can get it, natural gas) simply has no peers. Oil provides consumers with both high energy density and high power density. It burns cleanly. It's easily handled at atmospheric temperature and pressure, and the number of uses for it are essentially limitless.

My conversion to the idea that oil is green occurred in late 2007 when I was watching a *60 Minutes* report on the death of ten mountain gorillas in Virunga National Park in Congo. The mountain gorillas are among the most threatened species on the planet. After providing a few details about the deaths of the apes, reporter Anderson Cooper got to the motive. The gorillas hadn't been killed by poachers eager for a trophy, or even for food. They were killed because of a shortage of energy. Looking at the camera, Cooper held up a handful of black flakes and explained that the gorillas were killed "for this: charcoal."[6]

The problem, he said, was that "more than a million people in this area, practically everyone, use charcoal to cook their food. It's made by burning the trees in the gorillas' forest." And the gorillas in Virunga were in the way of the charcoal makers who were supplying fuel to the local population. Robert Muir of the Frankfurt Zoological Society told Cooper that solving the problem in Virunga was simple: "Provide alternative fuel, butane, for example." With a few thousand butane stoves and enough fuel to keep them running, Muir explained, the charcoal makers could be put out of business almost immediately.[7]

In short, the easiest and fastest way to preserve one of the world's most endangered species was not to forge more treaties banning trade in their skins, to create new regulations, or to fund more police to patrol the park. Instead, the simplest solution was to increase the use of oil products.

Only about seven hundred mountain gorillas remain on the planet— about half live in Virunga, one of the most important parks in Africa. The United Nations Educational, Scientific and Cultural Organization (UNESCO) has designated Virunga as a World Heritage Site, saying that it has "the highest biological diversity of any national park in Africa." The park straddles the backbone of the Albertine Rift, a region renowned for its biological diversity and richness. Extending through parts of Uganda, Rwanda, Congo, Burundi, and Tanzania, the Albertine Rift contains habitats ranging from active volcanoes to ice fields and from lowland forests to alpine vegetation.[8]

In April 2009, I corresponded via e-mail with Emmanuel de Merode, the chief warden for Virunga National Park.[9] "From a biodiversity perspective there is nothing more important in Africa than Virunga," de Merode told me. "The destruction of Virunga would be a loss to the whole of humanity, not just Africa, hence its status as a World Heritage Site and our efforts to globalize the effort to protect it."[10]

Hundreds of thousands of people live in the areas surrounding the park, but only a few of them have electricity. With few other sources of energy, the locals have little choice but to buy charcoal produced from wood that is cut illegally from the park. De Merode and his fellow park managers have declared that the charcoal trade "stands as the single biggest threat to the mountain gorillas and other flora and fauna in the park."[11]

Although the solution to the charcoal trade is obvious, the butane stoves, and the fuel they need, are simply too expensive for most of the people who live near the park. De Merode said supplying the stoves and the fuel would cost millions of dollars. But that cost must be compared with the value of keeping the mountain gorillas safe in the wild. Saving the gorillas and other wildlife and fauna in Virunga, de Merode told me, "is a test case, and a measure of our determination in addressing environmental issues of global proportions."[12] And what will happen if more energy services are not made available to the people who live in and around the park? De Merode replied that the deforestation will continue as the locals produce more charcoal. And that will mean destruction of the gorillas' habitat, "which would be the end for Congo's gorillas," he said. "It could take the overall population below the minimum threshold and threaten them with extinction."

Without butane as an option, de Merode and his team are trying to ramp up the production of biomass briquettes, which they hope will help cut demand for charcoal and therefore slow the rate of deforestation in Virunga. The biomass briquettes are made with grass, leaves, agricultural waste, sawdust, and other material that is fed into special presses and then dried in a greenhouse. He hopes that, eventually, the program will have 5,000 presses, enough, he believes, to supply briquettes for 800,000 people.[13] But by mid-2009, progress appeared to be slow.

The plight of the gorillas in Virunga is similar to the loss of tropical forest habitat in Indonesia, where the misguided quest to produce more biofuels has led to rapid deforestation. Since 1996, some 9.4 million acres of Indonesian forest have been destroyed to make way for palm oil plantations.[14] And much of that palm oil was aimed at producing biodiesel for export for the European market. The lowland tropical forests in Sumatra and Borneo have been decimated by the quest for palm oil. And as reporter Tom Knudson detailed in an early 2009 story for London's *Guardian* newspaper, as the forests have declined, so, too, have the numbers of rare endemic species such as tigers and orangutans. Sumatra, the sixth-largest island in the world, is home to the rare Sumatran tiger. Since the 1980s, the population of Sumatran tigers has fallen by more than 60 percent to fewer than 400 animals. "The tiger is going to go extinct, if we don't do something," one Indonesian biologist told Knudson. Indonesia's orangutans face similar threats. Found only on the islands of Sumatra and Borneo, the great apes are increasingly isolated because of the destruction of their rainforest habitats. On Sumatra, there are about 7,000 individuals.[15]

The destruction caused by palm oil production in Indonesia and elsewhere is a direct result of the misbegotten policies in Europe and elsewhere that assumed that biofuels were environmentally superior to petroleum. That wrongheaded thinking led to a huge rush to clear large swaths of the tropics in order to produce an energy alternative that has had the exact opposite effect of what was hoped. A 2008 World Bank report estimated that the deforestation in Indonesia—much of it associated with biofuels—had resulted in huge releases of carbon dioxide from the burning of the forest and the release of greenhouse gases from swamps and bogs. Those releases, amounting to some 2.6 billion tons per year, made Indonesia the third-largest emitter of carbon dioxide on the planet, behind China and the United States.[16]

Although the gorillas, tigers, and apes are important, the most important beneficiaries of more widespread use of oil would be humans. And, to be more specific, it would be the millions of young children and women who are sickened or who die prematurely every year from indoor air pollution caused by the burning of biomass.

In 2007, the World Health Organization estimated that indoor air pollution was killing about 500,000 people in India every year, most of them women and children. The agency also found that air pollution levels in some kitchens in rural India were thirty times higher than recommended and that the pollution was six times as bad as that found in New Delhi. Worldwide, as many as 1.6 million people per year are dying premature deaths due to indoor air pollution.[17]

About 37 percent of the world's population relies on solid fuels, such as straw, wood, dung, or coal, to cook their meals.[18] These low-quality fuels, combined with inadequate ventilation when the cooking is done inside, often results in the living areas being filled with a variety of noxious pollutants, including soot particles, carbon monoxide, benzene, formaldehyde, and even dioxin.[19] Continued exposure to polluted indoor air can result in numerous illnesses, ranging from relatively minor problems such as headaches and eye irritation to deadly conditions such as asthma, pneumonia, blindness, lung cancer, tuberculosis, and low birth weight in children born to mothers who were exposed to indoor air pollution during pregnancy.[20]

Despite these numbers, the problem of indoor air pollution doesn't get nearly as much attention as other public health issues, such as vaccination or safe drinking water. One of the most passionate voices proclaiming the need for more hydrocarbon use among the world's poor is that of Kirk R. Smith, a professor of global environmental health at the University of California, Berkeley.[21]

In 2002, Smith wrote a piece for *Science* magazine entitled "In Praise of Petroleum?" in which he challenged the notion that "for the poor as for everyone else, only renewable energy sources qualify as sustainable." He went on to say, "What possible better use for high-efficiency clean-burning fossil fuels such as LPG [liquefied petroleum gas] than providing high-quality energy services for poor households?" [22]

When I interviewed Smith in July 2009, he explained that "poor women in rural areas of developing countries are about as low on the totem pole, globally, as you can get. . . . They don't have anybody speaking for them. They don't have their own Sierra Club or whatever." Smith continues to advocate increased use of oil as a way to help the rural poor. "Even if you were to substitute LPG for all of the biomass used for cooking in the world, it would have very little impact on overall resources," he told me. "Why ask the poor to take on the need to use fancy, new, novel, untested renewable energy devices when we have something that's good for them? They have many other needs. And this is a great thing for them."[23]

Reducing the amount of biomass used for cooking in the developing world would also reduce black carbon emissions. Soot particles from burning biomass, as well as from other sources such as coal-fired power plants and diesel engines, are the second-largest manmade contributor to global climate change.[24] In the polar regions of the Arctic and Antarctic, and in other areas where snow and ice are prominent, deposition of black carbon particles causes the surfaces to absorb more solar radiation and therefore accelerates the melting process. Research has shown that these black carbon deposits may be causing more of the Arctic climatic change than all other manmade causes combined.[25]

About one-third of all the black carbon emissions on the planet come from inefficient cookstoves used in households in the developing world. By pushing more efficient biomass stoves as well as promoting increased use of stoves that use butane, propane, or other clean-burning fuels, Smith said, those releases could be cut dramatically. And even better, the new stoves would reduce the number of trees being cut to produce charcoal in developing countries like Congo, a process Smith calls "the most greenhouse-intensive fuel cycle in the world."[26]

Moving the world's rural poor away from charcoal to cleaner, denser fuels such as LPG and refined oil products will not only help save the world's forests and endangered animals, it will also dramatically improve their health and their standards of living. But to do that, we must move past the idea that oil is bad. The reality is that oil is greener than nearly everything else that might replace it. Natural gas is greener still. If we want to improve the lot of the world's poorest people, and women and girls in particular, we should be using more of both.

CHAPTER 18

Cellulosic Ethanol Can Scale Up and Cut U.S. Oil Imports

FOR YEARS, ETHANOL boosters have promised Americans that "cellulosic" ethanol lurks just ahead, right past the nearest service station. Once it becomes viable, this magic elixir—made from grass, wood chips, sawdust, or some other plant material—will deliver us from the evil clutches of foreign oil and make the United States "energy independent" while enriching farmers and strengthening small towns across the country.

Consider this claim: "From our cellulose waste products on the farm such as straw, corn-stalks, corn cobs and all similar sorts of material we throw away, we can get, by present known methods, enough alcohol to run our automotive equipment in the United States."

That sounds like something you've heard recently, right? Well, fasten your seatbelt because that claim was made way back in 1921. That's when Thomas Midgley, an American inventor, proclaimed the wonders of cellulosic ethanol to the Society of Automotive Engineers in Indianapolis. Though Midgley was excited about the prospect of cellulosic ethanol, he admitted that there was a significant hurdle to jump before his concept would be feasible: Producing the fuel would cost about $2 per gallon.[1] That's about $24 per gallon in current money.

Alas, what's old is new again. Over the past few years, cellulosic ethanol has been promoted by a Who's Who of American politics, including Iowa senator Tom Harkin,[2] President Barack Obama,[3] former vice president Al Gore, former Republican presidential nominee and U.S. Senator John McCain, former president Bill Clinton, former president George W. Bush, former CIA director James Woolsey,[4] and Rocky Mountain Institute cofounder Amory Lovins.[5] In August 2009, billionaires Ted Turner and T. Boone Pickens added their names to the list of cellulosic boosters when they co-wrote an opinion piece for the *Wall Street Journal*, in which they declared that "advanced biofuels from cellulosic material . . . can play a key role in reducing the vulnerabilities, emissions and costs associated with imported oil, while also providing new economic opportunities for America's farm communities."[6]

Of the people on that list, Lovins has been the longest-running—and the most consistently wrong—cheerleader for cellulosic fuels. His boosterism began with a 1976 article in *Foreign Affairs*, the piece that arguably made his career. In that article, called "Energy Strategy: The Road Not Taken?" Lovins declared that American energy policy was all wrong. What America needed was "soft" energy resources to replace the "hard" ones (namely, hydrocarbons and centralized power plants). Regarding biofuels, he wrote that "exciting developments in the conversion of agricultural, forestry and urban wastes to methanol and other liquid and gaseous fuels now offer practical, economically interesting technologies sufficient to run an efficient US transport sector."

Lovins went on: "Presently proved processes already offer sizable contributions without the inevitable climatic constraints of fossil-fuel combustion." And he claimed that, given better efficiency in automobiles and a large enough installation of cellulosic ethanol distilleries, "the whole of the transport needs could be met by organic conversion."[7] In other words, Lovins was making the exact same claim that Thomas Midgley had made fifty-five years earlier: Given enough money—that's always the catch, isn't it?—cellulosic ethanol could provide all of America's transportation fuel needs.

Almost thirty years after Lovins made his claims in *Foreign Affairs*, the United States still did not have a single biofuel company producing significant quantities of cellulosic ethanol for sale in the commercial market.[8] And yet, in 2004, Lovins and several coauthors wrote a book called *Winning the Oil Endgame* that, once again, said that advances in biotechnology would make cellulosic ethanol viable. And in doing so, claimed Lovins and his peers, it "will strengthen rural America, boost net farm income by tens of billions of dollars a year, and create more than 750,000 new jobs."[9] Two years later, Lovins was at it again. In 2006, while testifying before the U.S. Senate, he claimed that "advanced bio-fuels (chiefly cellulosic ethanol)" could be produced for an average cost of just $18 per barrel.[10]

Alas, Lovins isn't the only one drinking the cellulosic Kool-Aid. In his 2007 book, *Winning Our Energy Independence*, S. David Freeman, the former head of the Tennessee Valley Authority, said that to get away from our use of oil, "we must count on biofuels."[11] And a key part of Free-man's biofuel recipe was cellulosic ethanol. Freeman claimed that there was "huge potential to generate ethanol from the cellulose in organic wastes of agriculture and forestry."[12] Using some 368 million tons of "for-est wastes," he said, could provide about 18.4 billion gallons of ethanol per year, yielding "the equivalent of about 14 billion gallons gasoline [sic], or about 10% of current gasoline consumption."[13]

Cellulosic ethanol gained acolytes during the 2008 presidential cam-paign. In May 2008, Speaker of the House Nancy Pelosi touted the pas-sage of the subsidy-packed $307 billion farm bill, stating that it was an "investment in energy independence" because it provided "support for the transition to cellulosic ethanol."[14] Pelosi and her fellow members of Congress are such big believers in cellulosic ethanol that they have man-dated that U.S. fuel suppliers blend no less than 21 billion gallons of the nonexistent product into the American gasoline pool every year, starting no later than 2022.[15] (That volume of fuel is equal to about 1.37 million barrels per day, which is approximately equal to the volume of oil the United States imported from Venezuela in 2007.)[16]

Pelosi is just one of many Democrats who love the idea of cellulosic ethanol. In August 2008, Obama unveiled his "new" energy plan, which called for "advances in biofuels, including cellulosic ethanol."[17] In Janu-ary 2009, Senator Tom Harkin, the Iowa Democrat who has been a key backer of the corn ethanol industry for years, told PBS that "ethanol doesn't necessarily all have to come from corn. In the last farm bill, I put a lot of effort into supporting cellulose [sic] ethanol, and I think that's what you're going to see in the future."[18]

In April 2009, U.S. Energy Secretary Steven Chu wrote an article for *Newsweek* in which he said the United States "can develop new liquid biofuels that will be direct replacements for gasoline and diesel fuel." He claimed these fuels could be produced from grasses and "agricultural wastes" and that there was "an achievable strategy" for "using biomass to replace 30% of our transportation fuels." Once those fuels are ready, he declared, "the importance of oil as a strategic resource will plummet."[19] In late October 2009, Obama was still touting biofuels. But instead of singling out cellulosic ethanol for praise, he was talking about the need for "sustainably grown biofuels."[20] Exactly what that term means no one seems to know.

Although the hype continues unabated, cellulosic ethanol is scarcely closer to widespread commercial viability than it was when Midgley first began promoting it back in 1921. A September 2008 study on alternative automotive fuels done by Jan Kreider, a professor emeritus of engineering at the University of Colorado, and Peter S. Curtiss, a Boulder-based engineer, found that the production of cellulosic ethanol required about forty-two times as much water and emitted about 50 percent more carbon dioxide per unit of energy produced than standard gasoline. They also found that—as with corn ethanol—the amount of energy that could be gained by producing cellulosic ethanol was negligible.[21]

The underlying problem with cellulosic ethanol takes us back to the Four Imperatives. Cellulosic ethanol and other biomass-focused energy projects are plagued by their low power density. No matter how much the promoters want to talk about the merits of wood chips and switchgrass, they are fighting an uphill battle, because the power density of biomass production is simply too low: approximately 0.4 watts per square meter.[22] Even the best-managed tree plantations can only achieve power densities of about 1 watt per square meter.[23] For comparison, recall that even a marginal natural gas well has a power density of about 28 watts per square meter.

Fighting the inherently low power density of biomass production means that entrepreneurs must corral Bunyanesque quantities of the stuff in order to make even a small dent in America's motor fuel market. Let's assume that the United States wants to replace 10 percent of its oil use with cellulosic ethanol. That's a useful percentage, as it's approximately equal to the percentage of U.S. oil consumption that originates in the Persian Gulf.[24] Let's further assume that the United States decides that switchgrass is the most viable option for producing cellulosic ethanol.

Given those assumptions, here's the math: The United States consumes about 21 million barrels of oil per day, or about 320 billion gallons of oil per year.[25] Ten percent of that volume would be about 32 billion gallons of oil. But remember, ethanol's energy density is only about two-thirds that of gasoline. Thus, the United States would need to produce about 48.5 billion gallons of cellulosic ethanol in order to have the energy equivalent of 32 billion gallons of oil.

So how much biomass would be needed? Cellulosic ethanol companies like Coskata and Syntec have claimed that they can produce about 100 gallons of ethanol per ton of biomass. Therefore, producing 48.5 billion gallons of cellulosic ethanol would require about 485 million tons of biomass. How much is that? If we assume that a standard 48-foot trailer holds 15 tons of material, then we would need 32.3 million trailers to hold that 485 million tons of biomass. That's a lot of trailers. Arranged in a line, that column of trailers (not including any trucks attached to them) would stretch about 293,600 miles—long enough to reach from the Earth to the Moon and about one-fifth of the way back again.[26]

But let's continue driving down this road for another mile or two. Sure, it's possible to produce that much biomass, but how much land would be required to make it happen? A report from Oak Ridge National Laboratory suggests that 1 acre of switchgrass can produce 11.5 tons of biomass per year.[27]

Given those numbers, producing 485 million tons of biomass from switchgrass would require 42.1 million acres to be planted in nothing but switchgrass. That's equal to about 65,800 square miles, an area nearly equal to the size of Oklahoma.[28] Now, some wags might suggest that paving the Sooner State with nothing but switchgrass would be a significant improvement. But making room for all of that switchgrass would pinch America's ability to grow food. The 42.1 million acres needed for the switchgrass would be equal to about 10 percent of all the cropland now under cultivation in the United States.[29] Thus, for the United States to get 10 percent of its oil needs from cellulosic ethanol, it would need to plant an area equal to about 10 percent of its cropland in switchgrass.

And none of those calculations account for the fact that there's no infrastructure available to plant, harvest, and transport the switchgrass or other biomass source to the refinery. The U.S. farming sector has invested billions of dollars in all types of tractors, planters, and harvesters to grow and manage food crops. But none of that equipment—made by John Deere, Kubota, and other companies—has been designed to handle the gargantuan volumes of biomass that would be needed to make cellulosic ethanol viable.

Even if there is a breakthrough in ethanol production that allows the production of large quantities of alcohol fuel from biomass, it does not necessarily mean that the United States will see a corresponding drop in its imported oil needs. Why? The answer is simple: Ethanol only replaces one of the myriad products that refiners extract from a barrel of crude oil. And that leads to another heretical notion: Increased ethanol use won't cut oil imports.

▢▢▢

H. L. Mencken once remarked that there is always a "well-known solution to every human problem—neat, plausible, and wrong."

That quote comes to mind when considering the vocal group of politicos and neoconservatives who claim that the best way to cut American oil imports, and thereby impoverish the petrostates (and, in theory, reduce terrorism), is to require automakers to manufacture "flex-fuel" cars that can burn motor fuel containing 85 percent ethanol. (For more on this claim, and the people who are promoting it, see my book *Gusher of Lies*.)

Their rationale is that using more ethanol made from corn, switchgrass, or other biomass will create competition in the motor fuel market and thereby depose oil as the world's primary transportation fuel. Once that is done, they claim, oil will no longer be a strategic commodity; its price will fall, the petrostates will be bankrupted, and a newly energy-independent United States will race back to the head of the line as the world's undisputed sole superpower. The rhetoric put forward by these underinformed-but-persistent sophomores has proved so irresistible that several members of Congress have introduced legislation aimed at requiring automakers to produce flex-fuel cars. Fortunately, none of their proposed bills have passed.

The idea that ethanol provides a viable solution betrays a near-complete ignorance of the world petroleum business. The supposed solution fails because it only replaces part of the crude oil barrel. During the refining process, a barrel of crude yields several different "cuts," ranging from light products such as butane to heavy products such as asphalt. An average barrel of crude (42 gallons) yields about 20 gallons of gasoline.[30] And certain types of crude oil ("light sweet crude," for example) are better suited to gasoline or diesel production than others. The overall point is that refiners cannot produce just one product from a barrel of crude; they must produce several. And the market value of those various cuts is constantly changing.

But there's little growth in gasoline demand. Meanwhile, demand for other cuts of the crude barrel are booming. In short, the ethanol producers are making the wrong type of fuel at the wrong time. They are producing fuel that displaces gasoline at a time when gasoline demand, both in the United States and globally, is essentially flat. Meanwhile, demand for the segment of the crude barrel known as "middle distillates"—primarily diesel fuel and jet fuel—is growing rapidly. And ethanol cannot replace diesel or jet fuel, the liquids that propel the vast majority of our commercial transportation machinery.

In mid-2008, the Paris-based International Energy Agency released its medium-term oil market report, which said that while global gasoline demand is growing slowly, "distillates (jet fuel, kerosene, diesel, and other gasoil) have become—and will remain—the main growth drivers of world oil demand."[31] Between 2007 and 2013, the IEA expects distillate demand to nearly double, while global gasoline demand will grow only slightly.[32] Those projections were seconded by the Energy Information Administration in its 2009 Annual Energy Outlook, which expects demand for diesel fuel to grow by about 1.5 percent per year through 2030 while gasoline demand will fall by nearly 1 percent annually.[33]

The surge in diesel demand is due in large part to the ongoing "dieselization" of the European automobile market as well as continued economic growth in Asia and the United States. Ethanol is doing little, if anything, to reduce overall U.S. oil consumption or imports, because refiners are having to buy the same amount of crude (or more) in order to meet the demand for products other than gasoline—that is, jet fuel, diesel fuel, fuel oil, and asphalt. For instance, since 2000, the United States has consumed an average of 185 million barrels of asphalt and road oil per year,[34] and the vast majority of that asphalt is produced by domestic refiners who produce it from domestic and imported crude. Asphalt may not be as sexy as gasoline, but it is one of the myriad oil products that are essential elements of the U.S. economy.

The lesson here is obvious: Producing ethanol from corn or other substances cannot, and will not, significantly reduce oil use or oil imports, because it cannot replace the entire crude barrel. Unless inventors can come up with a substance (or substances) that can replace all of the products that are refined from a barrel of crude oil—from gasoline to naphtha and from diesel to asphalt—then the United States is going to continue using oil as a primary energy source for decades to come. And that will be true no matter how much corn gets burned up in America's delusional quest for "energy independence." But of course, ethanol isn't the only alternative fuel getting lots of hype these days.

Electric cars are all the rage. And the hype, once again, has lost all connection with reality.

CHAPTER 19

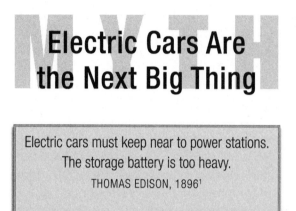

Electric Cars Are the Next Big Thing

Electric cars must keep near to power stations.
The storage battery is too heavy.
THOMAS EDISON, 1896[1]

There are not enough idiots who will buy it.
JOHAN DE NYSSCHEN, president of Audi of America,
talking about the Chevrolet Volt, September 2009[2]

ALL-ELECTRIC CARS are The Next Big Thing. And they always will be.

Okay, that's a bit harsh. But even a cursory look at the 100-year history of the electric car sector illustrates the need for a healthy dose of skepticism. And skepticism is a much-needed commodity today as the hype over electric cars slips into overdrive.

Consider the case of Shai Agassi, the founder of an electric car company called Better Place. In July 2008, Thomas Friedman of the *New York Times* dubbed Agassi "the Jewish Henry Ford" and said he was launching "an energy revolution" that would end the world's "oil addiction."[3] A few weeks later, *Wired* wrote a fawning article on Agassi saying that the entrepreneur had lined up $200 million in funding in his effort to launch "the fifth-largest startup of all time in less than a year." Agassi has convinced a lot of investors that his plan will work: "Everyone who meets him already believes he can see the future," gushed the author of the *Wired* piece.[4]

In May 2009, *Time* magazine declared that Agassi was going to "help the world end its addiction to oil by transforming cars from their climate-changing, lung-polluting, gas-guzzling design to one that's clean, afford-able and all-electric."[5] (The glowing article was written by a venture capitalist who had invested in Agassi's company.) All of that fantastic press, and yet only one or two of the stories bothered to mention the size of Agassi's electric car fleet, which consisted of exactly one prototype. Be-yond that, Agassi's company had no charging stations and no customers.[6]

The hype around Agassi was hardly unique. Two other startup car companies—Tesla Motors and Fisker Automotive—have been getting sim-ilar treatment. The all-electric Tesla Roadster retails for about $100,000, and the plug-in hybrid-electric car being sold by Fisker, called the Karma, sells for $87,500.[7] Despite that price, *Newsweek* explained that the Karma, a high-performance sports car, "will get 100 miles per gallon—and those who rarely travel more than 50 miles at a time will do even better."[8]

I am not disparaging hybrid-electric vehicles, such as the Toyota Prius, the Honda Insight, or the Ford Fusion. There is no question that the technologies behind those vehicles and others like them are durable, reliable, and functional. For those reasons, as well as their relative af-fordability, hybrid-electric drive cars, that is, the cars that don't come with an extension cord, are here to stay. And in the coming years, those vehicles, not their plug-in, or all-electric, peers, will likely claim an in-creasing share of the automotive market. That's the assessment of Mena-hem Anderman, a California-based battery expert who heads a firm called Total Battery Consulting. In 2007, Anderman told a U.S. Senate panel that conventional hybrid-electric technology was "the only one ma-ture enough for its market growth to have an impact on the nation's en-ergy usage in the next 10 years."[9]

Two recent studies have come to very similar conclusions. A 2008 study by the Center on Global Change at Duke University concluded that "gasoline prices would need to increase to around $6 per gallon to make plug-in hybrids cost effective; below $6 per gallon, regular hybrids are more cost effective than plug-in hybrids."[10] And an early 2009 study by a team of engineers at Carnegie Mellon University found that plug-in hybrid-electric vehicles that are "sized for 40 or more miles of electric-only travel do not offer the lowest lifetime cost in any scenario."[11]

And while sales of hybrids are growing, don't count out the old internal combustion engine just yet. Gasoline and diesel engines are seeing continual improvements as automakers wring yet more efficiency out of the designs that have ruled the terrestrial transportation market for more than a century. Nor is it a sure thing that hybrid-electric and all-electric cars will be able to surmount their challenges of material availability. The exotic components on many hybrid cars and all-electrics depend on the availability of elements such as lithium and neodymium. And the availability of those commodities is far from assured.

Despite the continuing advances being made to the internal combustion engine, that technology is being viewed by the media as rather stodgy. The sex, the sizzle—and of course, loads of media attention—are being given to all-electric and plug-in hybrid-electric vehicles. General Motors plans to start selling the Chevrolet Volt, a plug-in vehicle that is vying for the position of most-hyped car of the twenty-first century, in late 2010. The automaker has claimed that the Volt will get an astounding 230 miles per gallon.[12] (Shortly after that claim was made, *The Economist* calculated that 50 miles per gallon was more realistic.)[13] In mid-2009, Mitsubishi announced that it would begin selling the all-electric iMiEV car in Japan in 2010. The tiny, four-door car weighs just 2,400 pounds, has a 16-kilowatt-hour battery, and will sell for $44,700.[14]

Other automakers have jumped on the all-electric bandwagon, too. In early 2009, a Nissan official said the company saw "electrification of the transportation sector as our highest priority." The company claims that by 2020, 10 percent of the vehicles it produces will be all-electric.[15] In August 2009, Nissan announced that it would introduce an all-electric vehicle, the Leaf, which the company says will have a 100-mile range and cost about $30,000. Nissan expects to start selling the car in the United States sometime in 2010.[16] Honda is also planning an electric vehicle. The same month that Nissan announced the Leaf, Honda said it would begin selling an all-electric car in the United States as well, but did not specify a date.[17]

Thus, within the next year or two, nearly all of the world's major automakers are planning to introduce plug-in hybrid-electric or all-electric cars in the United States. But the hype isn't limited to the United States. In Britain, motorists may get government subsidies of up to 5,000 British pounds if they buy hybrid or all-electric cars. And the country's prime minister, Gordon Brown, has said that he wants all of the cars on the roads of the United Kingdom to be all-electric or hybrid-electric by 2020.[18] Some European automakers have announced plans to launch plug-in hybrid cars by 2012 or so, including PSA Peugeot Citroen, BMW, and Daimler.[19] Shortly after Nissan and Honda announced their plans to sell all-electric cars, the German government announced that it wanted to put 1 million electric cars on Germany's roads by 2020.[20]

There's no question that electric vehicles have many positive attributes: low refueling costs, no air pollutants at point of use, and quiet operation. But despite their promise, all-electric cars continue to be hampered by the same drawbacks that have haunted them for a century: limited range, slow recharge rates, lack of recharging stations, and high costs, particularly when compared to conventional cars. In short, the problems today are the same as they were back in Thomas Edison's day. And those problems can be summed up in one word: batteries. For more than a century, inventors have been trying to develop high-capacity batteries that will work well in the automotive market. And for the most part, they have failed. Since Edison, entrepreneurs and inventors have achieved a host of amazing feats, including putting a man on the Moon, building nuclear power plants, and figuring out how to deliver pornography to pocket-sized computer-phones. But even as other technologies have zoomed ahead in terms of cost and availability, batteries continue to be short-circuited by the same issues that hamstrung Edison. Sure, batteries have improved. But they haven't achieved the orders-of-magnitude improvements that are needed for them to compete effectively with other transportation fuels.

The problem is the second of the Four Imperatives: energy density. Automakers usually measure the energy density of batteries in watt-hours per kilogram—that is, the number of watt-hours of electricity stored per kilogram of battery weight.[21] Battery makers have been making significant improvements in energy density. Modern lithium-ion batteries provide a four-fold improvement in energy density when compared with their older lead-acid cousins. But neither source can hold a candle to gasoline, or for that matter, ethanol. Gasoline holds eighty times as many watt-hours per kilogram as a lithium-ion battery, and ethanol holds more than fifty times as many.

FIGURE 28 The Problem with Batteries: It's the Energy Density, Stupid!

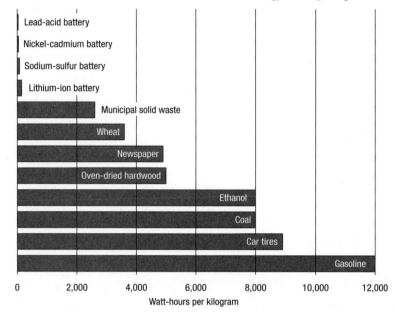

Sources: Donald Sadoway, "Power Storage: Batteries and Beyond," February 12, 2009; John Newman, University of California, Berkeley, personal communication with author, January 16, 2009; David J.C. MacKay, *Sustainable Energy—Without the Hot Air* (Cambridge: UIT Cambridge, 2009) (online at http://www.inference.phy.cam.ac.uk/sustainable/book/tex/cft.pdf), 284.

Given that batteries hold such small amounts of energy, designers of electric vehicles must, in effect, build their cars around the bulk and weight of the battery pack. For instance, on the Tesla Roadster, arguably the most famous electric car now available, the battery pack weighs 992 pounds, or slightly more than one-third of the curb weight of the vehicle, which tips the scales at 2,723 pounds.[22]

In 2009, Bill Reinert, one of the lead designers of the Toyota Prius, told me that existing battery technology simply "does not provide the cost, durability and energy storage attributes that allow for the development of mass-market products. We can get around some of these issues with niche products, or schemes like battery leasing, or subsidizing the products, but none of these are solutions for the mass market." And while Toyota is planning to develop an all-electric car, he said, the company is "looking at sales volumes of thousands, not millions. To produce an electric vehicle that's truly intended for a mass market, a replacement for your current gasoline car, we're going to need a battery chemistry that isn't currently available."[23]

Reinert's view was reinforced by a report on energy storage that was released by the U.S. Department of Energy's Office of Vehicle Technologies in early 2009. The report concludes that despite the enormous investments being made in plug-in hybrid-electric vehicles and lithium-ion batteries, four key barriers stand in the way of their commercialization: cost, performance, abuse tolerance, and life. The DOE's summary deserves to be read in its entirety:

> **Cost:** The current cost of Li-based [lithium-based] batteries (the most promising chemistry) is approximately a factor of three to five too high on a kWh [kilowatt-hour] basis. The main cost drivers being addressed are the high cost of raw materials and materials processing, the cost of cell and module packaging, and manufacturing costs.
>
> **Performance:** The performance barriers include the need for much higher energy densities to meet the volume/weight requirements, especially for the 40-mile system, and to reduce the number of cells in the battery (thus reducing system cost).
>
> **Abuse Tolerance:** Many Li batteries are not intrinsically tolerant to abusive conditions such as a short circuit (including an internal short circuit), overcharge, over-discharge, crush, or exposure to fire and/or other high temperature environments. The use of Li chemistry in these larger (energy) batteries increases the urgency to address these issues.
>
> **Life:** The ability to attain a 15-year life, or 300,000 HEV [hybrid-electric vehicle] cycles, or 5,000 EV [electric vehicle] cycles are unproven and are anticipated to be difficult. Specifically, the impact of combined EV/HEV cycling on battery life is unknown and extended time at high state of charge is predicted to limit battery life.[24]

Of course, the DOE's grim prognosis hasn't prevented the media from gushing over all-electrics. Consider this quote from the *Los Angeles Times*: "The electric automobile will quickly and easily take precedence over all other" types of motor vehicles. That declaration was published on

May 19, 1901, in a story headlined: "Edison's New Storage Battery." "If the claims which Mr. Edison makes for his new battery be not overstated, there is not much doubt that it will make a fortune for somebody," the article said.[25]

Today's electric cars—once again—may be on the verge of yet another spectacular breakthrough. But, just for fun, let's take a quick drive past a few newspaper stories:

- 1911: The *New York Times* declares that the electric car "has long been recognized as the ideal solution" because it "is cleaner and quieter" and "much more economical."[26]
- 1915: The *Washington Post* writes that "prices on electric cars will continue to drop until they are within reach of the average family."[27]
- 1959: The *New York Times* reports that the "Old electric may be the car of tomorrow." The story said that electric cars were making a comeback because "gasoline is expensive today, principally because it is so heavily taxed, while electricity is far cheaper" than it was back in the 1920s.[28]
- 1967: The *Los Angeles Times* says that American Motors Corporation is on the verge of producing an electric car, the Amitron, to be powered by lithium batteries capable of holding 330 watt-hours per kilogram. (That's more than two times as much as the energy density of modern lithium-ion batteries.) Backers of the Amitron said, "We don't see a major obstacle in technology. It's just a matter of time."[29]
- 1979: The *Washington Post* reports that General Motors has found "a breakthrough in batteries" that "now makes electric cars commercially practical." The new zinc-nickel oxide batteries will provide the "100-mile range that General Motors executives believe is necessary to successfully sell electric vehicles to the public."[30]
- 1980: In an opinion piece, the *Washington Post* avers that "practical electric cars can be built in the near future." By 2000, the average family would own cars, predicted the *Post*, "tailored for the purpose for which they are most often used." It went on to say that "in this new kind of car fleet, the electric vehicle could pay a big role— especially as delivery trucks and two-passenger urban commuter cars.

PHOTO 6 What's old is new again. In 1919, this Detroit Electric automobile stopped for a charge.
Source: Library of Congress, LC-USZ62-46285.

With an aggressive production effort, they might save 1 million barrels of oil a day by the turn of the century."[31]

But ignore the headlines of the past and consider the path of Tesla Motors, the outfit named for Nicola Tesla, the electricity pioneer who once worked with Thomas Edison and who went on to invent the induction electric motor. Although Tesla has been dead since 1943, the company named for him is producing the Roadster, which is snagging rave reviews. One writer called it "a frantic road dart on twisty roads."[32] In mid-2009, London's *Daily Telegraph* called the car "one of the stars of this year's London Motorexpo." The reviewer for the *Telegraph* praised the car's acceleration, saying that when one stomps on the accelerator, "the result is little short of astonishing. The Tesla belts away with the seamless surge of a catapult launch."[33] But given its high price, the Tesla is hardly a catapult for the common man,[34] and like its predecessors from a century ago, it faces the familiar issues of charging time, weight, and range. The car has a claimed range of 220 miles, and fully recharging the car's batteries takes at least four hours.[35]

That point leads to another critical challenge for all-electric cars: long refueling times. All energy sources are limited in how fast they can release energy and how fast they can be replenished. With batteries, the refiller is the killer, particularly when compared with the time needed to refuel with good old conventional gasoline.

In May 2009, with the fuel gauge on my 2000 Honda Odyssey minivan showing empty, I pulled into a Texaco station to fill up. Total elapsed time from inserting the nozzle into my tank until the automatic shut-off valve clicked off, signaling full: 1 minute and 59 seconds—about the time that it takes Michael Phelps to swim the 200-meter butterfly.[36] In less than 2 minutes, I pumped about 18.5 gallons of gasoline into the vehicle. That's the energy equivalent of more than 600 kilowatt-hours of electricity, or about eleven times as many kilowatt-hours as are contained in the Tesla Roadster's battery pack. Put another way, I loaded about eleven times as much energy as what is contained in the batteries in the Tesla—and I did it in 1/120th of the time that is needed to recharge the Tesla's battery system. (Recall that recharging the 53-kilowatt-hour battery pack in the Tesla takes about 4 hours, or 240 minutes.) The total cost of refueling my Honda van: $44.32.

Now, were I to buy 53 kilowatt-hours of electricity from the local utility, at an average cost of $0.10 per kWh, the total cost of the fuel would only be about $5.30—far less than the $44 I paid to refill my minivan. But then, my van doesn't need recharging every night—which leads to another key issue with all-electric or plug-in hybrid vehicles: the lack of recharging locations. According to a June 2009 report by the Government Accountability Office, "about 40% of consumers do not have access to an outlet near their vehicle at home." The report goes on to say that consumers who don't have access to electric power near their cars would need "public charging infrastructure, which manufacturers and others told us could be installed at a relatively low cost of perhaps a few thousand dollars for a new charging box."[37]

Despite the myriad challenges facing the electric car business, Congress and the Obama administration are hurling billions of dollars at it.

Among the most notable recipients of the government's largesse: Fisker Automotive. In September 2009, Fisker received a $529 million loan from the U.S. government to help finance its startup costs. One of Fisker's main financial backers is the venture capital firm Kleiner Perkins Caufield & Byers, a Silicon Valley firm where Al Gore is a partner.[38]

Fisker wasn't alone. Nissan got a $1.6 billion loan, and Tesla Motors got a $465 million loan.[39] Two Phoenix-based companies, Electric Transportation Engineering and ECOtality, were given $99.8 million in federal stimulus money to help roll out an electric vehicle pilot program in several U.S. cities.[40] Johnson Controls, one of America's biggest battery makers, got a federal grant for $299.2 million to help it build batteries for electric and hybrid cars. General Motors got $105.9 million to help it produce battery packs for the Chevy Volt. In all, about fifty different entities were given federal grants (all provided by the stimulus package passed by Congress) that totaled some $2.4 billion as part of an "electric drive vehicle battery and component manufacturing initiative."[41]

In announcing the initiative, President Obama said that the grants were "planting the seeds of progress for our country, and good-paying, private-sector jobs for the American people." He went on to say that the initiative would help in the "deployment of the next generation of clean-energy vehicles."[42] Obama may be right. All-electric cars may be on the verge of grabbing a significant percentage of the U.S. car fleet.

But history shows that skepticism is in order. I'm not saying there won't be electric vehicles. There are already millions of them. In 2008, Chinese manufacturers produced some 22 million electric two-wheelers. About 65 million electric scooters are now traveling on Chinese roads. And because most of those scooters use simple lead-acid batteries instead of more expensive lithium-ion units, consumers can buy them for as little as $250.[43]

Dozens of U.S. companies are selling electric scooters and motorcycles.[44] All-electric vehicles have become so commonplace that in early 2009, a Costco store in south Austin started selling an all-electric scooter for less than $1,000. Some of the all-electric cars now being developed will gain loyal customers, particularly among the rich. The ongoing advancements in battery technology will make electric vehicles more viable. And those improvements will be augmented by ultracapacitors.

Unlike batteries, ultracapacitors are not reliant on chemical reactions to store energy. Instead, they store electricity by physically separating the negative charge from the positive charge. (Batteries separate the two charges chemically.)[45] A key advantage of ultracapacitors is their ability to be charged and discharged very quickly, a process that tends to damage conventional batteries. By pairing batteries with ultracapacitors, automakers can assure high power delivery to the wheels and do so without wearing out the batteries.[46]

Though improvements in batteries and ultracapacitors will undoubtedly continue, and hybrid-electric cars will continue gaining in popularity and market share, it's worth questioning the environmental impacts of the all-electric car. An October 2009 analysis by the National Academy of Sciences looked at the environmental costs associated with various types of automobile fuel. The scientists at the National Academy looked at thirteen different fuel sources and analyzed their total impact on the environment, particularly those relating to what they called the "health and non-climate damages" for different combinations of fuels and vehicles. They then expressed those damages in cents per vehicle mile traveled (VMT) for two years: 2005, and estimates for 2030. The graph in Figure 29 combines the data from two charts published by the academy. For the sake of simplicity, I reduced the number of fuels displayed from thirteen to nine. The figure clearly shows that vehicles powered by politically favored fuels such as corn ethanol (E85) and electricity impose more "damages" on society than vehicles that are powered by gasoline or natural gas. The conventional fuels are also less costly to society than plug-in hybrid-electric vehicles are, though the plug-ins are another politically popular choice. Those vehicles are identified in the figure as "Grid Dependent SI HEV."

The findings of the academy provide more evidence that the era of the internal combustion engine will continue for decades to come. The life expectancy of the internal combustion engine continues to be extended by engineers, who are constantly making incremental efficiency gains. Those gains are particularly apparent in diesel-powered vehicles, which, by 2030, according to the Academy of Sciences, will impose the lowest total costs on society.

A number of automakers have been introducing diesel cars into the U.S. market of late. The 2009 Volkswagen Jetta TDI gets 41 miles per

FIGURE 29 National Academy of Sciences' Estimate of Total Life-Cycle Damages Imposed by Various Fuels Used in Light-Duty Vehicles, 2005 and 2030

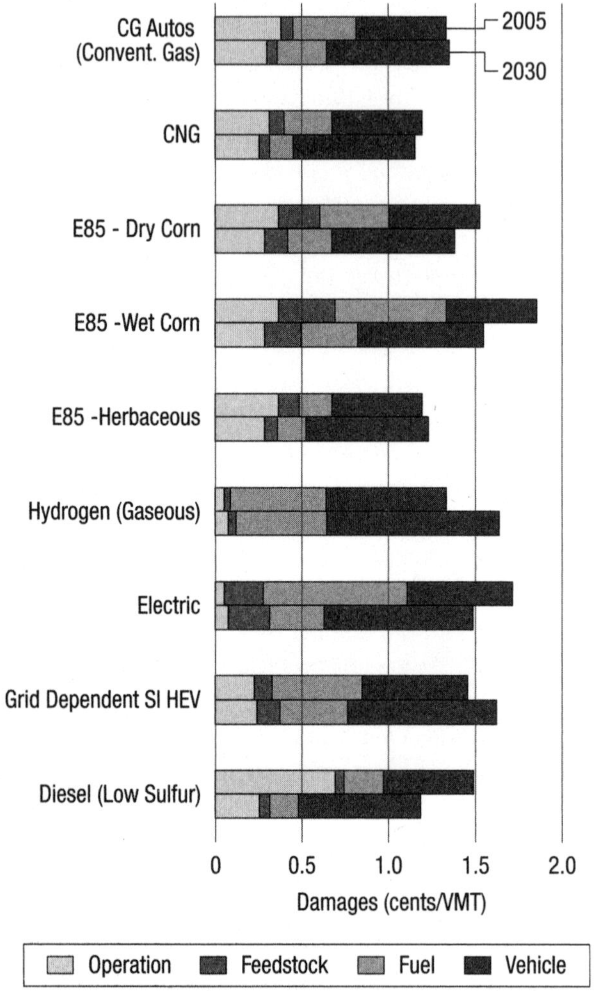

Note: CG stands for conventional gasoline. SI stands for spark ignition. CNG stands for compressed natural gas. HEV stands for hybrid-electric vehicle.

Source: National Academy of Sciences, "Hidden Costs of Energy: Unpriced Consequences of Energy Production and Use," 2009, executive summary, http://www.nap.edu/nap-cgi/report.cgi?record_id=12794&type=pdfxsum, 11.

gallon on the highway.[47] BMW now sells two diesel vehicles in the United States, one of which gets 36 miles per gallon on the highway.[48] Mercedes-Benz is selling three diesel vehicles that use their "BlueTEC" design.[49] Audi is selling two diesels, one of which gets more than 40 miles per gallon.[50] And although diesels are relatively rare in U.S. passenger cars, European car owners have been using diesels for decades. Consumers in Australia can buy the Hyundai i30 diesel wagon (cost: about $20,000), which gets about 40 miles per gallon and reportedly has a range of about 600 miles.[51] Automakers are also developing diesel hybrids, with Mercedes planning a sedan that could get 88 miles per gallon.[52]

Though diesels are more efficient than gasoline-fueled engines, the long-term prospects for gasoline are also good. The gasoline-powered 2009 Honda Fit gets 35 miles per gallon.[53] It has four doors and a roomy interior, and the base model sells for less than $15,000. Meanwhile, Ford has developed a turbocharged gasoline engine that will soon be the automaker's standard for its light-vehicle fleet. The new design, called EcoBoost, is smaller than conventionally aspirated engines while providing more power, and supplies fuel-economy improvements of up to 20 percent.[54]

Furthermore, diesel and gasoline vehicles are not overly reliant on rare earth elements such as neodymium and lanthanum—both of which are critical ingredients in the making of hybrid and electric vehicles. There is no way to know how the rare-earth-supply question will evolve. Perhaps China will adopt export policies that allow the goods containing rare earths to flow freely in world trade.

At the moment, it appears that thousands, perhaps millions, of hybrid-electric cars will be manufactured over the coming decades, and that those vehicles will continue to improve the overall efficiency of the U.S. auto sector. But remember that those high-tech vehicles are only part of the story. They will have to compete for market share with cheap, dependable vehicles powered by conventional internal combustion engines, engines that are still ubiquitous, cheap, and easy to maintain. Those hundreds of millions of internal combustion–powered vehicles can utilize a variety of fuels, including natural gas, propane, dimethyl ether, ethanol, methanol, used french-fry grease, soy diesel, and of course, conventional gasoline and diesel.[55] While all of those fuels will play a role, the key issue, as always, is the scale of the transition.

The introduction of the Prius about a decade ago marked the beginning of the electrification of the U.S. transportation sector. It's not the full electrification that many people dream of, but it is an important milestone in a process that will take decades. Remember, it took about ten years for Toyota to sell 1 million units of the Prius.[56] That sounds like a lot of cars until you remember that the global fleet now numbers about 1 billion vehicles. We can't be exactly sure how the electrification will proceed from here, but the decades-in-the-future transport system may include a fixed-guideway system that allows cars and trucks to be powered by the electric grid. The fixed guideways would allow vehicles to travel much closer together and at higher speeds than today's vehicles.

As the price of oil rises or falls in the coming decades, so will the acceleration/deceleration of the move toward using more electrons in transportation. Although opinions may differ regarding all-electric cars versus hybrids and other alt-fueled cars, we can agree on one basic theme: Electrons are good. The more electrons the better, because electricity is the basic commodity of modern life. And that leads to the last myth I want to debunk: the belief that we can create lots of electricity by burning biomass.

CHAPTER 20

We Can Replace Coal with Wood

EVERYONE LIKES WOOD. So when it comes to generating electricity, let's just replace coal with wood. Easy, right?

Some of the loudest voices on the Green/Left seem to think so. For instance, in March 2009, Joe Romm, a blogger on climate issues, wrote that "The best and cheapest near-term strategy for reducing coal plant CO_2 emissions without forcing utilities to simply walk away from their entire capital investment is to replace that coal with biomass." Romm cited a plan by Georgia Power, a subsidiary of utility giant Southern Company, to convert one of the company's plants so that it would burn wood rather than coal. The coal-fired power plant had 155 megawatts of capacity. After switching to wood, its output would be reduced to 96 megawatts. Romm praised the effort, saying that switching to biomass was "the most practical and affordable strategy for utilities with coal plants."[1]

In October 2009, the *New York Times* wrote about efforts to reduce carbon dioxide emissions from electric power plants, citing a Sierra Club effort to reduce coal consumption. The *Times* reporter, Matthew Wald, wrote that the promoters of non-coal sources "say that biomass fuels, derived from wood, waste, and alcohol, could offer an even better opportunity" for capturing the carbon dioxide that is generated during the combustion process. He went on to say that using trees could be advantageous, because "if a tree is cut down and burned in a boiler, a new tree can grow in its place, and absorb carbon dioxide from the atmosphere. That makes the process 'carbon negative;' for each ton burned, the amount of carbon dioxide in the atmosphere will decline."[2]

In November 2009, Romm was again singing the praises of using wood to produce electricity, with a blog post that cited a news story about a biomass power plant to be built in Ashland, Wisconsin.[3]

Those biomass-to-electricity pronouncements follow the unanimous 2008 vote by the Austin City Council to approve a plan put forward by the city's utility, Austin Energy, to spend $2.3 billion over twenty years to buy all of the power produced by a 100-megawatt wood-fired power plant to be built in East Texas.[4] Just before the Austin City Council voted on the wood-burning power plant, the city's mayor, Will Wynn, told the *Austin Chronicle* that the deal was a "strategic 'no brainer' that will keep our electric costs lower than the alternatives."[5]

While Romm, the Sierra Club, and Austin's environmentalists love the idea of biomass-fueled power plants, it appears that few of them have bothered to do the basic calculations that show just how much wood will be needed to replace even a small fraction of our coal needs. Here's the myth-busting reality: To replace just 10 percent of the coal-fired electricity capacity in the United States with wood-fired capacity would mean more than doubling overall U.S. wood consumption.

The math, as usual, is straightforward. The wood requirements for the Georgia Power facility and the East Texas generation project are about the same: 1 million tons of wood per year.[6] Thus, both projects will require 10,000 tons of wood per year to produce 1 megawatt of electricity.

The United States now has about 336,300 megawatts of coal-fired electricity generation capacity.[7] Let's assume that we want to replace just 10 percent of that coal-fired capacity—33,630 megawatts—with wood-burning power plants. Simple math shows that doing so would require about 336.3 million tons of wood per year.[8] How much wood is that? According to estimates from the United Nations Environmental Program, total U.S. wood consumption is now about 236.4 million tons per year.[9] Given those numbers, if the United States wants to continue using wood for building homes, bookshelves, and other uses—while also replacing 10 percent of its coal-fired generation capacity with wood-fired generators—

it will need to consume nearly 573 million tons of wood per year, or about 2.5 times its current consumption.

These numbers apparently don't bother Romm and other cheerleaders of this concept. In lauding the Georgia Power plant's move to burn wood rather than coal, he wrote that "the key, of course, [is] to make sure this is all done in the sustainable fashion. That will be the job of regulators and the Obama administration."[10]

But regulators, try as they might, can't overcome basic physics. The problems with biomass-to-electricity schemes are the same ones that haunt nearly every renewable energy idea: power density and energy density. Wood is a wonderful fuel for roasting marshmallows and keeping warm on a cold night. But its energy density is less than half that of coal's. That's why few people in the United States and in other developed countries use it for their cooking and heating needs. When you combine that low energy density with the low power density of wood and biomass production, the challenges become even more apparent. As discussed earlier, the power density of the best-managed forests is only about 1 watt per square meter.[11] And when a particular energy source, in this case, wood, has low power density and low energy density, that leads to problems with the other two elements of the Four Imperatives: cost and scale.

Alas, when it comes to discussions of "green" energy, too many advocates and politicos simply decide that it's a "no brainer." The result: Little critical thinking gets done about the issues of scale and the long-term impacts of the sources that are supposed to be making things better.

As I showed with just a bit of math, the myth that biomass provides a viable replacement for coal was easy to debunk. Now that many of the myths about green energy have been addressed, I will show, in the next section, why natural gas and nuclear hold so much promise for the future.

PART III
THE POWER
OF N2N

CHAPTER 21

Why N2N? And Why Now?
(The Megatrends Favoring Natural Gas and Nuclear)

IN AUGUST 2009, Fatih Birol, the chief economist for the International Energy Agency, did something that has become almost commonplace: He predicted a date for peak oil. In an interview with the British newspaper, the *Independent*, Birol said that his agency now believes "peak oil will come perhaps by 2020."[1]

Birol is one of dozens of prognosticators who have offered an opinion on the question of peak oil—the moment when the world reaches the limit of its ability to produce ever-increasing amounts of petroleum. Is Birol right? Maybe. We won't know for certain until about 2020 whether his prediction was right or wrong.

The importance of Birol's prediction is not that he picked a date for peak oil; rather, it's that he and so many other leading energy analysts are forecasting a peak. Indeed, when that peak is hit, it will be a milestone in our energy use—but it will not mean that we will quit using oil. As discussed in Part 1, the world will continue using oil for decades to come because no other fuel source provides such incredible versatility, ease of handling, and energy density. That said, it's also clear that oil's share of the primary energy market is declining. Back in 1973, oil provided 48 percent of the world's total energy. By 2008, that percentage had declined to 35 percent. And over the coming decades, the percentage will likely continue its slow decline.

It is true that this decline is part of a significant energy transition; it's just not the rapid move to the "green" sources that Al Gore and many other boosters have been hyping. The big challenge for today's policy-makers is to make the ongoing transition away from oil and coal to other energy sources as easy as possible; and in doing so, they should be encouraging energy sources that benefit both the environment and the economy. That brings us to the questions posed by the title of this chapter: Why N2N? And why now?

The answer to the first of those questions should, by now, be obvious. N2N means increasing our use of natural gas as we slowly transition to the use of more nuclear power over the next two to four decades. Because natural gas and nuclear power will have minimal negative impacts on the economy while providing significant environmental benefits, they provide the best no-regrets policy option. Indeed, natural gas and nuclear are far more environmentally friendly than the "green" energy sources that I debunked in Part 2.

Overhyped technologies such as wind power, cellulosic ethanol, and electric cars simply cannot provide the scale and reliability needed to meet global energy and power demands. Each one fails one or more of the Four Imperatives. Wind power has low power density, and without large-scale energy storage, it can't provide the always-on power that we demand. Cellulosic ethanol, too, is hamstrung by low power density, and despite decades of research, entrepreneurs still haven't found an economic way to turn wood chips and grass clippings into fuel. Meanwhile, the batteries used in today's electric cars continue to be limited by the same problem that flummoxed Thomas Edison when he wrestled with the battery challenge more than a century ago: low energy density. The density problem precludes wind, biomass, and batteries from meeting the final two of the Four Imperatives: cost and scale.

So why should we be pursuing N2N now? Again, the answer is apparent. If policymakers are serious about cutting carbon dioxide emissions and reducing air pollution, while minimizing land-use impacts and increasing the amount of energy available to their constituents, then they must embrace sources that can provide lots of power. Barring some magic solution to the energy storage problem, the incurable intermittency of wind and solar eliminates them from large-scale use. Of course, the world has plenty of coal, but coal's high carbon content and low hydrogen content is problematic. As I showed in Part 2, carbon capture and sequestration cannot, and will not, work on the scale that is needed to make a difference. The volumes of carbon dioxide are simply too large to be managed in an economic fashion.

All of those factors lead to the inevitable conclusion that the real fuels of the future are natural gas and nuclear. In fact, the future has already arrived. The world is readily embracing natural gas and nuclear power; policymakers need only provide proper encouragement. In 1973, natural gas and nuclear power combined to account for less than 20 percent of the world's primary energy consumption. By 2008, the two had a combined market share of nearly 30 percent.

For nearly four decades, natural gas and nuclear have been steadily stealing market share away from oil and coal. Between 1973 and 2008, worldwide consumption of natural gas jumped by 159 percent—faster than consumption of any other primary energy source with the exception of nuclear power, which grew by an amazing 1,253 percent. During that same time period, oil consumption rose by about 42.6 percent, and coal use increased by about 109 percent. In other words, since the 1973 Arab Oil Embargo—the tumultuous event that marked the beginning of the modern energy era—gas consumption has grown three times as fast, on a percentage basis, as oil consumption. Meanwhile, use of nuclear power grew nearly twelve times as fast, on a percentage basis, as coal.[2]

By using natural gas and nuclear power we will be able to meet the demands of the Four Imperatives while capitalizing on a number of megatrends. And those megatrends provide another set of reasons to embrace natural gas and nuclear: decarbonization, increasing use and availability of gaseous fuels, concerns about peak oil and peak coal, and increasing urbanization of the global population. The other key megatrend, which I have been discussing throughout this book, involves efforts to cut carbon dioxide emissions due to worries about climate change.

Before discussing those megatrends, it's worth looking at some of the countries that are already demonstrating a preference for natural gas and nuclear power. For instance, on August 4, 2009, the Italian utility Enel and Electricité de France agreed to build up to four nuclear reactors in Italy at a cost of up to $23 billion.[3] Two weeks later, China and Australia

FIGURE 30 Percentage Change in Global Primary Energy Consumption, 1973 to 2008

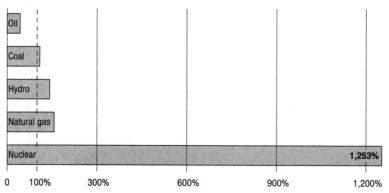

Source: BP Statistical Review of World Energy 2009, http://www.bp.com/liveassets/bp _internet/globalbp/globalbp_uk_english/reports_and_publications/statistical_energy_review _2008/STAGING/local_assets/2009_downloads/renewables_section_2009.pdf.

agreed to the largest-ever trade deal between the two countries, signing a natural gas supply agreement worth more than $40 billion that calls for Exxon Mobil to provide liquefied natural gas from Australia's giant offshore Gorgon field to PetroChina over twenty years.[4]

Italy, long opposed to nuclear power, has begun to realize that it must diversify its electricity-generation mix and reduce its reliance on imported energy, particularly natural gas from Russia. China, which has been snapping up natural resources of all kinds in recent years, is particularly interested in natural gas as a way to reduce some of the staggering pollution problems that are causing unrest among parts of its enormous population.

The deal to buy gas from Australia is just one example of China's growing appetite for natural gas. Although most energy analysts focus on China's hunger for oil and coal, the country's natural gas consumption is soaring. Between 1990 and 2008, China's consumption of natural gas jumped by 429 percent. On a percentage basis, China's gas consumption grew nearly twice as fast as its oil use (a 233 percent increase during that time frame) and nearly three times as fast as its coal consumption (165 percent).[5]

In addition to its huge hunger for natural gas, China has launched the most aggressive nuclear power program on the planet, with plans to add about 150 new nuclear reactors to its fleet. In September 2009, executives at Japan Steel Works announced that their company was spending about $1 billion on an expansion of its plant at Hokkaido. That facility is the only one that can manufacture the containment vessel needed for large reactors in a single piece. The expansion of the Hokkaido plant will allow Japan Steel Works to double its reactor vessel output to twelve per year.[6] And many of the containment vessels built there will go to China.

Of course, the Chinese are hardly alone. By late 2009, there were fifty-two reactors under construction around the globe, with total capacity of nearly 48,000 megawatts.[7] That amount of nuclear capacity is nearly equal to all of the electric-generating capacity of Mexico.[8] In addition to China, the countries now constructing new reactors include Argentina, Canada, Finland, France, India, Iran, Japan, Pakistan, Russia, Slovakia, South Korea, and the United States.

In addition, numerous other countries, including Belarus, Brazil, Bulgaria, Egypt, Indonesia, Kazakhstan, North Korea, Romania, South Africa, Thailand, Turkey, Ukraine, United Arab Emirates, United Kingdom, and Vietnam, are planning to add reactors. In all, some 150,000 megawatts of new nuclear power capacity is being planned.[9]

The power of N2N can be seen by looking at the countries that are simultaneously embracing both natural gas and nuclear power. For instance, between 1990 and 2008, gas demand in South Korea soared by nearly 1,100 percent—faster than that of any other country.[10] As for nuclear, by late 2009 South Korea had six new reactors under construction with a total capacity of 6,700 megawatts. Or look at Brazil, where natural gas use has increased by more than 700 percent since 1990, and where the government is now considering more than 5,000 megawatts of new nuclear power.[11] In China, gas consumption has soared, and more than 17,000 megawatts of new nuclear capacity are under construction.[12] Finally, consider India, where gas consumption has jumped by 243 percent since 1990, and where six new reactors are now under construction.[13] Natural gas and nuclear power are being embraced by countries all over the world because they provide clean, baseload, low-carbon or no-carbon power. And that's what every country wants. Although natural gas provides tremendous benefits, it cannot match nuclear power when it comes to power density. As discussed earlier in this book, nuclear reactors have very high power density, meaning they can produce large amounts of power from small amounts of real estate. That advantage is a direct result of the incredible amounts of energy that are produced by fission. The energy released from nuclear reactions are about 10 million times larger than those from chemical reactions. Or as one recent book, *A Cubic Mile of Oil*, explained it, about 2,000 tons of uranium-235 "can release as much energy as burning 4.2 billion tons of oil." That's the equivalent of about 30.7 billion barrels—or one cubic mile—of oil.[14]

The surging use of natural gas and nuclear power demonstrates and reinforces one of the most important energy megatrends of the modern era: decarbonization.

Decarbonization is the ongoing global trend toward consumption of fuels that contain less carbon. This megatrend was first identified by a group of scientists that included Nebosa Nakicenovic, Arnulf Grübler, Jesse Ausubel, and Cesare Marchetti,[15] who found that over the past two centuries, the process of decarbonization has been taking place in nearly every country around the world. Because consumers always want the cleanest, densest forms of energy and power that they can find, the trend will surely continue. The ratio of carbon to hydrogen atoms in the most common fuels tells the story.

From prehistory through, say, the 1700s and early 1800s, wood was the world's most common fuel. Wood has a carbon-to-burnable-hydrogen ratio (C:H) of about 10:1. That is, it contains about 10 carbon atoms for every 1 burnable hydrogen atom. But wood eventually lost its dominance to coal, which has far higher energy density and a C:H ratio of about 2:1. Coal lost out to oil, which has even higher energy density as well as easier handling characteristics. In addition, oil has a C:H ratio of about 1:2. Now we are seeing the rise of natural gas (methane), which, as its chemical formula (CH_4) suggests, has a C:H ratio of 1:4, or 1 carbon atom for every 4 hydrogens. In 2005, Marchetti, an Italian physicist, declared that for the next five decades, "methane is to be the dominant primary energy."[16]

The hunger that consumers have for cleaner sources fits perfectly with the goal that policymakers have decided should be a top global priority: reducing carbon dioxide emissions. In late 2008, Nobuo Tanaka, the executive director of the IEA, averred that "preventing irreversible damage to the global climate ultimately requires a major decarbonization of world energy sources."[17] Of course, not all countries are decarbonizing at the same rate. And some countries, including China and

India, are increasing, rather than decreasing, their coal consumption. But the long-term decarbonization of the global economy is continuing, and given concerns about climate change, that trend is likely to accelerate as countries around the world build more nuclear reactors and increase their consumption of natural gas.

Decarbonization favors natural gas and nuclear power at the same time that environmentalists and some politicians are working to impose country-by-country limits on carbon dioxide emissions. The efforts to limit those emissions go back to 1992 with the Earth Summit in Rio de Janeiro. They were formalized in 1997 with the Kyoto Protocol, which decreed that countries should cut their emissions by 5.2 percent below 1990 levels by 2012. But only a handful of countries have met their obligations under the protocol.[18] Nevertheless, in July 2009, UN Secretary-General Ban Ki-moon declared that global carbon dioxide emissions cuts are essential, saying they were "politically and morally imperative and a historic responsibility for the leaders for the future of humanity, even for the future of planet Earth."[19]

The efforts to impose carbon dioxide limits and decarbonize the world's energy sources are occurring at the same time that countries around the world are working to improve air quality. And air quality is an area where natural gas and nuclear power hold advantages over oil and coal. During combustion, natural gas emits about half as much carbon dioxide as coal and releases no particulates. Nor does it release significant quantities of sulfur dioxide or nitrogen oxides, two of the most problematic air pollutants. In 2005, the Environmental Protection Agency issued the Clean Air Interstate Rule, which aims to cut those two pollutants by as much as 70 percent by 2015. In addition, the agency has issued the Clean Air Mercury Rule, which aims to cut the releases of mercury from coal-fired power plants.[20] Those federal requirements will, in the coming years, favor natural gas and nuclear power over the use of coal.

The decarbonization trend is closely connected to another megatrend as well: the increasing use and availability of gaseous fuels.

The hydrogen economy remains decades away. But the increasing use of natural gas confirms the projections made by analysts who have predicted that the world's consumers will increasingly replace solid and liquid fuels with gaseous ones. Of course, the world will continue using coal, and lots of it, and oil, and lots of it, for many decades to come. But the trend toward gaseous fuels appears to be gathering speed. As Roberto F. Aguilera of the Vienna-based International Institute for Applied Systems Analysis put it in a March 2009 analysis, "we are on our way towards a methane economy that could pave the way to a hydrogen economy."[21] The trend toward gaseous fuels and the decarbonization trend are companions. About 95 percent of the hydrogen now being produced is derived from natural gas.[22] Thus, the long-anticipated and much-hyped "hydrogen economy," if it ever arrives, must begin with a methane economy. And over the past few years, thanks to improvements in drilling and recovery technologies, the estimated volumes of the world's recoverable natural gas resources have skyrocketed.

It's worth noting that, although the megatrend toward increased use and availability of gaseous fuels is now apparent, just a few years ago, some of the energy industry's top people were convinced that we were running out of gas.

□□□

In 2005, Lee Raymond, the famously combative CEO of Exxon Mobil, declared that "gas production has peaked in North America."[23] Raymond, who retired from the oil giant in 2006, said that his company was intent on building a new pipeline that would bring Arctic gas from Canada and Alaska south, and that more natural gas supplies would be needed "unless there's some huge find that nobody has any idea where it would be."[24] Raymond wasn't alone. In 2004, Julian Darley, a British writer, published *High Noon for Natural Gas: The New Energy Crisis*, a book documenting how a pending shortage of natural gas in the United States and Canada could "plunge" the two countries "into a carbon chasm, a hydrocarbon hole, from which they will be hard put to emerge unscathed."[25]

In 2003, Matthew Simmons, a Houston-based investment banker who had been issuing dire warnings about peak oil for years, predicted that natural gas supplies were about to fall off a "cliff." When asked about the future of natural gas supplies, Simmons said that "the solution is to pray. . . . Pray for no hurricanes and to stop [sic] the erosion of natural gas supplies. Under the best of circumstances, if all prayers are answered there will be no crisis for maybe two years. After that it's a certainty."[26]

The same year that Simmons petitioned for prayer, author Richard Heinberg published *The Party's Over*, a book filled with ominous warnings about the looming end of the hydrocarbon age. Regarding natural gas, he said there were "disturbing signs that rates of natural gas extraction in North America will soon start on an inexorable downhill slope—perhaps within a few months or at most a few years. When that happens, we may see a fairly rapid crash in production."[27]

Wrong, wrong, wrong, wrong.

In 2008, U.S. natural gas production hit 56.2 billion cubic feet per day, its highest level of production since 1974.[28] And given the abundance of new shale gas resources, some optimistic analysts are projecting that U.S. gas production could increase by 50 percent or more over the next two decades.

Of course, no one knows how much gas will be produced over the coming years. But the U.S. gas business has repeatedly shown that when it's not hampered by excessive governmental regulations, it can provide as much gas as the market needs. But to provide that gas, the U.S. gas industry, and more particularly, politicians and regulators, must once and for all overcome the notion that gas is scarce.

In less than five years, the global natural gas business has gone from shortage to surfeit. The industry has gone from concerns about having adequate available resources to a situation where a near-term glut—meaning the next two to five years, or perhaps even longer—is likely. That glut is due to several factors, including the global economic downturn, a surge in new natural gas liquefaction capacity, and, perhaps most important, the refinement of technologies that can unlock vast quantities of gas from shale deposits. Indeed, the shale gas revolution is what long-time gas analyst H. deForest Ralph has called the "black swan" in the gas business.[29]

The well-drilling and well-completion techniques that were perfected in the Barnett Shale in Texas proved that, when properly fractured under high pressure, shale—a rock source that has very low permeability—could yield enormous quantities of gas. But shale is only part of the story. Similar technological advances have resulted in increased gas production from other geologic formations with low permeability, such as coal beds and tight sands. These low-permeability reservoirs produce what the industry calls "unconventional" gas. And the potential volumes of unconventional gas resources are staggering.

In April 2009, Cambridge Energy Research Associates produced a study that estimated recoverable shale-gas resources outside of North America at between 5,000 and 16,000 trillion cubic feet.[30] Seven months later, the IEA estimated recoverable global gas resources—which includes both conventional and unconventional gas—at some 30,000 trillion cubic feet. That's the energy equivalent of about 5.4 trillion barrels of oil.[31]

Estimates of U.S. gas resources have escalated alongside those global estimates. Over the past two years, several reports have put potential U.S. gas resources on par with the gas reserves of Iran, Russia, and Qatar. In July 2008, Navigant Consulting put potential U.S. natural gas resources at 2,247 trillion cubic feet.[32] A few months later, another consulting firm, ICF International, estimated U.S. gas resources at 1,830 trillion cubic feet.[33]

Although those studies are important, a mid-2009 estimate by the Potential Gas Committee, a U.S. nonprofit organization, is the most credible. The Gas Committee, made up of experts from academia, the energy sector, and government, issues a report every two years. On June 18, 2009, it issued a report that estimated U.S. gas resources at 2,074 trillion cubic feet—the highest resource evaluation in the committee's forty-four-year history.[34] That quantity of gas is the energy equivalent of more than 350 billion barrels of crude oil, or about three times as much as the proved oil reserves of Iraq.[35] The 2009 estimate by the Gas Committee was a 35 percent increase over the estimate published in 2007, and like the reports from Navigant and ICF, it pointed to the boom in shale gas as a key reason for the big hike in the resource estimate.

The lead author of the report, John Curtis, a professor in geology and geological engineering at the Colorado School of Mines, cited the reasons for the big increase in resource estimates: "New and advanced exploration, well drilling and completion technologies are allowing us increasingly better access to domestic gas resources—especially 'unconventional' gas—which, not all that long ago, were considered impractical or uneconomical to pursue."[36]

Now, to be clear, resources are not reserves. In oil-field parlance, a "resource" is something that's probably out there. "Reserves" only applies to in-the-ground hydrocarbons that have been surveyed by drilling and

FIGURE 31 From Scarcity to Super-Abundance: U.S. Gas Resources
Compared to Proved Gas Reserves of Iran, Russia, and Other Countries

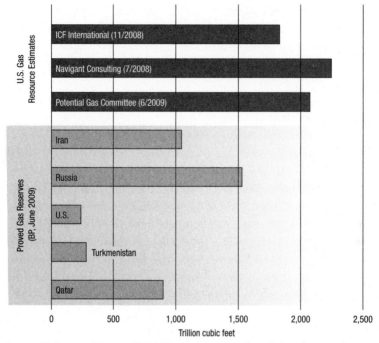

Sources: BP Statistical Review of World Energy 2009, http://www.bp.com/liveassets/
bp_internet/globalbp/globalbp_uk_english/reports_and_publications/statistical_energy
_review_2008/STAGING/local_assets/2009_downloads/renewables_section_2009.pdf; ICF
International, Table 7, "Availability, Economics, and Production Potential of North American
Unconventional Natural Gas Supplies," November 2008, http://www.ingaa.org/cms/31/
7306/7628/7833.aspx, 51; Navigant Consulting, "North American Natural Gas Supply
Assessment," July 4, 2008, http://www.cleanskies.org/upload/MediaFiles/Files/
Downloads2/finalncippt2.pdf, 14; Potential Gas Committee, "Potential Gas Committee
Reports Unprecedented Increase in Magnitude of US Natural Gas Resource Base," June 18,
2009, http://www.mines.edu/Potential-Gas-Committee-reports-unprecedented-increase-in
-magnitude-of-US-natural-gas-resource-base.

other agreed-upon techniques. In 2009, BP estimated proved global gas
reserves at about 6,534 trillion cubic feet, or less than a quarter of the
IEA's 2009 estimate of global gas *resources*.[37] The same distinctions apply
to the United States. BP puts proved U.S. gas *reserves* at about 238 trillion cubic feet, or about one-tenth of the Potential Gas Committee's
2009 estimate of U.S. gas *resources*.[38]

While the resources-versus-reserves caveat applies, the enormous gas estimates being put forward are part of a new gas paradigm that is based on abundant supplies of methane in a marketplace that is increasingly global. The growing globalization of gas can be seen by looking at the surge in trade of liquefied natural gas. Between 2000 and 2007, global trade in LNG increased by 67 percent.[39] And more LNG trade is on the way. By 2013 or so, the IEA expects global LNG production capacity to increase by about 50 percent.[40] And yet more gas liquefaction capacity is being planned.

In October 2009, the British-Dutch energy giant, Shell, announced plans to build a floating natural gas liquefaction facility that, when completed, will be deployed off the northwestern coast of Australia. The vessel, expected to cost about $5 billion, will be nearly 500 meters long and will be designed to monetize what the industry calls "stranded" gas, that is, gas that is in fields that are either too small or too far from commercial centers to justify the construction of pipelines or conventional gas liquefaction facilities. Australia alone has some 140 trillion cubic feet of stranded gas, and Shell may build as many as ten floating LNG production vessels, which could allow Australia and other countries to turn their stranded gas into cash. In addition to Shell, Japan's Inpex Holdings and Australia's Santos are also considering floating LNG projects.[41]

In November, Petrobras, the Brazilian energy giant, announced that it, too, is planning to build floating natural gas liquefaction facilities in order to bring ashore the massive amounts of gas that it has discovered in the Santos Basin, one of Brazil's huge offshore hydrocarbon reservoirs. The company said the floating platforms will operate about 190 miles offshore and will allow the offshore gas reserves "to be monetized, ensuring flexibility to supply the internal market and the possibility of exporting the product in the spot market."[42]

The new LNG production capacity is coming onstream at the same time that, thanks to the boom in shale gas production, America's need for LNG imports has largely disappeared. That point was made succinctly by Ian Cronshaw, a gas analyst at the IEA, who said that the United States was "now a virtual liquefied natural gas exporter because all the LNG that was supposed to be going there is now going somewhere else."[43]

The increasing globalization of the LNG market is occurring at the same time that the technologies for producing gas from shale are going global. In Canada, drillers have begun tapping two massive shale formations, Horn River and Montney.[44] Those resources are so big that there is even talk of building a gas liquefaction terminal in British Columbia that would use Canadian shale gas as a feedstock. The company proposing the idea, Calgary-based Kitimat LNG, claims that it already has potential buyers for the LNG in Asia.[45]

European countries see the shale gas revolution as an opportunity to reduce their dependence on the Kremlin-based kleptocrats who control the Russian gas business. The Europeans are justifiably worried about Russia. Gazprom, the Russian gas giant, has repeatedly shown its willingness to shut off gas flows to its neighbors. Those concerns have led European energy companies to invest in some of America's best shale-gas producers. In late 2008, Norwegian energy giant StatoilHydro paid Oklahoma City–based Chesapeake Energy nearly $3.4 billion to gain a stake in the massive Marcellus Shale play in Appalachia.[46] And in mid-2009, StatoilHydro and Chesapeake announced that they are assessing shale formations in China, India, Australia, and other countries.[47]

A host of companies have begun seeking permits to tap shale gas prospects in France.[48] In mid-2009, Exxon Mobil started preliminary exploration drilling in Germany, while ConocoPhillips started prospecting on a 1-million-acre section of Poland. Meanwhile, the Italian energy giant, Eni, paid $280 million to gain a stake in a Texas gas field operated by Fort Worth–based Quicksilver Resources so that Eni can learn some of Quicksilver's unconventional-gas production techniques.[49]

The surge in shale gas production is occurring at the same time that increasing numbers of countries are producing significant quantities of gas. Back in 1970, only about ten countries were producing more than 1 billion cubic feet of gas per day.[50] By 2008, there were forty-one countries producing at least 1 billion cubic feet of gas per day.[51]

One other important development in the megatrend toward increased use and availability of gaseous fuels is that new gas reserves are being found at a faster rate than new oil reserves are. Between 1988 and 2008, global proved natural gas reserves (remember, the numbers for reserves are always smaller than those for resources) jumped by more than 68 percent, reaching 6,534 trillion cubic feet.[52] During that same time period, global oil reserves increased by just 26 percent, to some 1.26 trillion barrels.[53] By any measure, that is a tremendous amount of petroleum. But the world's gas reserves are likely to last substantially longer than its oil reserves. In 2009, the reserves-to-production ratio for global proved oil reserves was 42.[54] That means that at current rates of extraction, the world's known oil reserves will be exhausted in 42 years. For comparison, the reserves-to-production ratio for natural gas is 60.4, meaning that at current extraction rates—and assuming no more gas is discovered—the world has more than 60 years of proved natural gas reserves left in the ground.[55] And those numbers will undoubtedly grow in the years ahead as more gas resources get moved into the reserves category.

The megatrend of increasing natural gas consumption and natural gas availability could scarcely be occurring at a better time. Over the past few years, worries about peak oil, and another possible peak—peak coal—have emerged as serious concerns. And that leads us to our next megatrend.

□□□

Peak oil is one of the most emotional and hotly debated issues in the energy business. There is widespread disagreement about when the world will hit peak production. For instance, former Princeton University geology professor Kenneth Deffeyes claims the peak was hit on Thanksgiving Day, 2005.[56] Veteran Houston-based energy analyst Henry Groppe says it occurred in 2006, and that the increases in production since then are due to production of natural gas liquids, not of crude.[57] Houston stock analyst Marshall Adkins, of the brokerage firm Raymond James, contends the peak was hit in 2008.[58] Other energy watchers claim that the global economic downturn has delayed any discussion about peak oil for the time being, perhaps until 2020 or so.[59]

What will reaching peak oil mean? Well, it will almost certainly mean higher prices. But how much higher? Groppe, the dean of the Houston energy analysts, with more than five decades of experience in the sector, provides a succinct prediction: "The price of oil will have to be whatever is required to cause total consumption to decline."[60] Put another way, oil prices will rise to whatever level is needed in order to make alternative energy sources more economic and thereby cut demand for oil products. As the old saying goes, the cure for high oil prices is high oil prices.

Of course, it's not known how much pain the economies of the United States and other countries will feel when the relatively high oil prices caused by a peak in production begin to kick in. Nor is it clear what will happen with oil demand. In fact, a peak in oil demand may be just as important as a peak in oil production.

The peaking of oil demand in the United States would be driven by a number of factors, including the economy, the age of the population, and the efficiency of the automotive fleet. The stagnating demand at present, due to the economic downturn, the aging of the population, and the increasing efficiency of the automotive fleet, can be seen in the numbers. In 2008, U.S. oil consumption fell to 19.5 million barrels per day, a level that's just 3.5 percent higher than it was back in 1978.[61]

In June 2009, Cambridge Energy Research Associates issued a report projecting that U.S. demand for automotive fuel would peak in 2014. That outlook agrees with that of the world's biggest publicly traded energy company, Exxon Mobil, which expects U.S. demand for transportation fuel to plateau by 2015 and then fall by about 10 percent by 2030.[62] Oil giant BP believes that U.S. demand has already peaked. In November 2009, the company's chief executive, Tony Hayward, declared "We will never sell more gasoline in the US than we sold in 2007."[63]

Meanwhile, some analysts are now predicting a similar peak in global oil demand. In October 2009, analysts at Deutsche Bank predicted that world petroleum consumption would peak in about 2015 at around 90 million barrels per day.[64] If those projections prove to be correct, then the United States, and much of the rest of the world, may soon have too much oil-refining capacity, and maybe too much oil-production capacity.

In fact, the United States already has too much refining capacity. In November 2009, Valero Energy, the biggest independent refiner in the country, announced that it was permanently shutting down its 210,000-barrel-per-day refinery in Delaware because of slack motor-fuel demand. The refinery had been losing $1 million per day. Three months earlier, Valero announced that it was halting production at a 235,000-barrel-per-day refinery in Aruba indefinitely. Valero had tried to sell the refineries but couldn't find buyers.[65]

Though we cannot predict the future, we can look backward and see that the beginning of the latest economic recession—like many recessions before it—coincided with a major spike in oil prices. History shows that sharp increases in oil prices are often followed by recessions. Those oil price spikes also lead to sharp decreases in oil demand. For instance, in 1978, U.S. oil consumption peaked at 18.8 million barrels per day. But the high prices that came with the 1979 oil shock, the second big price spike in six years, sent U.S. consumption tumbling. In fact, it took two decades for U.S. oil demand to recover after the price shocks of the 1970s.

It wasn't until 1998, when U.S. consumption hit 18.9 million barrels per day, that the 1978 level of consumption was surpassed.[66] And it took two decades for oil demand to recover, even though oil prices were remarkably low. From the mid-1980s through the early 2000s, prices largely stayed under $20 per barrel, and they even fell as low as $9.39 per barrel in December 1998.[67]

As we look forward, we can be sure that any peak in production, or spike in oil prices—regardless of the causes—will make consumers more judicious in their oil use. Groppe has pointed out that oil consumption in the OECD countries peaked way back in 1979 and has been flat or declining ever since. Future increases in oil prices will continue wringing less-efficient uses of oil—such as burning petroleum to make electricity— out of the system.

While peak oil discussions dominate the headlines, peak coal may be even more significant. Over the past two or three years, several analysts have been taking a close look at global reserves and production trends, and they are concluding that the peak in global coal output could come sooner than most people think.[68] Among the first to reach this conclusion was David Rutledge, an electrical engineering professor at the California Institute of Technology.[69] Although Rutledge spends much of his time working on radio and microwave circuits, he was intrigued by questions about peak oil and became interested in global coal resources. Rutledge looked at coal production histories and the reserves estimates in the United Kingdom, the United States, and other countries. His conclusion: Global coal production will peak at levels far lower than what many reports have projected.[70] Although Rutledge is somewhat reluctant to put an exact date on when he thinks global coal output will peak, during an interview in August 2009 he told me that it's reasonable to assume that coal production would peak within the next decade. Rutledge is particularly dubious about

China's ability to continue increasing its coal use. "If something can't go on forever, it won't," he told me.[71]

Other researchers are doing work that parallels Rutledge's. Tad Patzek, the head of the petroleum engineering department at the University of Texas at Austin, and Gregory Croft, a doctoral candidate in engineering at the University of California at Berkeley, have come to similar conclusions. Patzek and Croft have concluded that world coal production will peak in 2011. Furthermore, in a report that they completed in 2009, they projected that global coal production "will fall by 50% in the next 40 years" and that carbon dioxide emissions from coal combustion will fall by the same percentage.[72] For Patzek and Croft, the implications of the looming peak in coal production makes it apparent that the world must focus increasing effort on energy efficiency and that "new nuclear power stations should also be designed and put online fast, not because of the GHG [greenhouse gas] emissions issues, but because of the insufficient supply of coal."[73]

The physical production limits on oil and coal may keep carbon dioxide emissions far below the projections put forward by the Intergovernmental Panel on Climate Change, which has said that carbon dioxide concentrations could reach almost 1,000 parts per million by 2099.[74] In his analysis, Rutledge predicted that due to peak coal, global carbon dioxide concentrations will not rise much above 450 parts per million by 2065. If his predictions are correct, then some of the worry about future carbon dioxide emissions may be misplaced.

But N2N offers the most viable way to hedge our bets with regard to both peak oil and peak coal. To be sure, both oil and coal will continue to be key sources of primary energy throughout the world for the rest of the twenty-first century, but it is also apparent that the inevitable production plateaus of those fuels will force consumers to find alternatives. And natural gas and nuclear power are the only alternatives that can provide the scale of energy supplies needed to substitute for some of the expected declines in oil and coal production. In other words, natural gas and nuclear power can be used as a hedge against the looming "twin peaks."

Embracing N2N can also help the world deal with another megatrend: increasing urbanization. In his book *Whole Earth Discipline*, Stewart Brand wrote, "In 1800 the world was 3% urban; in 1900, 14% urban; in 2007,

50% urban. The world's population crossed that threshold—from a rural majority to an urban majority—at a sprint. We are now a city planet." According to Brand, "Every week there are 1.3 million new people in cities. That's 70 million a year, decade after decade. It is the largest movement of people in history."[75]

The migration to cities requires more clean energy sources that people can use in their homes. Coal and wood won't do; natural gas and nuclear power are the obvious choices to provide the cooking and heating fuel that city-dwellers need as well as the electricity they use to turn on their lights and keep their computers, entertainment centers, and appliances running.

Given these megatrends—decarbonization, increased use and availability of gaseous fuels, concerns about peak oil and peak coal, increasing urbanization, and continuing worries about carbon dioxide emissions—it makes sense for the United States to begin promoting N2N as a winning long-term strategy. Natural gas and nuclear power plants require far less land than wind and solar installations; both have lower carbon emissions than oil or coal; they emit no air pollutants; and both pass the challenges of cost and scale.

Although N2N is the obvious way forward, we must be realistic. Adding significant quantities of new nuclear power capacity in the United States will take decades. In the meantime, the United States should focus on the first N: natural gas. But emphasizing natural gas will be difficult, particularly given the strength of the coal lobby in Congress. It will also require Congress and federal regulators to overcome a decades-long parade of bad legislation that has had the perverse effect of stifling U.S. natural gas production and reducing gas's share of the U.S. primary energy market. Much of that legislation was based on a mistaken notion that America's natural gas resources were running out.

Thus, before looking forward to what should happen next with the U.S. gas sector, we need to take a quick look back at the history of an energy source that was once viewed as being nearly worthless.

CHAPTER 22

A Very Short History of American Natural Gas and Regulatory Stupidity

> The choice is not between cheap and expensive
> natural gas, because there is no such thing
> as a plentiful supply of cheap gas.
>
> Editorial, *Los Angeles Times*, May 26, 1975[1]

FOR DECADES, the calculus in the oil field was simple: Oil was cash. Natural gas was trash.

Oil was easily transported—by pipeline, rail, or truck—and it had a huge and growing market in the burgeoning fleet of trucks and automobiles. It could also be used for manufacturing, lubrication, heating, cooking, and lighting. Oil was relatively easy to handle. If a well blew out, the petroleum could be diverted into a hastily dug pit. Longer-term storage could be handled with relatively inexpensive wooden tanks, which would suffice until the oil could be transported.

Gas, on the other hand, was difficult to transport and had a nasty habit of exploding. The fuel could only be profitably moved by pipeline, and other than heating and municipal lighting systems—which required good pipelines—it had few apparent uses. That meant the fuel was nearly worthless to the oil companies—and in the early days of the industry, that's what they were, oil companies. Few companies targeted natural gas when drilling. Instead, they were forced to deal with the natural gas that was mixed with the oil. In the industry, that type of gas is known as "associated gas." And given that it was mixed with oil and was hard to get rid of, it was viewed as more of a problem than an asset. Thus, during the first few decades of the U.S. oil industry, there was no "gas industry" to speak of. The idea of actually drilling for natural gas on purpose—or, in industry parlance, looking for "unassociated gas"—was rare.

A few entrepreneurs did see value in natural gas. In 1891, America's first significant high-pressure gas pipeline was built by Indiana Gas and Oil to carry natural gas from a field in Indiana to customers in Chicago, a distance of about 120 miles. However, the pipe was a leaky mess, and by 1907, it was shut down. Two decades later, there were only a handful of natural gas pipelines in the entire country, and many of them were plagued by leaks and inefficient operations. It wasn't until the early 1920s—when manufacturers began making significant quantities of seamless steel pipe, which could be joined using new technologies such as oxyacetylene and electric welding—that pipeline builders were able to string together longer sections of high-strength steel pipe.[2] But even with those technologies, by the mid-1920s the longest gas pipeline in the United States was only about 300 miles long.

By 1930, when oil was selling for about $1 per barrel, natural gas was selling in some locales for as little as $0.03 per thousand cubic feet—if a market for the gas was even available.[3] With scant pipeline capacity and little monetary interest in selling the gas, most producers looked for ways to simply get rid of the fuel. During the 1940s, gas was so abundant in the areas around Amarillo, Texas, that the city offered companies free natural gas for five years if they agreed to set up shop in the city and employ at least fifty people.[4] The city got no takers. With few uses for the gas, huge quantities of the fuel were simply vented into the atmosphere. That worked, but it also came with the risk of explosion. (Furthermore, we now know that methane is a greenhouse gas that is about twenty times more effective at trapping heat in the atmosphere than carbon dioxide.)

It was both safer and better, from an environmental standpoint, to burn the gas near the wellhead. Companies with too much gas on their hands put up tall pipes near their producing wells and set the gas aflame.

During the 1930s and 1940s in Texas, gas flares were so common, and so bright, that in parts of the Lone Star State motorists could drive for hours at night without ever needing to flip on their headlights. According to historian David Prindle, "Miles away from any major oil field, newspapers could be read at night by the light" from the flares. In 1934 alone, about 1 billion cubic feet of gas per day were being flared in the Texas Panhandle.[5]

Flaring might have continued but for the intervention of the Texas Railroad Commission, the Austin-based agency that from the 1930s through 1973 determined production levels for key U.S. oil fields, which had the effect of largely determining global oil prices. In 1947, after years of wrangling with the industry, the agency passed a rule that prohibited the flaring of gas. The agency determined that gas was a valuable natural resource and therefore should be conserved.[6] Some producers fought the rule, but the courts sided with the agency. The Railroad Commission's move forced Texas producers to either reinject the natural gas into their wells—a move that helped to maintain the oil reservoir's natural pressure, and therefore its long-term productivity—or put it into a pipeline.[7] The ban on flaring was later adopted by other oil-producing states in the United States, and the domestic gas industry began to mature.

Between 1949 and 1957, U.S. gas consumption doubled, and in 1958, gas surpassed coal to become the second-largest source of primary energy in the country.[8] By the late 1950s, gas looked ready to rob even greater market share away from coal. But just as that energy transition was beginning, natural gas became a favored target for federal regulators. And the hodgepodge of regulations that resulted would hamstring the U.S. gas industry for decades. The result was predictable: U.S. gas consumption fell, coal consumption rose, and along with increased coal use came increased emissions of carbon dioxide.

␣␣␣

The case was known as *Phillips Petroleum Co. v. Wisconsin*. In 1954, just four years before natural gas eclipsed coal as the second-biggest source of primary energy in the United States, the U.S. Supreme Court issued a ruling in the Phillips case that gave federal authorities the power to set prices on gas that was sold into the interstate market. Federal regulators were empowered to set prices based on the gas producers' costs, plus a "fair" profit.[9] But the prices deemed "fair" by federal regulators were often uneconomic for gas producers. The result was a downward spiral in the amount of gas destined for sale across state lines. Gas producers could sell their production to customers inside their own state (intrastate) for a mutually agreed-on price. But if those same producers—say, from gas-rich states such as Oklahoma or Texas—wanted to sell their gas to a pipeline for delivery to customers in, say, Ohio or Illinois, then that interstate gas was subject to federal price controls. In short, the price controls effectively prevented producers from sending their gas to the most lucrative markets, and that, predictably, slowed the growth of the U.S. gas industry.

By 1971, a shortage of natural gas was looming. That year, Monty Hoyt, a reporter for the *Christian Science Monitor*, summed up the situation: "The wellhead price for interstate natural gas has been kept artificially low for more than a decade" by federal regulators, and "this has stimulated demand for the 'Mr. Clean' of fuels, while at the same time discouraging exploration and drilling of new wells. Reserves have dwindled as a result." He went on to say that "cheap natural gas has caused utilities and major industries to switch to gas-burning boilers," a move that depressed demand for other boiler fuels such as coal and oil.[10]

In January 1973, the *Washington Post* reported that the United States was in the midst of a "chronic shortage of natural gas," and when a cold snap hit the nation that month, energy shortages were widespread. Schools and factories in eleven states, from Colorado to Ohio, were forced to close. Homeowners in several states, including Michigan, Minnesota, and Illinois, were running short of natural gas. The havoc wasn't limited to natural gas. Price controls imposed by the Nixon administration on refined oil products led to widespread shortages.[11]

But despite the recurring shortages, some of the energy industry's loudest critics simply couldn't fathom the concept that government price controls might be the problem rather than the solution. In 1974, S. David Freeman, a lawyer who later became the head of the Tennessee Valley Authority, wrote an opinion piece for the *Los Angeles Times*, in which he insisted that price controls on natural gas were essential and must be continued. His solution to America's energy crisis was more governmental intervention to the point where the government, he said, would "provide everyone with a smaller ration of gasoline."[12]

Fortunately, Freeman's grand plan for rationing motor fuel didn't gain any traction. But his belief in the need to keep energy scarce and expensive gained plenty of traction in California. About seven months after Freeman wrote his piece for the *Los Angeles Times*, the paper's editorial board published its own piece, arguing that natural gas prices should be deregulated—but only for "new gas." That is, any gas discoveries that were made from 1975 onward should be allowed to sell their gas "in a fundamentally free market." But existing gas supplies would continue to be sold at prices set by federal regulators. The *Times* went on to contend that a "windfall profits tax" should be imposed on the gas producers who "failed to plow most of the profits back into the hunt for new supplies." The editors concluded their May 26, 1975, opinion piece by declaring that "the choice is not between cheap and expensive natural gas, because there is no such thing as a plentiful supply of cheap gas."[13]

That belief—that there's no such thing as a plentiful supply of cheap gas—continued to be the dominant mindset throughout the mid- to late 1970s. And that mindset assured that the federal price controls stayed in place. In January 1977, when the United States was hit by another blast of nasty winter weather, natural gas shortages came roaring back. The shortages were so severe that a utility in Buffalo, New York, asked its business, industrial, and school customers to shut down for two days and suggested that residential customers turn their thermostats down to 55 degrees. In Ohio, gas utilities were forced to cut off supplies to 4,500 industrial customers, and automobile plants in Michigan and Ohio were closed, putting 56,000 people out of work.[14] By late January, some 300,000 workers had been laid off as a result of the shortages of natural gas. The crisis was so acute that on January 26, 1977, President Jimmy Carter, who'd just been sworn into office about a week earlier, asked Congress for emergency legislation that would temporarily lift the federal price controls on interstate gas shipments.[15]

Throughout the late 1970s, ideologues such as Freeman and other Carter-era energy bureaucrats were convinced that there was a geological shortage of natural gas and therefore the United States had to embark on an alternative path away from the fuel. In 1977, John O'Leary, the administrator of the Federal Energy Administration, told Congress that "it must be assumed that domestic natural gas supplies will continue to decline" and that the United States should "convert to other fuels just as rapidly as we can." That same year, Gordon Zareski of the Federal Power Commission testified before Congress and stated that U.S. policies "should be based on the expectation of decreasing gas availability." He went on to say that annual production of natural gas would "continue to decline, even assuming successful exploration and development of the frontier areas."[16]

What O'Leary and Zareski didn't tell Congress was that within gas-rich states such as Texas, there was a surfeit of gas. Why? The answer was simple: Gas producers could get good prices for their product by keeping it within the state's borders. In mid-1973, when federal inter-state prices were about $0.25 per thousand cubic feet, some Texas producers were selling their production intrastate for as much as $0.82.[17] By 1978, Texas had so much gas in its intrastate market that the Texas Rail-road Commission began acting as a type of referee for the intrastate sales so that it could more closely match output with demand.[18] The havoc caused by federal meddling in the interstate gas market was recognized by energy-industry historian Ruth Sheldon Knowles, who, in a 1977 opinion piece for the *Los Angeles Times*, wrote that "the nation's natural gas shortage was created by interstate controls. By contrast, the free-market approach has actually increased supplies for intrastate use." The reason federal price controls on interstate gas persisted was that "keep-ing the price artificially low—that is, below its true market value—has been politically popular."[19]

The mistaken belief that the United States was running short of nat-ural gas was politically advantageous for the coal producers, who were eager to limit competition from the oil and gas business. And those coal producers had powerful friends in Congress, including, but not limited to, Senator Robert Byrd, the powerful Democrat from West Virginia.

In 1978, Byrd and his allies convinced Congress to pass two bills that would haunt the gas industry for years: the Powerplant and Indus-trial Fuel Use Act and the Natural Gas Policy Act. The most important provision of the Fuel Use Act was that it prohibited the use of gas for electricity generation. Meanwhile, the Natural Gas Policy Act created a briar patch of categories for gas pricing based on whether it was sold in interstate or intrastate commerce, what type of wells were involved, and even how deep the wells were.[20] In the wake of the legislation, gas consumption plummeted. By 1986, natural gas consumption had fallen to about 44 billion cubic feet per day—a level not seen in the United States since 1965.[21]

In 1987, Congress finally reversed course and repealed the Powerplant and Industrial Fuel Use Act. Although the law was in effect for less than a decade, it did plenty of damage. The ban on the use of natural gas for power generation led many electric utilities—which were seeing booming demand for electricity—to build more coal plants.[22] In 1978, natural gas was generating 13.8 percent of U.S. electricity. By 1988—a decade after the Powerplant and Industrial Fuel Use Act was passed—natural gas's share of the U.S. electricity business had fallen to a modern low of just 9.3 percent.

The winner of all this federal intervention was coal. Between 1973 and 2008, coal's share of the primary energy market jumped from 18 percent to 25 percent.[23] Much of that coal was used for electricity generation. Between 1978—the year that Congress passed the Powerplant and Industrial Fuel Use Act—and 1988, coal's share of the U.S. electricity generation market soared, going from 44.2 percent to 56.9 percent, the highest level of the modern era.[24]

The irony here is almost too great. Today, Congress is working mightily to impose a cap-and-trade system or some similar plan aimed at reducing carbon dioxide emissions, and in particular to reduce coal consumption. Had the natural gas sector that was booming back in the 1950s been allowed to continue without excessive federal regulation, U.S. carbon emissions would undoubtedly be lower today, because more gas would be employed for electricity generation and other uses. That would have meant less coal consumption, and therefore lower carbon emissions and less coal-related pollution of mercury, soot, and sulfur dioxide.

In other words, Congress is now trying to find a cure for some of the very same maladies it helped to create. That can be seen by comparing the history of the global gas market with that of the U.S. gas sector. Between 1973 and 2008, natural gas's share of the global primary energy market

FIGURE 32 U.S. Electricity Generation, by Fuel Shares, 1973 to 2008

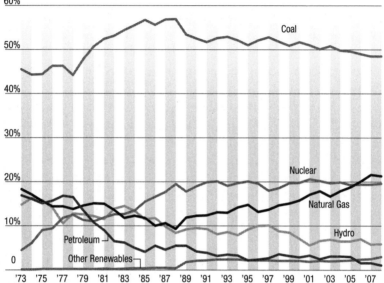

Source: Nuclear Energy Institute, "Resources and Stats: Generation Statistics,"
http://www.nei.org/resourcesandstats/graphicsandcharts/generationstatistics/.

rose from 18 percent to 24 percent. Meanwhile—thanks to Congress's ill-advised intervention in the domestic energy market—natural gas's share of the U.S. primary energy market fell from 32 percent to 26 percent.[25]

In his 1996 book *Oil, Gas & Government*, an exhaustive history of the regulation of the U.S. oil and gas business, author Robert L. Bradley Jr. wrote that the result of federal regulatory forays into the natural gas market was that the electricity industry had to substitute "the most pollutive fossil fuel (coal) for the cleanest fossil fuel (gas)."[26]

The thicket of regulations on the gas sector eventually became so onerous that Congress finally had no choice but to repeal most of them. And much of that deregulation occurred during the administrations of Ronald Reagan and George H.W. Bush. But even as the natural gas business was being gradually deregulated, many analysts continued to claim that America was running out of natural gas.

For instance, in 1983, the U.S. Office of Technology Assessment (a now-defunct arm of Congress) predicted that by 2000, U.S. gas output

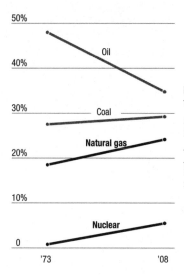

FIGURE 33
World Primary Energy Mix, 1973 to 2008

Source: BP Statistical Review of World Energy
2009, http://www.bp.com/liveassets/bp
_internet/globalbp/globalbp_uk_english/reports
_and_publications/statistical_energy_review_2008/
STAGING/local_assets/2009_downloads/
renewables_section_2009.pdf.

FIGURE 34
U.S. Primary Energy Mix, 1973 to 2008

Source: BP Statistical Review of World Energy
2009, http://www.bp.com/liveassets/bp
_internet/globalbp/globalbp_uk_english/reports
_and_publications/statistical_energy_review_2008/
STAGING/local_assets/2009_downloads/
renewables_section_2009.pdf.

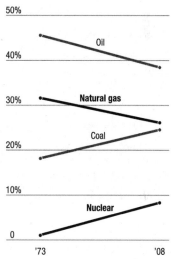

would likely be no more than about 19 trillion cubic feet per year.[27] The reality was quite different. By 2000, total production was almost 24.1 trillion cubic feet.[28] After declaring that existing proved gas reserves would only provide marginal amounts of gas after the year 2000, the analysts at the Office of Technology Assessment said, "All other domestic production must come from gas which has not yet been identified by drilling."[29] Or, as former defense secretary Donald Rumsfeld might have put it, that yet-to-be-found gas was an "unknown unknown."[30]

Federal analysts weren't the only ones grappling with the unknown unknowns of U.S. gas resources. Top executives at Exxon were also convinced that the United States was running out of natural gas. In 1984, Charles B. Wheeler, a senior vice president at the oil giant, told a Senate committee that the yet-to-be-discovered gas resources in the United States likely amounted to about 300 trillion cubic feet. Given that small resource base, Wheeler said, the United States should "conserve our scarce gas resources."[31]

But that mindset ignores one of the great paradoxes of the past century of natural gas exploration and production in the United States: The more gas the nation produces, the more gas it finds. It sounds odd, but the numbers tell the story: In 1989, the United States had about 168 trillion cubic feet of proved gas reserves.[32] By the end of 2008, proved gas reserves had increased by 41 percent, to some 237 trillion cubic feet.[33] But here's the amazing thing: Over that twenty-year period, U.S. gas wells produced more than 390 trillion cubic feet of gas—that's more than two times as much gas as was foreseen in the proved reserve estimates put forward back in 1989.[34]

Despite those huge gas withdrawals, U.S. natural gas production, and natural gas reserves, have increased. Why? The answer is that technological advances in the oil and gas sector are continually unlocking resources that were unavailable just a few years earlier. And the keys needed to unlock natural gas from shale were cut in the state that has been on the forefront of oil and gas technology for more than a century: Texas.

Stripper Power!

Strippers don't get much respect.

In the energy business, the sexy projects nearly always involve expensive wells, huge dams, giant electricity plants, or sprawling arrays of wind turbines and solar panels. Those headline-making projects are invariably designed to produce big quantities of oil, natural gas, or electricity, and do so quickly. That leaves marginal oil and gas wells—commonly referred to as "stripper" wells—as something of an afterthought in the U.S. energy sector.

But when it comes to the Four Imperatives, stripper wells are awfully impressive, particularly when they are compared to wind and solar. Let's consider the first of the Four Imperatives: power density. It ranks first in the Four Imperatives because energy sources with high power density nearly always prevail over those with low power density as they require less land and fewer resources. And that usually translates into advantages when it comes to two of the other imperatives: cost and scale. So what qualifies as a stripper? The industry defines a gas stripper well as one that produces 60,000 cubic feet of gas or less, and an oil stripper well as one that produces 10 barrels of oil or less per day.[35]

Basic math shows that a gas stripper well, covering 2 acres, has a power density of about 150 horsepower per acre, or about 28 watts per square meter, which is about twenty-three times as much as the average wind turbine.[36] (Recall that the average wind turbine has a power density of about 1.2 watts per square meter.)

Meanwhile, an oil stripper well producing 10 barrels of oil per day has a power output of about 150 horsepower per acre, or 27 watts per square meter.[37] Again, that's about twenty-three times as much power as is produced by a wind turbine. Even if we reduce the stripper well's output to just 2 barrels of oil per day—the average output for that type of well, according to the Interstate Oil & Gas Compact Commission—it will still produce about 30 horsepower per acre, or 5.5 watts per square meter. That's about four times as much power density as the average wind turbine.[38]

Stripper wells are a critical source of U.S. energy production, accounting for about 9 percent of domestic gas production and about 16 percent of oil production.[39] In 2007, U.S. oil stripper wells produced about 800,000 barrels of oil per day.[40] That's more oil than was produced that year by a number of other oil-producing countries, including Argentina, Colombia, Ecuador (which is a member of OPEC), Oman, Syria, Egypt, Australia, India, and Malaysia.[41]

The high power density of oil and gas wells—and in particular, the marginal production that comes from stripper wells—provides yet more evidence in favor of hydrocarbons. The real-estate footprint of oil and gas—particularly when compared with those of wind power—is small. And, as E. F. Schumacher made clear nearly four decades ago, small is beautiful. But future unconventional gas production may require an even smaller footprint than modern wells. Gas producers are now drilling multiple wells from one location. For instance, in 2009, the Denver-based Bill Barrett Corporation announced that it was planning to drill forty-seven gas wells on a single 8-acre site in western Colorado.

As those Colorado wells go into decline and become, by definition, stripper wells, their power density will make wind turbines look positively puny by comparison. If we assume six gas wells per acre, all producing 60,000 cubic feet per day, their collective power output will total 168 watts per square meter, or about 140 times as much as that of a wind turbine.

Call it stripper power.

CHAPTER 23

It's a Gas, Gas, Gas

Welcome to the "Gas Factory"

> You will not find a single cubic foot of natural gas unless you drill more wells.
>
> TEXAS WILDCATTER FRANK PITTS, 1977[1]

YOU CAN'T COAX NATURAL GAS out of shale deposits unless you have a lot of sand. And you can't manage that sand without a sand chief, or, better yet, a couple of sand chiefs.

That was one of the first lessons I learned during a visit to a frac spread located a few miles west of Hillsboro, Texas, on a soggy day in March 2009. The term "frac spread" is oil-field lingo for the collection of trucks, trailers, pumps, hoses, pipes, personnel, sand chiefs, and tanks that are needed for the hydraulic fracturing ("frac" or "frac job") of a particular subsurface geologic zone. On this particular day, a crew of about two dozen men backed by dozens of trucks and more than 10,000 diesel-fueled horsepower were working on two wells, the Padgett #1-H and the Greenhill #1-H. The wells were operated by Houston-based EOG Resources, one of the most aggressive drillers in the Barnett Shale. Randy Hulme, an affable EOG petroleum engineer, explained the layout. The company had drilled the two wells within a few yards of each other. Both were drilled to a depth of about 8,000 feet, and when that level was achieved, the roughnecks and tool pusher working on the drilling rig had angled the drill bit sideways and sent it clawing into the shale on a horizontal tangent for another half-mile or so, all of it, the company hoped, in the sweet spot of the Barnett Shale. That half-mile section, called the lateral, was the objective of the frac job.

Over the roar of the diesel pumps, Hulme pointed to an 18-wheeler loaded with sand that had just pulled into the frac spread and parked next to the sand chief: a Winnebago-sized metal box with a big conveyor slung under its belly. "We blow the sand from the trucks into the chief, from there it goes into the blender, and then gets pumped into the well," he said. A few minutes later Hulme led me into a cramped trailer loaded with electronics, video readouts, and gauges, where several technicians monitored the equipment involved in the fracturing process. Pointing to one of the dozen flat-panel video screens, Hulme said, "We're now pumping 45 barrels of water per minute into the well at a pressure of 6,500 pounds per square inch."

EOG was doing what's called a "multistage frac" on both of the wells. The idea was to create multiple 250-foot-long segments on the laterals of the wells. Each segment, or stage, would be individually fractured to allow the maximum recovery of gas from the lateral. One of the laterals was going to have twelve stages, the other, fifteen. And for each of those stages, the massive pumps—powered by five roaring V-8 diesel engines, each rated at 2,000 horsepower—were going to slam about 300,000 pounds of sand into the well. By the time they were finished with the two wells, EOG was going to pump a total of 8 million pounds (4,000 tons) of sand into the two holes.

Hulme explained that the goal of all that pressure and sand was simple: to pulverize a large section of the shale and create a network of fractures. The sand gets pumped into the fractures and holds them open to provide a system of channels that are similar to the air-conditioning ducts in a house. After those sandy channels are opened, tiny pores in the shale—some of them just a few nanometers (a nanometer is 1 billionth of a meter) in diameter—release the methane molecules that have been trapped inside. The gas then flows through the sand, into the well bore, and to the surface, where it is put into a local pipeline. It then goes to a processing plant where any water or natural gas liquids—such as propane, butane, or ethane—are removed. The gas is then fed into an intrastate or interstate pipeline.

By themselves, the wells near Hillsboro were not that important; they were just two of the thousands of wells that have been drilled in the sprawling Barnett Shale. But the Padgett and Greenhill wells are emblematic of the shale gas revolution that has changed the global natural gas business. The key technologies used on the two wells near Hillsboro— horizontal drilling and multistage hydraulic fracturing of wells with long laterals—have become standard for producing gas from shale. But it took years to perfect them. Over the past decade or so, the Barnett has been one of the most intensively drilled pieces of real estate on the planet. Between 1997 and 2005, the number of producing wells in the region jumped tenfold. And with that increase in drilling came huge increases in production. By the end of 2008, there were more than 12,000 producing gas wells in the Barnett Shale with a total production of over 4.8 billion cubic feet per day, the equivalent of about 875,000 barrels of oil per day.[2]

The intensive drilling and huge gas production from the Barnett has vaulted it from obscurity into one of the ten most prolific gas fields on the planet, ranking on par with Iran's giant South Pars field.[3] And all of that gas has been produced from the Barnett because a Texas energy baron named George Mitchell, who owned a lot of leases in the region, kept looking for ways to wring gas out of the shale. During the 1990s, Mitchell spent millions of dollars before his drilling crews discovered the right recipe for hydraulic fracturing. In 1997, they found that water injected under extremely high pressure was the winning formula, and that technique quickly spread. In 2003, horizontal drilling became widespread in the Barnett. The combination of the two techniques broke the code. A frenzy of leasing and drilling began that continues to this day.

And that leads me to an essential point: The Barnett Shale is the single most important hydrocarbon development in North America since the discovery of the East Texas Field in the 1930s. The East Texas Field was a big deal. It led to the empowerment of the Texas Railroad Commission, which would go on to control world oil prices for the next four decades.[4] The massive oil field helped transform Texas from a provincial backwater into what Republican political strategist Karl Rove has dubbed "America's Superstate."[5] The money from the East Texas Field turned Dallas, the nearest major city, into a thriving metropolis. The

FIGURE 35 Barnett Shale Producing Wells, 1982 to 2008

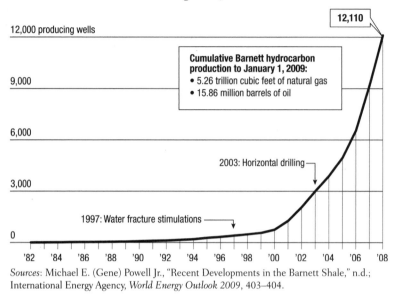

Sources: Michael E. (Gene) Powell Jr., "Recent Developments in the Barnett Shale," n.d.; International Energy Agency, *World Energy Outlook 2009*, 403–404.

gusher of money that came out of the field allowed arch-conservative wingnuts such as H. L. Hunt to fund Joe McCarthy, the John Birch Society, and other right-wing politicians and groups.

The oil money that came out of East Texas and other fields in the Lone Star State played a critical role in the rise of numerous Texas politicians, including Lyndon Johnson, George H. W. Bush, and George W. Bush.[6] The East Texas Field will always stand as a key marker in the history of U.S. energy and politics. But the East Texas Field didn't have any children—that is, its discovery did not immediately lead to the discovery and development of other, similar-sized fields. The Barnett Shale has already had children, and those children have changed the global gas business.

In April 2009, the U.S. Department of Energy estimated the total amount of recoverable shale gas in the United States at 649.2 trillion cubic feet. That's the energy equivalent of about 118 billion barrels of oil—or about four times America's proved oil reserves.[7] And though that

TABLE 4 The Barnett Shale and Its Children: Recoverable Gas in Major U.S. Shale Gas Basins

Shale Basin	Recoverable Gas (in trillion cubic feet)	Recoverable Gas (in billion barrels of oil equivalent)
Barnett	44	8
Fayetteville	41.6	7.5
Haynesville	251	45.7
Marcellus	262	47.7
Woodford	11.4	2
Antrim	20	3.6
New Albany	19.2	3.5
Total	649.2	118

Source: Energy Information Administration, "Modern Shale Gas Development in the United States: A Primer," April 2009, http://www.fe.doe.gov/programs/oilgas/publications/naturalgas _general/Shale_Gas_Primer_2009.pdf, 17; author calculations.

quantity of energy is impressive, the even better news is that much of the available shale gas lies close to major population centers: The Marcellus Shale is relatively close to New York City and Philadelphia. The Barnett, as well as the Haynesville Shale in Louisiana, are relatively close to existing gas pipeline infrastructure. The Department of Energy estimated that the Barnett Shale alone contains some 44 trillion cubic feet of recoverable natural gas. That's a huge volume of methane, but it's a mere fraction of the size of the Marcellus Shale.

The Marcellus Shale underlies an area covering about 50,000 square miles in a swath that runs from New York through Pennsylvania and southward almost to North Carolina.[8] If the Department of Energy's estimate of the Marcellus resource is correct—262 trillion cubic feet of gas—then the shale contains the equivalent of 47.7 billion barrels of oil. The potential of the Marcellus is so great that Rick Smead, a gas analyst at Navigant Consulting in Houston, has dubbed it "Prudhoe Bay under Pittsburgh."

Right behind the Marcellus is the Haynesville, with an estimated 251 trillion cubic feet of gas, the energy equivalent of 45.7 billion barrels of oil. For reference, the biggest conventional hydrocarbon discovery of the past decade is Tupi, the giant field located offshore Brazil. Tupi is estimated to hold about 8 billion barrels of oil equivalent, and most of it is in the form of oil, not gas.[9] The Haynesville formation may hold more than five times as much raw energy as Tupi. But remember, Tupi is located offshore in about 7,000 feet of water. The oil itself is located another 17,000 feet below the muddy floor of the Atlantic Ocean.[10] Developing Tupi will cost tens of billions of dollars. And that field is located in Brazil—which, given its enormous oil resources, is frequently rumored to be thinking of joining OPEC.

Meanwhile, the energy locked up in the Haynesville—as well as that in the Barnett, the Marcellus, and the other shale formations—is on land, right here in the United States, where it can be put into the existing pipeline network. Of course, there's no way that all of the gas that's in the Marcellus, the Haynesville, or any of the other shale plays will be produced. Much of that gas will be too deep, or too far from pipelines, or some unforeseen obstacle (or obstacles) will prevent it from being profitably developed. But even with that caveat, it's abundantly clear that the United States has enormous quantities of natural gas—more gas than was ever thought possible. The gas resources are so big that some companies in the industry are now calling the business of drilling in shale the "gas factory."[11]

For decades, the oil and gas business depended on "E&P"—that is, exploration and production. But thanks to the revolution in shale gas, the "E" no longer applies. Companies no longer have to look hard to find gas. When the United States and the rest of the world needs more gas, drillers will simply dial up the number of shale gas wells that they drill and fracture. When demand slackens, they will dial it down.

The surge in unconventional gas production in the United States provides a prime example of how new technologies have overhauled the energy sector. Indeed, the U.S. natural gas business is now at an inflection point. For decades, American policymakers have been convinced that the United States was running out of natural gas. Today, the exploitable quantities of natural gas known to be in the country could easily last a century. And while technology has played a key role in the shale gas revolution, another factor—one that's rarely discussed—was also a key driver in the push to exploit America's vast natural gas wealth: mineral rights.

FIGURE 36 From Unconventional to Conventional: U.S. Natural Gas Production, 1990 to 2030

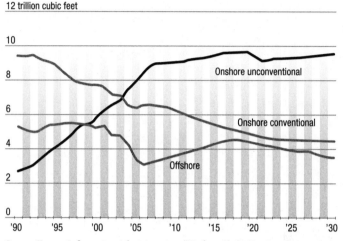

12 trillion cubic feet

Source: Energy Information Administration, "Modern Shale Gas Development in the United States: A Primer," April 2009, http://www.fe.doe.gov/programs/oilgas/ publications/naturalgas_general/Shale_Gas_Primer_2009.pdf, 7.

The private ownership of mineral rights has been essential to the development of the U.S. oil and gas industry. And the continued development and production of the nation's natural gas resources will further strengthen the U.S. economy, because the payments related to those minerals rights provide a significant, but largely unseen, economic stimulus.

Elephant Hunting:
Comparing the Barnett Shale and the East Texas Field

Once every few decades, the energy industry is forced to overhaul its thinking. New technologies or new discoveries replace the old ways, and the industry must quickly adapt to the new paradigm. In the 1930s, the East Texas Field turned the U.S. oil sector on its head. In the 2000s, the Barnett Shale did the same for natural gas.

Both fields proved the doubters wrong. Both fields were huge and both were located in northern Texas. And the discoveries of both coincided with a price collapse in their respective commodities and the commencement of worldwide financial calamities. Of course, there's an obvious difference between the Barnett and East Texas Field: The shale is a natural gas play, whereas East Texas was almost exclusively about oil. And yet, even accounting for that difference, the two fields are remarkably similar.

The two fields are geographically close to each other: The East Texas Field was centered around Kilgore, about 125 miles east of Dallas; the Barnett Shale pivots to the north, west, and south of Fort Worth, about 30 miles west of Dallas. Both fields shattered claims that the world was running out of hydrocarbons. In 1914, a U.S. government agency, the Bureau of Mines, predicted that world oil supplies would be depleted within ten years.[12] In the 1980s, federal officials and some of the top people in the energy sector believed that the United States was running out of natural gas.[13]

Both fields are huge. The East Texas Field dwarfed all of the oil fields that came before it. Historian Lawrence Goodwyn described it as "one vast lake of oil 43 miles long and 3 to 10 miles wide. It was so big it was hard to think about."[14] It contained some 6 billion barrels of oil and was the largest oil field in the world known at that time.[15] By comparison, the sprawling Barnett Shale makes the East Texas Field look almost dainty. The Barnett deposit sprawls across at least seventeen counties and covers some 5,000 square miles—that's nearly the size of the state of Connecticut.[16]

The size of the two fields led to a flurry of drilling activity unlike anything that had come before. In 1930, Columbus Marion "Dad" Joiner, a stubborn, Bible-quoting promoter from Alabama, drilled his way into history by bringing in a massive gusher with the Daisy Bradford No. 3. About a year later, in October

1931, one well was being completed in the East Texas Field every hour. In the first half of 1931 alone, 1,100 wells were drilled. In 1932 in Kilgore, one city block contained 44 different oil wells. By the end of 1933, nearly 12,000 wells were sucking oil out of the East Texas Field.[17] The activity in the Barnett Shale has not been quite that intense, but the drilling activity has been remarkable. In 2003, as the outlines of the shale play began to be understood, 47 drilling rigs were working in the Barnett Shale during an average week. By 2008, that number had soared to 182 rigs.[18] Put another way, during any given week in 2008, about 10 percent of the wells being drilled in the United States were being drilled in the Barnett Shale.[19] In 2003, the Barnett had about 3,000 producing wells. By the end of 2008, it had more than 12,000.[20]

The doubters had insisted that the Barnett Shale would never pay off. The shale was just too dense, too impermeable. Houston oilman George Mitchell's persistence (and sizable wallet) proved them wrong. He kept his drilling crews working on the Barnett Shale for years even though his own personnel were telling him it was a waste of money. Six decades earlier, Dad Joiner had silenced the critics who claimed that East Texas was devoid of hydrocarbons when his rank wildcat yielded a gusher in Rusk County.[21]

Both fields resulted in a surge of drilling, which quickly led to a flood of production and a collapse in prices. In 1930, the year that Joiner found oil, crude prices in the United States averaged $1.19 per barrel.[22] By August 1931, oil was selling for 13 cents per barrel, and in parts of East Texas it was selling for as little as 3 cents per barrel.[23]

The development of the Barnett Shale and other major shale gas plays in the United States contributed to a similar price collapse. During the summer of 2008, the Barnett was producing about 8 percent of all the gas produced in the country.[24] And that gas was being fed into a booming global economy that was reaching the tail-end of a long bull market. On July 3, 2008, near-month natural-gas futures prices hit a near-record $13.58 per thousand cubic feet.[25] By April 2009, gas futures had fallen to about $3.50,[26] and numerous forecasters were estimating the U.S. natural gas prices would stay under $4 for months, perhaps even years to come.

Just as the East Texas Field changed the American oil business, the Barnett Shale has fundamentally changed the U.S. gas business—and that change is just getting started.

CHAPTER 24

America's Secret Google

J. PAUL GETTY, one of the world's first billionaires, once declared that "the meek shall inherit the Earth, but not its mineral rights."[1] Getty—who made his first million dollars in the Oklahoma oil fields—was on to something.

Although dozens of economists have written about the critical role that private-property rights play in building wealth in developing countries (among the more notable: Peruvian economist Hernando de Soto), few, if any, have considered the importance of private ownership of mineral rights.[2] And fewer still have written about America's anomalous status as the only country on the planet that allows individuals, rather than the state or the crown, to own the minerals beneath their feet.

The private ownership of mineral rights in the United States is a key—but perennially overlooked—reason why the nation has become so prosperous, and why the American oil and gas industry continues to lead the world in developing new technologies to extract hydrocarbons.

In the rest of the world, the king or the state decides which resources get drilled and when. The individuals who own or occupy the land where the drilling is occurring have little or no financial interest in the outcome. By contrast, in the United States, mineral owners are motivated to exploit their minerals. That helps to explain why the United States—despite the fact that it is the most-drilled country on the planet—has the most dynamic and innovative industry. Over the past century, about 1.7 million oil and gas wells have been drilled in

America.[3] No other country comes anywhere close to that level of prospecting. The oceans of natural gas in the Barnett Shale, the Haynesville Shale, and other formations would likely never have been developed if that gas had been owned by the federal or state government. Individuals and entrepreneurs persist in developing new drilling and completion technologies because it's in their interest to do so. The more they produce from the properties on which they lease or own the minerals, the more money they make.

When it comes to getting oil and gas out of the ground, private ownership of mineral rights provides proof positive that greed is good. Without greed—the greed that springs from individuals owning the minerals beneath their feet—the vast oil and natural-gas fields in Texas, California, Oklahoma, Louisiana, and other states would likely remain underdeveloped.

Increased domestic gas production—as well as more domestic oil production—will mean more royalty payments to average Americans. In 2007, the total mineral interest payments to individual Americans totaled about $21.5 billion. If the U.S. mineral rights payments to private individuals were a stand-alone company, its revenues would nearly equal those of Google, whose 2008 revenues were $21.8 billion.[4]

Despite the enormous value and importance of mineral rights to middle-class Americans, the importance of mineral rights rarely gets mentioned during discussions of energy policy. That may be due to the fact that obtaining royalty information is notoriously difficult. There are no central sources of data for royalty payments on oil and gas production. Therefore, I did my own calculations to come up with that $21.5 billion figure.[5]

Whatever the exact amount of U.S. royalty payments from oil and gas, it's not all going to the J. Paul Gettys of the world. The boom in U.S. natural gas production has provided windfall lease and royalty income for ordinary people. About 1 million Americans own mineral interests.[6] And while those mineral royalties help middle-class individuals pay their bills, educate their children, and stimulate local economies, the fuel being produced from privately held mineral deposits will be used for manufacturing, home heating, transportation, and, of course, electricity generation.

In the areas around the Barnett Shale in Texas, one neighborhood group was paid $25,000 per acre for the lease rights.[7] In Louisiana, in the area around the Haynesville Shale gas play, energy companies paid De-Soto Parish $27 million just for the right to drill on parish land.[8] *Haynesville*, an insightful documentary about the effect that the new shale play is having on residents of the Bayou State, profiles one of the big winners of the natural gas boom: a middle-class real-estate appraiser named Mike Smith. A resident of DeSoto Parish, Smith owns about 300 acres in the heart of the Haynesville Shale. For the rights to drill on his land, one natural gas driller paid Smith nearly $1.3 million. In addition, Smith will earn a royalty of up to 25 percent on the gas produced from beneath his property.[9]

There are many reasons to develop America's abundant natural gas resources. Increasing the amount of royalty payments—money that will stay near the communities where the gas is being extracted—provides direct cash payments to a huge number of citizens.

Moreover, the domestic oil and gas business creates jobs, lots of them. In 2007, the Colorado Energy Research Institute released a study on the economic impact that the oil and gas industry has on Colorado. It found that in 2005, nearly 71,000 jobs in the state were directly attributable to the oil and gas industry. That same year, there were about 74 drilling rigs operating continuously in the state; thus, simple arithmetic shows that each rig correlates to about 1,000 jobs.[10] A 2008 study by Oklahoma State University estimated that the oil and gas industry in the Sooner State employs about 76,000 people, and those workers are paid about $8.9 billion per year. When accounting for direct and indirect economic impacts, the oil and gas sector accounts for more than 14 percent of Oklahoma's employment and 18 percent of the state's labor income.[11] The Independent Petroleum Association of America estimates that the entire U.S. oil and gas sector—from drilling and refining to transportation and retailing—employs about 1.8 million people.[12]

Are those the type of "green" jobs that are being hyped by various politicians and environmental groups? Probably not. But there is no question that the oil and gas industry is employing lots of people in good-paying jobs. And maintaining a strong domestic oil and gas sector will help assure that those jobs don't go to China or another foreign country.

Although royalties and jobs are critical to the U.S. economy, it's also true that increased drilling activity comes with significant costs. A handful of landowners, like Mike Smith, will get rich off of their mineral rights. But many others will only see the industrial side of the business. Producing natural gas from shale beds and other unconventional sources requires lots of drilling—and drilling rigs are not always welcome.

CHAPTER 25

Gas Pains

THE OLD ADAGE—"There ain't no such thing as a free lunch"—applies to nearly everything. But it is particularly true with energy. Every form of energy production comes with costs to humans and the environment.

Hydropower requires the flooding of rivers and streams, thereby ruining habitats for aquatic life. Oil and gas production requires significant amounts of land for drilling and pipelines. Oil spills during transportation can harm all types of wildlife. Coal mining—particularly strip mining and mountain-top removal—results in huge swaths of denuded land. And, of course, the combustion of hydrocarbons emits enormous quantities of carbon dioxide.[1]

Although natural gas production has many favorable attributes, it still comes with significant environmental costs. Producing gas from coal beds, tight sands, and shale deposits requires high well densities. That is, in order to produce large quantities of gas, companies have to drill large numbers of wells in relatively close proximity—and they have to keep drilling. And those wells can pose problems for neighbors in rural and urban areas. During the well drilling and completion process, traffic—from both heavy and light trucks—to and from the well site can be heavy.

In addition, there are concerns about water and hydraulic fracturing. Properly fracturing a shale gas well requires the use of 1 to 3 million gallons of water.[2] To obtain optimum results, oil-field service companies add various constituents to the water, including small amounts of surfactants, antibacterial agents, and friction reducers. After the water is used, it must be disposed of properly—which usually means injecting it into a disposal well. Regulation of the fracturing process is generally left to the states—and some states that are seeing big increases in gas drilling are under pressure from environmental groups to ramp up regulation of the fracturing process. There is also pressure at the federal level.

In June 2009, shortly before the Potential Gas Committee estimated U.S. gas resources at more than 2,000 trillion cubic feet, U.S. Representative Diana DeGette, a Colorado Democrat, along with representatives from New York and Pennsylvania, introduced a bill called the "FRAC Act" that would put regulation of hydraulic fracturing under the purview of federal regulators.[3] If the bill becomes law, the oil and gas industry estimates that it could increase the cost of fracturing each well by about $150,000.[4] Representative Maurice Hinchey, a New York Democrat who is among the bill's authors, said the legislation would protect drinking water and lift "the veil of secrecy currently shrouding this industry practice."[5]

The push for more regulation comes, in part, from articles written by Abrahm Lustgarten, a reporter for ProPublica, an independent news outlet. In a November 2008 article, Lustgarten reported on water contamination in Sublette County, Wyoming, near the huge Pinedale natural gas field, a region that has been extensively drilled over the past few years. He claimed that a contaminated water well in Sublette County was among "more than 1,000 other cases of contamination" that have been "documented by courts and state and local governments in Colorado, New Mexico, Alabama, Ohio, and Pennsylvania."[6]

While there's no doubt that the oil and gas industry has caused isolated cases of groundwater contamination, the extent of that contamination is not known. Industry officials contend there is no need for additional regulation. A 2004 study by the EPA found that there was no evidence that fracturing was a danger to drinking water. The Independent Petroleum Association of America claims that more than 1 million wells have been fractured over the past fifty years and that there have been no documented cases of contaminated drinking water during that time.[7] But as a reporter with a long history of writing about the oilfield, I find that claim disingenuous.

Lustgarten's reporting on the problems in Wyoming helped to bring about an inquiry by the EPA. In 2009, the agency found traces of toxins—

including arsenic, barium, cobalt, and copper—in eleven of the thirty-nine drinking-water wells it tested near the town of Pavillion. One of the wells that tested positive was owned by a local farmer named John Fenton, who said the water in his well was fine until EnCana, the big Canadian gas company, began drilling near his home.[8]

The EPA's investigation into the problems in Wyoming are continuing. Regardless of what happens in Wyoming, some industry opponents want more federal oversight on the oil and gas industry in general and the hydraulic fracturing process in particular. Calls for more regulation will almost certainly grow as drilling ramps up in the Marcellus Shale, which underlies large swaths of New York, Pennsylvania, and other eastern states. Gas producers have begun responding to the pressure. In late October 2009, Chesapeake Energy announced that it would not do any drilling in the upstate New York watershed, a region that provides drinking water for 8.2 million people in New York City and surrounding areas. The CEO of Chesapeake, Aubrey McClendon, has called on the industry to reveal all of the chemicals that are used during the hydraulic fracturing process.[9]

While environmentalists lauded Chesapeake's announcement that it wouldn't drill in the upstate New York watershed, the U.S. gas industry will still need lots of new wells in order to keep gas production in line with gas demand. The new shale gas wells being developed in the Barnett, Haynesville, and Marcellus regions have steep decline curves, meaning that output from some wells may fall by 80 to 90 percent during the first year of production. Those steep declines are part of a continuing drop in average well productivity. In 1971, the average U.S. gas well was producing 435,000 cubic feet of gas per day. By 2008, that number had decreased to 113,000 cubic feet per day.[10] That reduction in productivity, combined with the steep decline rates in the new shale gas plays, leaves the U.S. energy industry with no choice but to continue drilling tens of thousands of new wells per year. In 2008 alone, more than 60,000 wells were drilled in the United States, with gas wells outnumbering oil wells by 2 to 1.[11]

Exactly how many new wells will be needed is not yet clear, and it will likely take several years before the industry figures out how to manage the huge quantities of gas available in the various shale deposits.

FIGURE 37 U.S. Natural Gas Wells, Average Productivity

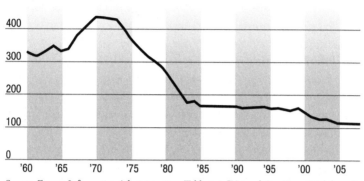

Source: Energy Information Administration, Table 6.4, "Natural Gas Gross Withdrawals and Natural Gas Well Productivity, 1960–2008," n.d., http://www.eia.doe.gov/emeu/aer/txt/ptb0604.html.

But as the number of gas wells increases, and those production facilities get closer to people's homes and businesses, the costs of developing U.S. natural gas will be felt by more people. Concerns about water—both in terms of the quantity of water used during the fracturing process, and the allegations that fracturing may have negative effects on surface and groundwater supplies—will almost surely increase.

Although some of those concerns may be valid, many of the complaints about water usage are clearly overblown. For instance, if the gas industry is able to ramp up its operations in the Marcellus Shale in Pennsylvania to the point where it is drilling and fracturing 3,000 wells per year, the industry expects water consumption to total about 30 million gallons per day. For comparison, the Pennsylvania electric utility sector uses about 5.9 *billion* gallons of water per day, or about two hundred times more than the projected needs of the natural gas sector.[12]

Furthermore, the gas industry is continually improving both its seismic monitoring and its ability to perform drilling that has a small footprint. The industry's ability to extract hydrocarbons from reservoirs that are miles away can be seen by the November 2009 news that Chesapeake Energy had drilled a gas well inside the city limits of Fort Worth. That's not overly uncommon, given that the Barnett Shale underlies much of the city. What was newsworthy was that the target zone for the well was underneath the north end-zone of Amon G. Carter Stadium, the home venue of the Texas Christian University Horned Frogs football team. But visitors to the stadium, and students, faculty, and university administrators, will never notice the well, because the drilling site for the well is about 1.5 miles southeast of the stadium.[13]

Though critics of the industry will claim that the costs of natural gas development are significant, in reality those costs are not new, and they aren't unusual. For decades, drillers have been sinking tens of thousands of wells all over the United States in their search for more hydrocarbons. And they have found them in huge quantities. The development of shale gas causes some disruption—in the form of traffic, noise, big drilling rigs, temporary fluid tanks, and generators—no question about it. But once the gas wells are in place, there is virtually no more disruption, and those wells can remain productive for decades.

Natural gas is not a perfect fuel. But natural gas is the greenest of the hydrocarbons. Once we move beyond hydrocarbons in the search for something that is yet greener, there is only one choice that can provide the scale of energy we need and provide it in a way that is both affordable and environmentally friendly. That choice, of course, is nuclear power.

CHAPTER 26

Nuclear Goes Beyond Green

Nuclear power is beyond green.

Of course, that's not the message you're going to hear from the media darlings and mainstream environmental groups. For them, nuclear power has become a rallying point around which they can raise money and continue pushing their message that the only options for the future are renewable energy and efficiency. They insist that nuclear power is too expensive—and too dangerous—for use in the modern world.

That message, particularly the part about danger, evokes a strong response among the population. Some of the fear is understandable. The enduring image that marks the beginning of the nuclear age is, of course, the mushroom cloud. By unlocking the forces inside the atom, humans unleashed the most fearsome weapons the world has ever known, and the United States has used that knowledge twice, at Hiroshima and Nagasaki, to devastating effect. More recently, the accidents at Three Mile Island and Chernobyl, along with movies such as *Silkwood* and *The China Syndrome*, have stoked fears about what might happen in the case of a nuclear accident. And environmental groups continue to use fears about nuclear proliferation as the reason to fight nuclear power.

In short, for many people, nuclear power's future has not yet overcome its past. But progress on the issue is being made. Perhaps the best single rebuttal to these fears comes from James Lovelock, the British scientist who proposed the Gaia theory, which posits that the Earth is a self-regulating organism. In 2004, Lovelock wrote an opinion piece for the *Independent* in which he made it clear that nuclear power is the only viable option for large-scale reductions in carbon dioxide emissions. "By all means," he wrote, "let us use the small input from renewables sensibly, but only one immediately available source does not cause global warming and that is nuclear energy." Lovelock went on, writing,

> Opposition to nuclear energy is based on irrational fear fed by Hollywood-style fiction, the Green lobbies and the media. These fears are unjustified, and nuclear energy from its start in 1952 has proved to be the safest of all energy sources. . . . I am a Green and I entreat my friends in the movement to drop their wrongheaded objection to nuclear energy. . . . We have no time to experiment with visionary energy sources.[1]

Other leading environmentalists have also endorsed nuclear, including Patrick Moore, who was a founder of Greenpeace, and the late Anglican bishop Hugh Montefiore, who was a trustee of the United Kingdom's Friends of the Earth for two decades. Despite the growing support for nuclear power, some of the most established members of the Green/Left continue their opposition. Among the most strident—and consistently wrong—of the nuclear opponents: Amory Lovins.

In 1986, when asked about the future of nuclear power, Lovins declared flatly, "There isn't one. . . . No more will be built. The only question is whether the plants already operating will continue to operate during their lifetime or whether they will be shut down prematurely." Since then, Lovins has repeated one of his favorite lines: "Nuclear is dying of an incurable attack of market forces."[2]

In 2007, when I interviewed Lovins, he declared that "a huge and capable propaganda campaign by the [nuclear] industry and its political allies is spinning an illusion of a renaissance that deceives credulous journalists but not hard-nosed investors."[3]

How did Lovins do on his prediction back in 1986? According to data from the International Atomic Energy Agency (IAEA), about 130 new reactors with nearly 123,000 megawatts of generating capacity have been brought online over the past two decades or so. Those reactors represent nearly one-third of global nuclear capacity, which in late 2009 included 436 reactors with 370,000 megawatts of capacity.[4] As for his 2007 claim about the "illusion of a renaissance," the numbers, once again, have proven him wrong. By the end of 2009, more than four dozen new reactor projects, representing nearly 48,000 megawatts of new nuclear capacity, were under construction.[5] And many more were on the way. Japan, the third-biggest producer of nuclear power (after the United States and France), plans to construct 11 new reactors over the next decade or so.[6] And the country plans to be getting 60 percent of its electricity from nuclear power by 2050—double the current percentage.[7]

The International Energy Agency (IEA) sees nuclear power as an essential part of the effort to stabilize global carbon dioxide levels. In its 2009 World Energy Outlook, the agency said it expected global investment in nuclear power to total some $1.3 trillion over the next two decades.[8] More importantly, the IEA's latest report makes it clear that nuclear power is competitive with conventional power plants. "New nuclear power plants can generate electricity at a cost of between $55 and $80 per MWh [megawatt-hour], which places them in a strong competitive position against coal- or gas-fired power plants, particularly when fossil-fuel plants carry the burden of the carbon cost associated with the cap-and-trade system" that is in place in Europe, and proposed for the United States.[9]

The IEA projects that for power plants that begin operations between 2015 and 2020, nuclear will be among the cheapest options, even when compared to wind power and coal-fired power plants that use high-efficiency ultra-supercritical combustion. The agency estimates that nuclear power plants will be able to produce electricity for about $72 per megawatt-hour, whereas onshore wind costs will be about $94 per megawatt-hour.[10]

Despite the data, Lovins continues singing from his same tired hymnal. In 2009, he said that nuclear is "continuing its decades-long collapse in the global marketplace because it's grossly uncompetitive, unneeded and obsolete."[11] Lovins may be wrong, but at least he's been consistently wrong—for nearly three decades. The same can be said of the major environmental groups. Consider this line from Greenpeace International: Nuclear power is "an unacceptable risk to the environment and to humanity. The only solution is to halt the expansion of all nuclear power," and, says Greenpeace, to begin "the shutdown of existing plants."[12]

Here's the Sierra Club's position on nuclear, a position it has held since 1974: "The Sierra Club opposes the licensing, construction and

FIGURE 38 International Energy Agency's Projected Costs for Commercial Electricity Generation Plants That Begin Operations from 2015 to 2020, in Dollars Per Megawatt-Hour

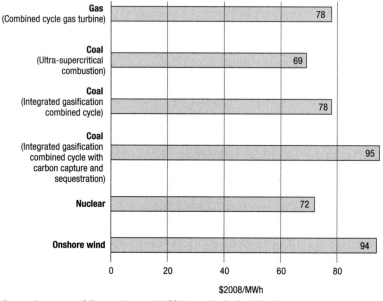

$2008/MWh

Source: International Energy Agency, *World Energy Outlook 2009,* 381.

operation of new nuclear reactors utilizing the fission process." The club plans to continue its opposition, pending "development of adequate national and global policies to curb energy over-use and unnecessary economic growth."[13]

Unfortunately, neither Greenpeace nor the Sierra Club explains how they plan to replace nuclear power, which now provides about 15 percent of the world's electricity needs and about 5 percent of its total primary energy.[14] And while the Sierra Club may be opposed to "energy over-use and unnecessary economic growth," there haven't been many countries in Africa—or anywhere else—that have expressed concern about using too much energy or about too much economic growth.

Nuclear power is the only always-on, no-carbon source that can replace significant amounts of coal in our electricity generation portfolio. If the United States is serious about cutting carbon dioxide emissions and reducing the harmful environmental side effects of coal-fired power while keeping the lights on and the beer cold, nuclear has to be an integral part of the plan. Indeed, when all factors are considered, nuclear power may be the most environmentally friendly form of electricity generation.

That's not to say that nuclear doesn't come with environmental costs. It does. But then, so do renewable sources such as wind and solar, which require hundreds—or thousands—of square miles of land for power generation and transmission. The same problems of energy sprawl hamper the development of hydropower and biofuels.

Thanks to their super-high power density, nuclear reactors require small amounts of land. When operational, they emit no carbon dioxide, and the volume of their solid waste production is minuscule. For instance, a 1,000-megawatt nuclear reactor produces about 20 cubic meters of solid waste per year.[15] Every year, the entire fleet of U.S. nuclear reactors produces about 2,000 tons of spent fuel. Over the entire history of the U.S. nuclear power industry, it has produced about 60,000 tons of high-level waste.[16] That volume of material, if stacked to a depth of about 15 feet, would cover an area the size of a single football field.[17]

The key, of course, is proper waste management. Other countries, including Russia, Japan, and France, are actively and responsibly handling the nuclear waste produced by their reactors. The same can be done in the United States. And it can be done because the volumes of waste being produced are relatively small, particularly when compared with the amounts being produced by the coal industry. In 2007 alone, coal-fired power plants in the United States generated 131 million tons of coal ash—and much of that material is contaminated with heavy metals.[18] Thus, in one year, *the U.S. coal industry produces nearly 2,200 times as much solid waste as the U.S. nuclear industry has produced in more than four decades.*

By nearly any metric that the environmental groups choose—footprint, solid waste production, neurotoxin releases, or carbon dioxide emissions— nuclear power is beyond green. The main problem facing nuclear power is the environmental groups themselves. Mainstream environmental groups continue to oppose nuclear energy despite the fact that the existing global fleet of reactors prevents the emission of about 2 billion

TABLE 5 Estimated Construction Cost of Various Electric Generation Plants

Source	Construction cost per kilowatt of capacity
Nuclear	$4,000 to $6,700
Offshore wind	$5,000
Coal	$2,300
Natural gas	$850

Sources: Rebecca Smith, "The New Nukes," Wall Street Journal, September 8, 2009, http://online.wsj.com/article/SB10001424052970204409904574350342705855178.html. The wind figure is for the Sheringham Shoal offshore wind farm in the United Kingdom. Estimated cost is 1 billion British pounds. In mid-September 2009, that was equal to about $1.7 billion. See "Onshore Construction Begins for Sheringham Shoal Wind farm," NewEnergyFocus.com, September 7, 2009, http://www.newenergyfocus.com/do/ecco.py/view_item?listid=1&listcatid=32&listitemid=2978&section=Wind.

tons of carbon dioxide per year—that's about 7 percent of global carbon dioxide emissions.[19]

While nuclear power's green credentials are obvious, critics bring up valid concerns about its cost. Building a large nuclear plant in the United States will cost billions of dollars. Utilities are understandably nervous about committing $10 billion or more to a project that could be delayed and cost more than expected. And over the past decade or so, the costs associated with building new reactors have increased substantially. For instance, a reactor being built at Olkiluoto, Finland, by the French nuclear giant, Areva, has been hampered by repeated delays, with the price tag for the project reportedly increasing by about 50 percent. Another Areva reactor project in Flamanville, France, is also over budget and running behind schedule.[20] In May 2009, Areva officials in Paris admitted to me that the company was having problems with the deployment of its latest reactor design, the European Pressured Reactor, in Olkiluoto and Flamanville. But Areva is confident the problems will be worked out over time.

Some U.S. utilities are shying away from nuclear power over cost concerns. In April 2009, resistance from Missouri legislators led the state's biggest electric utility, Ameren UE, to drop its plan to hire Areva to build a $6 billion copy of the European Pressured Reactor.[21] A few months later, in October 2009, the San Antonio city council delayed a vote on $400 million in bonds that were to be sold to support the municipal utility's plan to invest in two additional reactors at the South Texas Project. The vote was suspended after reports surfaced that the two new reactors, with a total capacity of 2,700 megawatts, were going to cost a total of $13 billion, or about $3 billion more than previously expected.[22] At the $13 billion price, that works out to about $4,800 per kilowatt.

Obviously, the relatively high cost of nuclear power presents a barrier for both utilities and consumers. But comparing the initial construction costs of a nuclear power plant with those of coal and natural gas is misleading, because the long-term operating costs of the nuclear plant are lower than those for coal and natural gas plants. The reason: The fuel for nuclear reactors costs a fraction of what utilities pay to fuel their coal- and gas-fired plants. But the higher operating costs of coal- and gas-fired plants appear to be acceptable in the current marketplace, particularly given the recent declines in U.S. electricity consumption and the general nervousness about the future health of the economy.

Although the high initial costs of nuclear power are substantial, the per-kilowatt construction costs of nuclear power plants are similar to the per-kilowatt costs of constructing offshore wind projects. In 2009, Norwegian energy giant StatoilHydro began building the 315-megawatt Sheringham Shoal offshore wind farm. That project, located in British territorial waters about 120 miles northeast of London, carries a price tag of about $1.7 billion, which works out to about $5,000 per kilowatt of installed capacity—a sum that puts it in the same ballpark as a nuclear power plant.[23] And unlike nuclear plants, which usually have a capacity factor of 90 percent, those offshore generators will likely only produce power about 30 to 40 percent of the time.

The high capital costs and low power density of offshore wind means higher costs for consumers. In November 2009, a Rhode Island electric utility rejected a proposal from a company called Deepwater Wind, which wanted to sell electricity. Deepwater Wind had proposed a $200 million array of wind turbines off the Rhode Island coast. According to the *Providence Journal*, Deepwater wanted to sell electricity to the utility at a cost of $0.253 per kilowatt-hour. And, as the newspaper reported, "that price would increase by 3.5% annually." The *Journal* went on to report that the local utility currently pays an average of $0.092 per kilowatt-hour for electricity produced from conventional generators. Thus, for the luxury of buying wind generated by offshore wind turbines, Deepwater Wind wanted Rhode Island consumers to pay more than twice as much for their electricity as they would from conventional generators. Nevertheless, the company's chief development officer, Paul Rich, was unabashed, declaring that the expected price of electricity from Deepwater Wind's offshore project was "in line with major European wind farms in an established market with an established supply chain."[24]

Solar power is even more expensive than offshore wind. In early 2009, Austin's utility, Austin Energy, agreed to spend $180 million on a 30-megawatt solar facility. At that price, the solar plant will cost about $6,000 per kilowatt. And according to Austin Energy officials, the solar farm will run at a capacity factor of about 23 percent.[25] Thus, Austin Energy has agreed to build a solar plant that will operate about one-fourth as often as a nuclear plant and cost about 25 percent more on a per-kilowatt basis.[26] And the cost of the Austin solar project appears to be about average. In late October 2009, Barack Obama gave a speech at a new 25-megawatt solar photovoltaic facility in south Florida owned by Florida Power & Light. The cost of that project: $152 million, or about $6,000 per kilowatt.[27]

To summarize, officials in San Antonio are worried, rightly, about the $13 billion cost of a new nuclear plant at the South Texas Project. But if they wanted to build a solar plant with the same output—2,700 megawatts—as the new reactor, it would cost about $16.2 billion. And the energy production—measured in kilowatt-hours—from the solar facility would be one-third, or less, of the output from the nuclear reactors.

Obviously, the price tag for nuclear plants presents a significant obstacle, no question about it. But when compared to offshore wind or onshore solar, the costs of nuclear are comparable—and unlike wind and solar, nuclear plants can provide the always-on electricity that our society demands. The reality is that every form of power generation exacts costs. But when considering all of the costs, there's no greener choice than nuclear.

Antinuclear ideologues remain opposed to nuclear power even though the nuclear industry has developed new reactor designs—both big and small—that can be used to meet almost any need. But before I discuss the merits of the new class of reactors, we must deal with the issue of nuclear waste.

The Real Story on Subsidies

Critics complain that the nuclear sector gets too much in the way of federal subsidies. That claim doesn't square with the data. According to a 2008 study by the Energy Information Administration, nuclear power now gets federal subsidies and support worth about $1.59 per megawatt-hour of electricity produced. For comparison, wind power gets $23.37 and solar gets $24.34 per megawatt-hour.[28]

But those numbers don't tell the full story. In 2007, the U.S. nuclear sector produced 794 million megawatt-hours of electricity. The wind and solar sectors combined to produce 32 million megawatt-hours. That means that when measured on per-unit-of-output basis, wind and solar are getting about 15 times as much in federal subsidies as nuclear even though nuclear is producing about 25 times as much energy as wind and solar *combined*.[29] And while those numbers are important, remember that nuclear plants are providing electric power that is available 24/7/365. In a society where constant, reliable power is an essential commodity, Congress is lavishing subsidies on wind and solar even though they cannot, and will not, be able to provide the type of always-available electricity that consumers demand.

While additional nuclear generation capacity makes sense, a substantial expansion of the industry will take twenty to thirty years of sustained investment, and that will mean increased government support. The U.S. nuclear industry has already received substantial federal loan guarantees— and it wants more.[30] Furthermore, the nuclear industry gets blanket liability insurance from the federal government in case of a major accident.[31] But recent federal support for the nuclear sector appears to be far less than what is being extended to the renewables sector. According to one report, in 2009, the total federal loan guarantees available through the Department of Energy for renewable energy totaled $78.5 billion, whereas available guarantees for nuclear power totaled $18.5 billion. (Loans for "clean coal" got $8 billion.)[32]

By comparison, natural gas subsidies are a pittance. According to the EIA, federal subsidies for natural gas–fired electricity production totaled just $227 million in 2007. Meanwhile, coal-fired electricity subsidies totaled $3 billion. Nuclear power subsidies totaled $1.26 billion, and wind and solar power combined received subsidies totaling $738 million.

But the real story, again, comes from looking at the subsidies on a per-unit-of-output basis. In 2007, natural gas–fired electricity got just $0.25 per megawatt-hour in federal subsidies. And it got that tiny subsidy even though natural gas was used to produce about 900 million megawatt-hours of electricity in 2007.[33]

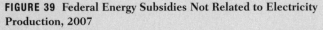

FIGURE 39 Federal Energy Subsidies Not Related to Electricity Production, 2007

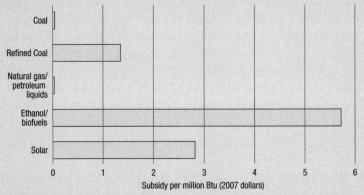

Subsidy per million Btu (2007 dollars)

Source: Energy Information Administration, "Federal Financial Interventions and Subsidies in Energy Markets 2007," April 2008, http://www.eia.doe.gov/oiaf/servicerpt/subsidy2/pdf/subsidy08.pdf.

How does that compare with wind and solar? Well, in 2007, the wind power sector got 93 times as much in federal subsidies as the natural gas sector even though the gas sector produced 28 times more electricity than wind.[34] Solar is even worse. It received 97 times as much in subsidies per megawatt-hour produced as gas, even though the gas-fired electric sector produced 900 times as much electricity as solar.

Alas, there's more subsidy madness. And predictably, it involves the corn ethanol scam, the black hole of federal taxpayer dollars. The EIA report shows that in 2007, ethanol and biofuels got $5.72 per million Btu of energy produced. That's 190 times as much subsidy as was provided to the entire U.S. oil and gas business, which received just $0.03 per million Btu. And the biofuel scammers got those fat subsidies even though the oil and gas business provided about 98 times as much energy as the bio-fuels sector.[35]

FIGURE 40 Federal Energy Subsidies for Electricity Production, 2007

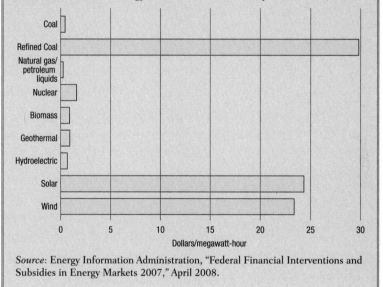

Source: Energy Information Administration, "Federal Financial Interventions and Subsidies in Energy Markets 2007," April 2008.

CHAPTER 27

A Smashing Idea
for Nuclear Waste

THE BIGGEST PROBLEM facing the future of nuclear power isn't science, cost, or how to handle the issue of radioactive waste. Instead, it's the equivocators. And those equivocators invariably seize on the issue of nuclear waste as the reason why the United States cannot pursue nuclear power. For instance, during a February 2009 speech in Houston, Fred Krupp, the president of the Environmental Defense Fund, said that "all of us need to be open-minded about nuclear power as part of the solution."[1]

Krupp's equivocation came immediately afterward: "Having said that, I'm not ready to call for a new wave of construction because there are some very legitimate concerns not only about costs but also about nuclear waste. . . . It's partly technical and it's partly political as to what we do with the waste. I think we need to come up with those answers."[2]

The answers are here. What's lacking aren't answers, but political will. That's not to say the challenge of handling nuclear waste can be solved easily or cheaply. Coming up with a long-term solution will take years of work, lots of money, and sustained support from Congress. And that's the crux of the problem: Nuclear power requires strong governmental involvement. One analyst summed it up well when he told me, "The Republicans like nuclear, but they hate government. The Democrats like government but they hate nuclear power." And those conflicting views

269

have contributed to the stalemate on nuclear power development in the United States.

That stalemate is most obvious when it comes to dealing with nuclear waste. In 2009, the Obama administration—bowing to pressure from Senate Majority Leader Harry Reid, who hails from Nevada—decimated funding for the waste disposal site at Yucca Mountain. Administration officials said they were abandoning the project and would begin looking for other waste sites.[3] Reid's political power play has left the United States without a long-term program—or even the beginnings of one—to deal with its spent nuclear fuel.[4] Reid's NIMBY posturing may be handy politics for Nevada, but it effectively renders moot a two-decade-old federal law that requires the federal government to take possession of the high-level waste produced by the country's nuclear power plants. It also means that the two decades and $13.5 billion of taxpayers' money that has been spent researching and developing the site at Yucca Mountain (which is ready for use and only awaits licensing) has effectively gone up in smoke—thereby adding just a bit more carbon dioxide to the atmosphere.[5]

In their 2008 political platform the Republicans called nuclear energy the "most reliable zero-carbon-emissions source of energy that we have." The GOP went on, saying that "unwarranted fear mongering" has prevented the country from starting a new reactor for more than three decades.[6]

Meanwhile, in the Democrats' 2008 political platform, the phrase "nuclear power" occurs exactly one time—and refers only to the spread of nuclear weapons. The Democrats' discussion of nuclear power focuses almost exclusively on the issues of proliferation and the threat of Iran and North Korea having nuclear weapons. Furthermore, the Democrats made their opposition to the waste repository at Yucca Mountain, Nevada, clear, saying "We will protect Nevada and its communities from the high-level nuclear waste dump at Yucca Mountain, which has not been proven to be safe by sound science."[7]

The Democratic platform does two things that have become common in the fight over nuclear waste disposal: It conflates the issue of nuclear power with nuclear weapons proliferation, and it restates the canard that nuclear waste cannot be handled safely. This practice of equating nuclear power with weapons proliferation extends back to the 1970s, when Jimmy Carter halted the recycling of spent fuel rods from nuclear power plants. Carter claimed that recycling the fuel would lead to the spread of plutonium. But there is no connection whatsoever between the reprocessing of spent nuclear fuel in the United States with the proliferation of nuclear weapons.

Although Carter killed America's waste reprocessing plant (then being built at Barnwell, South Carolina), France, Russia, England, Canada, and Japan all continued with their waste reprocessing operations. By doing so, those countries reduced the volume of their high-level nuclear waste. In addition, reprocessing recovers valuable fuel.[8] About 95 percent of the material contained in the spent fuel rods that come out of nuclear reactors can be reused. During reprocessing, that material—about 1 percent of which is plutonium—is captured and used to produce mixed oxide fuel (called MOX) that can then be burned in nuclear reactors to generate more electricity.[9]

Of course, the plutonium recovered during reprocessing could, in theory, be used to make a weapon, and some antinuclear groups have used that fact as a reason to oppose the use of nuclear power. For instance, the Rocky Mountain Institute has declared that commercial nuclear power is the "biggest driving force" behind nuclear proliferation and that it provides "do-it-yourself bomb kits—nearly all the needed materials, skills, knowledge, and equipment—in innocent-looking civilian disguise."[10]

Associating nuclear power with nuclear weapons requires ignoring the obvious: Several countries have developed nuclear weapons without even thinking about producing electricity. If a country wants an atomic bomb, they can build one with or without a nuclear power program. During World War II, the United States developed the atom bomb. It took another decade or so for the nation to begin using fission to produce electricity. Other countries, including France, Israel, North Korea, and Pakistan, all developed nuclear weapons before they developed nuclear power. And of those four, only France is now producing significant amounts of electricity with fission. Since 1945, a new nation has been able to build or acquire a nuclear weapon about every five years.[11] The world's leading countries may not want Iran to have such a weapon, but the unfortunate reality is that the knowledge that allows countries to build nuclear bombs has diffused to the point that nuclear proliferation has become a fact of the modern world. The problem of nuclear proliferation has been a key challenge in global geopolitics since the end of World War II. Properly addressing it will take concerted international effort, and a central part of that effort must be U.S. leadership on the issues of nuclear fuel and proper handling and disposal of nuclear waste.

The United States has been stockpiling its nuclear waste for decades while debating what to do, but the debate continues to ignore one of the best options: "integrated energy parks." The idea, now being promoted by some officials inside the Department of Energy, would allow the nation to use its own real estate for interim nuclear waste storage. The parks could also be used as locations for new nuclear reactors and nuclear fuel reprocessing.

The U.S. Department of Energy's nuclear-focused national laboratories—Los Alamos, Idaho, Savannah River, Sandia, and Oak Ridge—have decades of experience with nuclear materials and technologies. The communities near the labs are familiar with nuclear issues and are interested in keeping the jobs that the labs provide. Perhaps most important, the nuclear-focused national labs already have the essential security and safety systems in place. Using the integrated energy parks to store nuclear waste would allow the federal government to fulfill its obligations under the Nuclear Waste Policy Act, which requires the DOE to take and dispose of the waste from nuclear power plants.[12]

Today, the United States has thousands of tons of nuclear waste spread among 121 different locations across the country. Consolidating that waste at the national laboratories would improve overall security, allow the federal government to comply with the Nuclear Waste Policy Act, and legitimize the concept of using those sites as integrated energy parks, where the waste could be stored indefinitely in dry casks. When the nation finally musters the political will to do something with the waste, it could then be reprocessed at those same sites. After reprocessing, the fuel could be used in reactors built on the very same properties.

That brings us to another key point: Nuclear reactors are not the problem when it comes to nuclear weapons proliferation, they are part of the solution. In June 2009, Stewart Brand, the longtime environmental activist—as well as an ardent backer of nuclear power—declared that "nuclear energy has done more to dismantle weapons than any other activity."[13] Since the mid-1990s, Russia has been shipping bomb-grade nuclear material from dismantled warheads to the United States, where it is being converted into fuel for use in commercial reactors. By mid-2009, the program, called "megatons to megawatts," had eliminated the equivalent of nearly 15,000 nuclear warheads.[14] That program will continue for years to come. As it continues producing electricity from the world's nuclear weapons stockpiles, it will be making the world a safer place.

The megatons to megawatts program is laudable, and yet the United States must begin developing a long-term solution for nuclear waste disposal. One intriguing option involves using transmutation to destroy the most worrisome segments of the radioactive waste stream. Transmutation—a process that involves transforming long-lived radioactive isotopes into ones with shorter half-lives—reduces the toxicity of the materials that come out of nuclear reactors.[15] Using transmutation has at least three advantages: The most dangerous materials, such as plutonium, are burned rather than stored, thereby minimizing the proliferation risk; the process reduces the amount of dangerous wastes produced by fission reactors, thereby reducing the number of long-term repositories needed to handle the waste; and finally, it reduces the long-term toxicity of the final waste products from millennia down to centuries or decades.

Before discussing transmutation, I need to explain the basics of nuclear waste. Today, all of the waste that comes out of America's nuclear reactors uses the "open" or "once through" fuel cycle. The cheapest method of handling waste, the open fuel cycle takes the spent fuel rods from nuclear reactors and stores them in spent-fuel pools or sealed metal casks for several years. From there, the waste is supposed to go into a permanent central waste repository such as Yucca Mountain, where it will remain, in theory, forever.[16]

Meanwhile, France and several other countries take a different tack, reprocessing the waste that comes out of their reactors. Doing so is more expensive than employing the open fuel cycle but makes better use of the energy potential in the spent fuel rods. Reprocessing, which involves capturing the uranium and other materials so that they can be sent through the reactors again, reduces the volume of waste by a factor of two or three.

Regardless of whether the spent fuel rods are reprocessed or not, the final result is a stream of highly radioactive waste that must be buried somewhere and kept safe for millennia. If that waste is buried without further treatment, then the relatively short-term radioactivity of the fission products, such as cesium and strontium, is greatly reduced after about a hundred years. At that time, says Swadesh Mahajan, a thermonuclear physicist at the Institute for Fusion Studies at the University of Texas, the repository will be, in effect, "a plutonium mine," because the spent fuel rods will still contain significant quantities of the bomb-making material as well as other isotopes that can be made to undergo fission relatively easily. And that's why Mahajan and several other nuclear scientists around the world are rallying around the concept of transmutation.

"Burying the waste isn't enough," says Mahajan. "We must reduce the quantity and therefore the long-term biohazards associated with the toxic waste. Doing so cuts the number of final repositories for the waste, which is critical. But it also cuts the long-term risk of proliferation."[17]

The key ingredient needed to make transmutation viable: neutrons—the uncharged particles that make up half or more of the mass in an atom. Given a good source of "fast" or high-energy neutrons, the most troublesome nuclear waste could be effectively burned. And the process of burning the waste also produces significant quantities of energy that can be captured to produce electricity. There are several methods for producing the fast neutrons needed for transmutation, but one promising technology appears to be a hybrid reactor that combines both fusion and fission.

Fusion's reputation has been tarnished by an excess of hype. Just like fuel cell–powered cars and the hydrogen economy, producing electricity from fusion has been touted as the Next Big Thing for decades. But the hybrid fusion reactors now being discussed are not designed for electricity production. Instead, their main purpose would be the production of fast neutrons, a process that is far less technically demanding.

Those fast neutrons would be used to bombard the most problematic wastes, including plutonium, americium, curium, and neptunium, which have very long half-lives. (Neptunium's half-life is 2 million years.)[18] That bombardment forces the radioactive atoms to undergo fission, which turns them into lighter elements and isotopes that decay much more rapidly.[19]

The idea of using a fusion-fission system has been promoted by two Nobel Prize–winning-scientists: Andrei Sakharov and Hans Bethe. Sakharov, a Soviet-era physicist who is best-known for his brave advocacy for human rights and freedom, proposed a fusion-fission system back in 1950.[20] Bethe, the 1967 recipient of the Nobel Prize in Physics, was German American and worked on the Manhattan Project during World War II. He advocated the use of the fusion-fission hybrid as a way to create additional nuclear fuel and also saw that model as a method of dealing with high-level nuclear waste, saying it could provide "a great advantage for the prevention of proliferation of nuclear weapons."[21]

Here in the United States, scientists from the University of Texas, Georgia Institute of Technology, and Princeton University are leading the push to make the hybrid fusion-fission reactor idea into a reality. Scientists in China are also pursuing the hybrid concept.[22] But although the concept shows promise, it's still just a concept, and it would have to undergo years of testing and development before it could be used in an actual working reactor. Advances in materials science will be needed to make certain that the walls of the hybrid could withstand the extreme heat that would be generated by the new type of reactor. All of that research and development would require Congress to appropriate billions of dollars to multiple research teams over several decades. And that would require bipartisan political support.

And though transmutation offers one option, it doesn't get rid of all the waste. Even if the fast neutrons produced by the fusion reactor are able to burn the majority of the waste, some radioactive materials will remain, and that means the United States would still need a long-term waste repository. Given all the work that's been done at Yucca Mountain, that site continues to be a viable candidate. But the government has other options, including the Waste Isolation Pilot Plant in New Mexico. Located near Carlsbad, it began accepting defense-related nuclear waste in 1999.[23] The site—which utilizes a salt formation located half a mile below the surface that has been geologically stable for some 250 million years—could also be used to dispose of the waste coming out of nuclear reactors. The only thing standing in the way of that concept: Congress.

Alas, Congress continues to dither. Meanwhile, the United States is missing an opportunity to lead the world in the development of the technologies that could revolutionize the nuclear power sector by finally resolving the problem of nuclear waste disposal. Congress has effectively dictated that U.S. electric utilities must continue piling up spent nuclear waste at the sites where it is generated. But congressional inaction should not hinder the development of new nuclear plants. Fully addressing the challenge of nuclear waste will take decades. The issue is not so urgent that it must be resolved before new plants are built.

Indeed, the new designs for nuclear power plants are so attractive that building more of them makes sense, both economically and environmentally. They have been made safer, more efficient, and, in some cases, smaller. And those new reactor designs may provide a key breakthrough, both in terms of cost and scale.

CHAPTER 28

Future Nukes

FOR DECADES, there was one constant in the U.S. electricity business: growing demand. After the invention of the incandescent lightbulb, America's booming economy meant an ever-growing need for more electricity. For instance, during the 1950s, electricity demand grew by about 9 percent per year. But thanks to steady increases in the efficiency of the U.S. economy, that demand growth steadily fell during the second half of the twentieth century, though annual power demand growth was still averaging about 1.1 percent by the mid-2000s.[1] That growth stopped altogether with the global economic crisis that hit in late 2008, and during the first six months of 2009, U.S. electricity demand actually fell by 5 percent.[2]

Nevertheless, there is every reason to expect that U.S. electricity consumption will rebound as the nation's economy and population continue to grow. The Energy Information Administration projects that by 2030, U.S. electricity use will rise by at least 14 percent and perhaps by as much as 34 percent when compared to 2007 levels.[3]

Nuclear power can, and should, be used to meet some of that new demand—and it can do so by using reactors that are both large and small. It can also meet some of that new demand by using thorium, rather than uranium, to fuel those reactors. More on thorium in a moment.

Al Gore and other critics contend that nuclear power plants only come in large sizes with large price tags.[4] Although it's true that most of the commercial reactors now being used to generate electricity have capacities of about 1,000 megawatts or more, some of the most exciting developments in the nuclear power sector are happening around reactors with outputs of 125 megawatts or less. And those smaller reactors could be used in a variety of locations and applications, revolutionizing the electricity generation business.

Modular reactors could be used in series to displace larger coal- or natural gas–fired generators and could be particularly appealing to remote towns and cities that currently rely on diesel-fired generators to supply their electricity. They could also be used at locations that need highly reliable electric power, such as military bases or large industrial facilities. Modular reactors could be used to help stabilize the electric power grid, as dispersing small reactors over a large power grid could help a city or a utility assure more reliable power delivery.

Small reactors have a long history aboard military vessels. The U.S. Navy has long been one of the world's biggest users of nuclear power, relying on small reactors to propel their submarines and surface ships. The USS *Virginia*, for instance, one of the newest attack submarines in America's nuclear-powered fleet, uses a pressurized water reactor with a total output of about 37 megawatts.[5]

Using smaller reactors for land-based applications offers a number of advantages, particularly when these small plants are compared with their larger cousins. First, they would cost a fraction of the cost of the larger plants. Second, they could be used as single or multiple units. Thus, if a utility needs, say 800 megawatts of generation capacity, it could buy as many small reactors as it needed to meet that demand and add more in stages. Third, small reactors could be manufactured in a central location. One of the reasons that large reactors are so expensive is that they are, well, big. They must be built at the final location. By contrast, the companies proposing to build modular reactors are planning to use a factory-based approach where the reactors are manufactured and then shipped to their final destination. This approach, which is not quite mass-production, should result in dramatically lower costs. Fourth, the modular reactors are designed to be buried in the ground, which makes them more resistant to any weather- or terrorism-related event.

Now, before you start shopping for a modular reactor at your nearest Home Depot, understand that these designs are still conceptual. None of the companies that are proposing modular reactors have received approval from the Nuclear Regulatory Commission for their designs.

Among the first modular reactor designs to make headlines was a 10-megawatt idea put forward by the Japanese firm Toshiba. In 2006, it began discussing the possibility of locating a small reactor in the remote town of Galena, Alaska, which relies on diesel-fired generators for its electricity. Toshiba, which owns Westinghouse, claimed that the reactor could produce power for about $0.13 per kilowatt-hour, which would be far cheaper than what the town is paying for power from its diesel generators. Toshiba has dubbed its design "4S" (for super-safe, small, and simple). The reactor would produce enough electricity for about 10,000 homes, and it would be cooled by liquid sodium instead of water. Toshiba has called its design a "nuclear battery" that could operate for up to thirty years without refueling.[6]

Three other companies—Hyperion Power Generation, NuScale Power, and Babcock & Wilcox—are also vying to build the first modular reactor for the U.S. market. Hyperion, a Santa Fe–based company, is using technology developed and licensed by Los Alamos National Laboratory. Hyperion's 25-megawatt reactor would be about the size of an average hot tub, making it small enough to be transported via tractor-trailer. Hyperion says that it wants to build "about 4,000" of its reactors within the first ten years of production. The units would be encased in concrete, buried underground, and refueled every five to seven years. Hyperion is backed by venture capital money.[7]

Oregon-based NuScale is also backed by venture capitalists. The NuScale design would produce 45 megawatts of power, a size that the company says would give utilities and other power providers "a way to add and finance new generating capacity in a manner and on a time scale similar to gas turbines."[8] Like many of the large reactors now in use, the NuScale reactor is a pressurized water reactor, but its design is simpler. The water in the system is cooled by natural circulation, thus eliminating many pumps, pipes, and other parts that could fail.

Perhaps the most credible bid to build modular reactors comes from Babcock & Wilcox. In June 2009, the company announced plans to build a modular reactor capable of generating 125 megawatts of electricity. Babcock & Wilcox has a key advantage over Hyperion and NuScale because it already has a long history of manufacturing components for the electricity sector. When Thomas Edison established the first central power plant in the United States, on Pearl Street in Manhattan in 1882, he relied on boilers made by Babcock & Wilcox.[9] Furthermore, Babcock & Wilcox is a subsidiary of Houston-based McDermott International, a publicly traded engineering and construction company whose 2008 revenues totaled $6.5 billion.[10] Thus, it has plenty of capital available to back what will clearly be a multiyear licensing, designing, and manufacturing process. Babcock & Wilcox already has factories and trained employees with decades of experience building nuclear power components for the U.S. Navy. Hyperion and NuScale do not yet have any manufacturing facilities; nor do they have prior experience in building nuclear components.

Though the designs for modular reactors are promising, they are likely several years away from being licensed by the Nuclear Regulatory Commission, a process that is notoriously expensive and slow. It will likely be at least five years, perhaps even ten years, before any of the modular reactor companies are able to begin manufacturing their units.

In the meantime, several new designs for large reactors are coming to the market, and they are safer and more powerful than their predecessors. General Electric, Westinghouse, Areva, and Mitsubishi have all submitted design certification applications to the Nuclear Regulatory Commission.[11] The agency has already approved a Westinghouse reactor, the AP1000, that is proving popular in other countries because of its simplified design.

Other designs are also showing promise. Engineers at Massachusetts Institute of Technology, along with companies in South Africa, China, and the Netherlands, are working on "pebble-bed" reactor technology that may be even safer than other designs. The reactor uses fuel pellets that are difficult to reprocess, a feature that makes the fuel cycle safer as it reduces the possibility that the plutonium that is left in the spent fuel might be diverted for nefarious purposes.[12]

In addition to the pebble-bed design, some engineers are working on reactor designs that would be fueled with thorium, an element that sits near uranium on the periodic table. But unlike uranium, when used in reactors thorium would not produce any plutonium, a characteristic that makes thorium more attractive with regard to potential weapons prolif-eration.[13] Thorium is about four times more abundant than uranium and is easier to mine. Better still, the United States holds about 20 percent of the world's known supply of thorium.[14]

The history of thorium-fueled reactors goes back to the days of Pres-ident Dwight D. Eisenhower's Atoms for Peace program. In fact, the first commercial power plant developed under that program, a reactor in Shippingport, Pennsylvania, was initially fueled with thorium.[15] India, which holds about 25 percent of the world's known thorium deposits, has been pursuing thorium-fueled reactors for many years.[16] Lately, sev-eral other countries have begun looking more closely at thorium, includ-ing Canada, China, the United States, France, Japan, Norway, and Russia. Here in the United States, Virginia-based Lightbridge Corporation, a small company that specializes in thorium fuel, claims that it will begin testing of thorium fuel rods in commercial reactors by 2012 or 2013. The CEO of Lightbridge says that his company's thorium fuel rods can be used in existing reactors without any modifications and that the tho-rium fuel would be about 5 to 15 percent cheaper than comparable amounts of uranium.[17] Lightbridge also claims that the thorium fuel cycle produces far less radioactive waste than uranium.[18]

Of course, substituting thorium fuel for uranium will take time. Lightbridge and other companies must continue their development work and obtain licensing from nuclear regulators here in the United States and in other countries. But the potential offered by thorium, modular reactors, pebble-bed reactors, and other reactor designs is obvious.[19] Fur-thermore, it's clear that nuclear power must be part of the energy mix if the world is to achieve any significant progress in cutting the growth of carbon dioxide emissions. In its 2009 World Energy Outlook, the Inter-national Energy Agency declared that "nuclear technology is the only large-scale, baseload, electricity-generation technology with a near-zero carbon footprint."[20]

If policymakers are going to agree that carbon dioxide is a problem, then, as the Pulitzer Prize–winning author Richard Rhodes has put it, "nuclear power should be central." In 2000, Rhodes—who has probably written more about nuclear weapons and nuclear power than any other author—along with Denis Beller of Los Alamos National Laboratory, writing in an article in *Foreign Affairs* called "The Need for Nuclear Power," concluded that "despite its outstanding record, [nuclear power] has . . . been relegated by its opponents to the same twilight zone of contentious ideological conflict as abortion and evolution. It deserves better. Nuclear power is environmentally safe, practical, and affordable. It is not the problem—it is one of the best solutions."[21]

Rhodes and Beller made an essential point: The discussion about nuclear power has devolved to the point where it is akin to the debates over some of the most controversial issues of our time. The abortion and evolution debates are ruled by emotion and faith, not rationality. When it comes to nuclear power, the United States must undertake a relentlessly logical approach, one that depends on facts. It must also embrace the development of emerging nuclear technologies such as modular reactors, thorium fuel, and other reactor concepts. If it does so, then nuclear power, as Rhodes and Beller concluded, will undoubtedly be seen as one of the best available long-term solutions to our energy challenges.

PART IV
MOVING
FORWARD

CHAPTER 29

Rethinking "Green" and a Few Other Suggestions

> Doubt is not a pleasant condition,
> but certainty is absurd.
>
> VOLTAIRE

I STILL REMEMBER the first time I mixed blue paint with yellow paint. The result, of course, was a revelation to my young eyes: green.

Today, more than four and a half decades later, I'm eager to return to simpler ideas about what is, and isn't, green. Over the past few years, the concept of "green-ness" has become so overused as to become devoid of meaning. As I hope this book has helped to make clear, most, or perhaps all, of the renewable energy push, and in particular, the push for more wind power, is based on the bogus notion that those sources are "greener" than hydrocarbons such as oil and natural gas. That's simply not true.

All the blather about "green" has fostered the delusion that we can get our energy on the cheap, without any environmental impacts at all. Again, that's just not true. Sure, the idea of wind turbines might have a certain charm, and arrays of solar panels might make our cities and towns look like settings for science-fiction films. And if only we could just get a few coal-fired power plants to belch a rainbow every once in a while, they might look kind of pretty, too. But the hard truth is that energy production is not pretty, cheap, or easy.

Although I have attacked many of the claims about alternative energy, it's clear that the push for renewable energy has lots of momentum. The industry has captured much of the public's imagination, and that means that sources such as wind and solar will continue their rapid growth. Between now and 2030, the International Energy Agency expects that some $5.5 trillion will be spent on renewable energy projects,[1] and by the end of that period, renewables could be providing 10 percent of the world's primary energy needs.[2] Significant strides are being made in reducing the cost of solar power. In early 2009, First Solar, one of America's biggest producers of photovoltaic cells, said it had reduced its manufacturing costs to about $1 per watt, a key threshold for economic viability.[3] And in August 2009, eSolar, a thermal-solar company, christened a facility in the California desert that the company claims has higher power densities than similar solar projects and does so at lower cost.[4]

In mid-2009, the desire to find alternative motor fuels led Exxon Mobil to team up with California-based Synthetic Genomics to study photosynthetic algae. The deal calls for the oil giant to invest up to $600 million in the project.[5] Although algae-based fuels are many years away from being commercially viable, other alternative energy technologies are making progress. In November 2009, a California-based company, SolarReserve, announced plans to build a concentrated solar farm that will use molten salt to store energy. The company claims that it will be able to store up to seven hours of the project's solar energy in the form of molten salt.[6] Meanwhile, Dow Chemical has developed a solar roof shingle for the residential market that the company claims can be installed just like asphalt shingles to form an array. They are cheaper than conventional photovoltaic panels, Dow says, and could "offset between 40 percent and 80 percent of a home's electricity consumption."[7] Another intriguing possibility: spray-on solar cells. Researchers in the United States, Australia, Canada, and Switzerland are working on plastic coatings that contain tiny particles of titanium, copper, gallium, and indium. The coatings could be far cheaper than today's solar panels and could be applied to both vertical and horizontal surfaces.[8]

All of those possibilities are exciting, and they should find a ready market. While I believe that natural gas and nuclear power offer the best short- and long-term energy options for the future, I'm also bullish on solar. That said, I'm leery of making sweeping pronouncements. My wariness is, in part, a product of seeing how the shale gas revolution has swept away much of the conventional wisdom about the future of the U.S. gas industry. In the span of about three years, from 2005 to 2008, the industry swung from fears of gas shortages to a glut of gas. Looking back even further provides evidence of another disruptive technology. In 1882, Thomas Edison built the first central power station on Pearl Street in New York City, and within eight years, there were 1,000 similar stations operating all across the country.

Similar technological disruptions may lie ahead. They could include a breakthrough in energy storage technology or perhaps the discovery of a massive new oil field. A drop in oil demand combined with excess oil production capacity would result in a major drop in the price of oil—and that price decrease would immediately undermine the push for alternative energy efforts. I've made clear my position that hydrocarbons will persist for many decades to come, and yet I know that I could be proven wrong. As Voltaire said, "certainty is absurd."

At the end of my last book *Gusher of Lies*, I offered a few suggestions about energy policy. As I look at some of those suggestions today, roughly two years later, I find that my positions haven't changed.* But given those suggestions, I need to outline the gist of the N2N Plan.

When he announced the Pickens Plan, Dallas oil baron T. Boone Pickens said that "an idiot with a plan is better than a genius with no plan." Unlike Pickens, I don't have $60 million to launch a media drive to promote N2N, but perhaps that's okay, because my plan isn't as complicated as his. In fact, the N2N Plan has just four concepts:

1. Promote natural gas and nuclear power through targeted use of tax incentives.
2. Encourage oil and gas production in the United States.

---

*Those suggestions: Get the government the hell out of the energy business; stop obsessing over prices and reduce the number of fuel blends; reject the culture of fear and engage the Arab and Islamic worlds; redefine energy security; accept increasing energy use and adapt to a changing global climate; embrace solar and nuclear; pursue new technologies and efficiency; create the superbattery prize; increase domestic oil production; and embrace natural gas.

3. Continue promoting energy efficiency.
4. Continue working on renewables and energy storage technologies such as batteries and compressed-air energy storage.

That's it. The N2N Plan doesn't make any promises about reductions in foreign oil, or carbon dioxide emissions, or anything else. Further note that the last two items don't even really need to be stated. The United States, and the other countries of the world, will keep pursuing efficiency, renewables, and energy storage because those areas have always attracted capital. We don't need to tell entrepreneurs and engineers to make more efficient machines. They do it on their own because they are interested in making money. Huge amounts of capital—including billions of dollars courtesy of U.S. taxpayers—are going into the energy storage business because batteries, hybrid cars, and electric cars all have some potential for displacing some hydrocarbons.

Although the United States should be encouraging more production of oil—and of natural gas—within the United States and in its offshore waters, the Obama administration and leading congressional Democrats have threatened to repeal a pair of tax breaks that oil and gas industry officials believe are essential. Obama's 2010 budget called for the elimination of the expensing of "intangible drilling costs," which allows energy companies to deduct the bulk of their expenses for drilling new wells; it also called for ending "percentage depletion," which allows well owners to deduct a certain amount of the value of their production in a given year. In May 2009, Obama called the tax treatments "unjustifiable loopholes" that do "little to incentivize production or reduce energy prices."[9] But getting rid of those tax breaks now, just months after the U.S. natural gas sector has unlocked enormous quantities of shale gas, makes no sense whatsoever. Fortunately, common sense appears to have triumphed and the tax breaks remain in place—at least for the time being.

Given that it's in America's long-term interest to promote nuclear power at home, it must also take a leadership role globally on the issues of nuclear safety, regulation, and proliferation. Therefore, my first suggestion, and perhaps the most important one, is for the United States to support the one organization that is already set up to do that on the international level.

Vigorously Support the IAEA

The damage done by George W. Bush, Dick Cheney, and their cronies to the reputation of the United States in the international arena will last for decades. Bush and Cheney tarnished America's image on a variety of issues, including the use of torture against prisoners and adherence to human rights principles and the rule of law. But when it comes to energy issues, few things have hurt America's long-term credibility more than the Bush administration's steamrolling of the International Atomic Energy Agency in its headlong rush to unleash the dogs of war on Saddam Hussein's Iraq.

Prior to the invasion of Iraq, top Bush administration officials repeatedly claimed that Iraq was trying to build a nuclear weapon. On September 7, 2002, Bush himself falsely claimed that the IAEA, the lead global agency in dealing with proliferation issues, had issued a report saying that Iraq was just six months away from developing a nuclear weapon.[10] A few months later, on January 27, 2003, the director general of the IAEA, Mohamed El-Baradei, told the United Nations Security Council that there was no evidence of a nuclear weapons program in Iraq and that the aluminum tubes that were a focal point of the U.S. disinformation program "would not be suitable for manufacturing centrifuges" needed to enrich uranium.[11]

ElBaradei, of course, was right. After the invasion, the U.S. military never found any nuclear weapons or weapons of mass destruction in Iraq. But that didn't stop the Bush administration from belittling the agency and ElBaradei. In 2005, John Bolton, whom Bush had nominated to be America's ambassador to the UN, said that the IAEA's declaration that Iraq didn't have a nuclear weapons program was "simply impossible to believe." That same year, the Bush administration tried to have El-Baradei replaced.[12]

In 2005, ElBaradei was awarded the Nobel Peace Prize,[13] and in 2009 he told *Time* magazine that the "most dissatisfying moment of my life, of course, was when the Iraq war was launched. That hundreds of thousands of people lost their lives on the basis of fiction, not facts, makes me shudder."[14] Although ElBaradei left the IAEA in late 2009, the Vienna-based agency remains the essential international agency for nuclear issues and is focused solely on those issues. ElBaradei played a critical role in the international negotiations aimed at heading off a military confrontation with Iran over its nuclear aspirations, and his replacement will have to take a similar high-profile role.[15]

For the United States to embrace N2N, it must be committed to vigorous international regulation and policing of the nuclear sector. That means closer monitoring of the fuel being used by the growing number of nuclear reactors as well as increased efforts at ports and other locations to detect any radioactive materials that could be used for nefarious purposes.

Created in 1957, the IAEA was set up as a global "atoms for peace" agency under the aegis of the UN.[16] It now operates on a budget of about $400 million per year.[17] But it needs more money, political support, and technology. The United States should help to provide all three.

President Obama has indicated that he wants more international nuclear cooperation. In his April 2009 speech in Prague, he said that he wanted to "build a new framework for civil nuclear cooperation, including an international fuel bank, so that countries can access peaceful power without increasing the risks of proliferation."[18] That's an easy speech to make. But making that kind of program into a reality requires having a strong, credible, forceful IAEA—and no country is more important in making the IAEA credible than the United States. That is nothing new. Supporting the IAEA has been in America's long-term interests since the end of World War II.

In 1946, U.S. Undersecretary of State Dean Acheson (who became secretary of state in 1949) asked David Lilienthal, the head of the Tennessee Valley Authority, to chair a panel to advise President Harry Truman about nuclear weapons.[19] That same year, Lilienthal headed the production of a document known as the Acheson-Lilienthal Report, which concluded that the world had entered a new era in which nuclear technology would be widely known and understood. It said flatly that "there will no longer be secrets about atomic energy," and it declared that, given the potential destructiveness of nuclear weapons, there must be "international control of atomic energy" coupled with "a system of inspection."[20]

Today, more than six decades after that document was written, the need for international control of nuclear materials, along with reliable systems of inspection of nuclear facilities, remains essential. In fact, the need for a strong IAEA has never been more obvious. The United States must—repeat, must—be a leader in giving the IAEA all of the authority it needs to finally make the objectives of the Acheson-Lilienthal Report into a reality.

End Iowa's Monopoly on the Presidential Primaries

Iowa has unilaterally insisted that it must have the first presidential primary. As a result, a tiny percentage of a tiny percentage of people in a relatively small, rural, agricultural state have undue influence over the selection of the person who will become president of the United States. And that also makes Iowa ground zero for the corn ethanol boondoggle, a farm subsidy program that masquerades as an energy program. (Iowa calls its primary a "caucus," but it is, in reality, a primary.)[21]

Barack Obama was pro-ethanol when he went to Iowa, the biggest corn ethanol producer in the United States, and his win in the Iowa primary was key to his winning the White House. His key contenders—John McCain and Hillary Clinton—were both ardently opposed to the corn ethanol boondoggle until they decided to run for president. Once they launched their campaigns for the presidency, both of them became ethanol evangelicals because they understood the need to win in Iowa.

The United States must reform its presidential primary system. The first step in doing so is to reduce Iowa's importance in the selection process. The state's powerful agriculture lobby has corrupted our presidential selection process and in doing so has made support for corn ethanol a litmus test for any candidate aiming for the White House. One of the first steps in reforming our energy policy must be the elimination of the corn ethanol outrage. But that won't happen as long as Iowa maintains its stranglehold on the presidential primary system.

Elect More Engineers and Push Science, Technology, Engineering, and Math

France is run by engineers. The United States is run by lawyers. And that difference goes a long way in explaining why France has a forward-looking energy policy and the United States has, well, no stated energy policy at all.

Engineers build things. Lawyers sue people who build things. One of the greatest challenges in the making of a smart, forward-looking, no-regrets energy policy in America is the paucity of knowledgeable people in positions of power on Capitol Hill and in Washington who truly understand energy.

Congress is dominated by lawyers who want to make policy and write super-long, super-complex bills. The 2005 Energy Policy Act, for instance, was 550 pages,[22] and the Energy Independence and Security Act of 2007 was 310 pages.[23] In 2009, the House passed another energy bill, the American Clean Energy and Security Act, that amounted to 1,428 pages.[24] When printed out on standard paper, the 2009 bill (also known as Waxman-Markey) creates a stack nearly 7 inches tall. These mammoth bills are written by lawyers, for lawyers (and of course, lawyer-lobbyists). The result is an ever more fragmented and complicated U.S. energy policy that has little effect on overall energy-consumption patterns.

The over-abundance of lawyers in American government can be seen by looking at the Senate. In 2007, 60 of the 100 senators had law degrees,[25] and in 2009, only 3 had engineering degrees.[26] The House had exactly 1 registered professional engineer in 2009: Joe Barton, a Republican from the Dallas–Fort Worth area.[27] (Just for reference, the House had 13 medical doctors, and the Senate had 2 doctors.)[28] Barack Obama is a lawyer, and so was Bill Clinton. In France, the most prestigious school is probably the École Polytechnique. In the United States, it's Harvard Law. France has about 46,000 lawyers (about 1 for every 1,300 citizens),[29] whereas the United States has about 1.1 million (about 1 for every 280 citizens).[30]

If the United States is to move forward in a constructive way on energy policy, it must begin putting more people into positions of power who not only like to build things, but who are also scientifically literate and numerate. Unless or until there is greater numeracy among policy-makers, the nation will continue to get lousy energy policy promulgated by lawyers who have no concept of physics, numbers, scale, or, most essentially, the difference between energy and horsepower. As Stan Jakuba has told me many times: "People will take years to debate an energy issue but not a semester of physics."

Of course, electing more engineers and fewer lawyers will require us to have an education system that is capable of producing more students who are savvy in math and science. And given the parlous state of the American education system, particularly when it comes to mathematics, this challenge is particularly daunting. The United States must get more serious about teaching science, technology, engineering, and mathematics. America's future competitiveness depends on it.

Emulate Iran and France

If the United States is going to embrace N2N, it must emulate some of the energy policies of Iran and France.

Though that statement will anger many flag-waving Americans, the simple truth is that those two countries are embracing natural gas and nuclear power. Of course, Iran's pursuit of nuclear power has led to widespread controversy, with the true intentions of the country's leaders under suspicion—but that only further underscores the need for a strong IAEA, the only international agency with the capability of assuring that Iran does not use its nuclear facilities to build weapons, as well as the credibility to make its findings heard in the international arena. And although some critics argue that Iran doesn't need nuclear power, the country's electricity demand is soaring. Between 1990 and 2008, Iran ranked eighth in the world in terms of the speed of growth in electricity demand.[31] Over that time span, Iran's electricity generation nearly quadrupled—and given the country's young population (the median age is twenty-seven), that electricity demand will almost certainly continue its rapid growth.[32]

Regardless of Iran's nuclear plans, the data show that both Iran and France have dramatically increased their consumption of natural gas. At the time of the Arab Oil Embargo, France was getting just 8 percent of its total primary energy from gas. By 2008, that number had risen to 15 percent. Meanwhile, gas accounted for 27 percent of Iran's primary energy back in 1973, and thirty-five years later gas was providing 55 percent of the country's primary energy.

Iran has launched the world's most aggressive natural-gas-vehicle adoption plan. Between 2007 and 2008, the country tripled the number of NGVs on its roads. By mid-2009, Iran, the biggest auto producer in the Middle East, had about 1.5 million NGVs on its streets and was adding new NGVs at a rate of about 100,000 a month. Between the summer of 2008 and mid-2009, Iran more than doubled the number of NGV-refueling stations on its roads, bringing the total to about 1,000 locations— approximately the same number as now exist in the United States.[33] At this rate, Iran will surpass Pakistan as the country with the most NGVs before the end of 2010. (Pakistan has about 2 million NGVs.)[34]

Before going on, let me state the obvious: It will take years to add significant numbers of new NGVs to the U.S. auto fleet. Moreover, even a hundredfold increase in the number of NGVs in the United States will not result in "energy independence" or put much of a dent in foreign oil imports. But just as Iran is putting its copious quantities of natural gas to use in its transportation sector, it makes sense for the United States to increase utilization of natural gas as a way to hedge against potential oil price increases.

One promising technology that could make NGVs more viable is the use of "adsorbed gas tanks." Adsorption occurs when a gas or liquid accumulates on the surface of a solid, and adsorption technologies facilitate the storage of natural gas at far lower pressures, about 500 pounds per square inch, than those required in standard natural gas tanks, which generally store gas at more than 3,000 pounds per square inch. Two universities—the University of Missouri and the University of California, Los Angeles—have developed adsorption technologies that are awaiting commercialization.[35] Substituting adsorbed natural gas tanks for high-pressure tanks (which are akin to scuba tanks) would lower costs and allow carmakers to conform gas storage tanks to the shape of the vehicle.

Regarding nuclear power, no other country has embraced the atom as tightly as France. In 1973, nuclear power was providing just 2 percent of France's primary energy. By 2008, that percentage had risen to 39 percent, the highest rate of any country. Today, France is the world's most nuclear-dependent nation, getting nearly 80 percent of its electricity from its fleet of fifty-nine nuclear reactors—the majority of which have a standardized size and design.[36]

Along with their embrace of fission, the French have taken a serious approach to nuclear waste. And they have a place to put it: After decades of operating the world's second-largest fleet of fission reactors (the U.S. fleet is the biggest), France compacts or vitrifies all of its high-level radioactive waste and safely stores it in an area covering about 1.75 acres, which is about the size of one soccer field, at a site near La Hague.[37] The United States, meanwhile, continues its feckless, reckless, decades-long bender of denial, the recurring theme of which appears to be the notion that if only we ignore the nuclear waste problem long enough, maybe it'll go away.

In the years to come, if the United States wants to encourage nuclear power production, Congress will have to take a more dirigiste attitude. Although some free-market critics of nuclear power, including Jerry Taylor of the Cato Institute, believe that such intervention makes for bad policy, the reality is that the U.S. government has already made huge commitments to the nuclear power industry in the form of insurance, waste disposal, and loan guarantees.[38]

Over the past two years, Congress has extended massive handouts to the renewable energy sector for wind power, solar power, and batteries. Extending more federal support—funding, oversight, and research grants—to the nuclear sector will provide far bigger paybacks than the funds that are being given to the wind and solar sectors. That's not to argue that the United States should aim to exactly replicate the French model. That won't happen. The U.S. electric utility sector is highly fragmented, with electricity being provided by a mish-mash of cooperatives, investor-owned companies, and municipally owned utilities, and that diffused ownership necessitates a different approach than what we see in France. But that should not prevent the United States from moving forward with nuclear power initiatives.

The French have shown that nuclear power can be managed and managed well. In particular, France provides a workable model for dealing with nuclear waste. The United States should learn from that example.

Ban Mountaintop-Removal Mining

I have spent a fair amount of time in and around coal mines. Thirty years ago, while living on the Navajo Reservation, I spent many days digging live piñon pine trees from land that was soon to be strip mined. (I sold the salvaged trees to nurseries and homeowners.) While living on the reservation, I stayed warm on winter nights by burning coal in the cast-iron stove that sat in the middle of a one-room hogan near Fort Defiance that I was renting from a Navajo family. Years later, while living in Denver, I used my pickup truck to deliver half-ton loads of coal to homeowners who wanted low-cost fuel for their stoves and fireplaces.

I get it. I understand why we use coal. Coal consumption and coal mining are not going away. Coal-fired power generators provide about half of U.S. electricity, and their share of the power market will remain sizable for decades to come. But that doesn't mean that I have to like watching (albeit from a great distance) while big swaths of Appalachia are flattened. For years, Congress, along with the past few administrations, including the Obama administration, has winked while the coal industry razes Appalachia.

More than 1 million acres of Appalachian mountains and forests have been leveled since the mid-1990s.[39] The coal industry insists that this type of mining is essential because it is cheap. But although it may be cheaper to get coal out of the ground by using mountaintop removal than by using underground mining, the environmental effects are horrendous.

Smart energy policy goes beyond fuel mix, carbon output, and import levels. It's also about land use. That is evident in the myriad fights over high-voltage transmission lines and wind farms—and those fights are going to continue for a long time. But the siting of wind turbines, electric transmission lines, and oil wells is fundamentally different from the leveling of wide swaths of mountains.[40]

The EPA has finally begun taking a harder look at some of the federal permits needed by coal companies eager to increase their mountaintop-removal operations.[41] In August 2009, Ken Ward Jr., a hard-nosed writer for the *Charleston Gazette* who has documented the years-long battles over mountaintop removal, reported that coal-industry officials have admitted that much of the coal they are now producing via mountaintop removal could also be mined economically using shafts.[42]

I want energy sources that disturb the smallest possible amounts of real estate. That necessarily excludes mountaintop removal. Simply put, mountaintop removal should be prohibited.

Quit Wasting Natural Gas!

The world is awash in natural gas. And while that reality slowly sinks in, the countries of the world continue to waste vast quantities of it. In 2007 alone, about 5.3 trillion cubic feet of gas were burned off as waste, or in industry parlance, flared. That quantity of gas is the equivalent of about 30 percent of the European Union's total annual gas needs.[43] Put another way, on an average day the world flares about 14.5 billion cubic feet of gas, the energy equivalent of 2.6 million barrels of oil per day.[44]

The worst offenders when it comes to gas flaring are the oil producers in Russia and the countries around the Caspian Sea. That region flares some 2.1 trillion cubic feet of gas per year. Running a close second are the petrostates of the Middle East and North Africa, which flare about 1.6 trillion cubic feet per year.[45] Iraq alone is flaring about 1 billion cubic feet of natural gas per day, enough fuel to generate most—or perhaps all—of Iraq's electricity.[46] Or consider this: The 1 billion cubic feet of gas wasted every day in Iraq could supply nearly all of the natural gas needs of Taiwan.[47]

Of course, capturing all of the gas now being flared would cost tens of billions of dollars. And that's the key issue: Many producers flare the gas because saving the gas costs more than burning it. But some new technologies are emerging that could allow companies to turn that gas into a salable product. For instance, R3 Sciences, a Louisiana-based startup company, is developing a small system that can convert gas into methanol near the wellhead.[48] Although methanol is a relatively low-value product, R3's technology includes a process that can upgrade the methanol into dimethyl ether, which, like propane, can be used as motor fuel. Dimethyl ether can also be turned into gasoline.[49] Given those technologies, and the relative abundance of natural gas around the world, there is enormous potential for converting natural gas into motor fuel that can be used with the existing infrastructure.

The point here is obvious: Natural gas is too valuable for it to simply be wasted. The United States began eliminating the flaring of natural gas back in the 1940s. It's time for the rest of the world to follow suit.

Remember That Energy Demagoguery Will Continue Because No One Is in Charge

Energy policy is like the weather: Everybody complains about it, but nobody does anything about it. The reason: Nobody is really in charge.

Sure, Congress writes laws—it has passed two major energy bills since 2005—but the process of implementing energy policy is diffused across huge bureaucracies without any consideration of how to best coordinate substantive changes. And in that regard, very little has changed in the past forty years. In 1971, Monty Hoyt of the *Christian Science Monitor* estimated that federal energy policy was determined by forty-eight federal agencies and fourteen congressional committees.[50] By 2009, the crazy quilt of regulatory oversight was different but just as convoluted. That year, the U.S. Chamber of Commerce's Institute for 21st Century Energy estimated that there were twenty-four federal agencies and twenty-five congressional committees with a hand in shaping U.S. energy policy.[51] (See Appendix C for the chamber's list.)

This phalanx of regulators and overseers assures that demagogues on energy policy will continue to kvetch about what should, could, and might happen. And the ranks of energy posers will continue to swell, because the possibility of effecting significant change over a short period of time is nil, while the potential for outrage is infinite.

Reorienting energy diets takes time even in countries ruled by authoritarian regimes, and it will take even longer in the United States because U.S. energy policy decisions are so diffused among so many different entities. Creating a more cohesive, forward-looking energy policy will require each of those entities to give up some of their control. That might happen. But it will take a long time.

CHAPTER 30

Toward Cheap, Abundant Energy

IN THEIR 2005 book on the energy business, *The Bottomless Well*, authors Peter Huber and Mark Mills made the case for increased energy supplies, writing that, "over the long term, societies that expand and improve their energy supplies overwhelm those that don't. The paramount objective of US energy policy should be to promote abundant supplies of cheap energy and to facilitate their distribution and consumption."[1]

That's it exactly. More energy—and more power—equals more wealth. Period. End of story. Leave. Go home. Elvis has left the building.

The United States must—repeat, *must*—pursue cheap and abundant energy as a primary goal. But it must do more than that. It must also promote cheap and abundant energy abroad. By promoting cheap and abundant energy abroad, the United States can help stimulate its own economy.

Unfortunately, rather than focusing on cheap, abundant energy, policymakers in Congress and in the federal bureaucracy are launching mandates and subsidies for programs that will likely do just the opposite. And the results of those overly intrusive policies could be disastrous.

In August 2009, the U.S. Commerce Department announced that about 12.7 percent of the U.S. population—some 38.8 million Americans—were living in poverty. The agency estimated that between 2008 and 2009, more than 1.5 million Americans fell below the income level that qualifies them as "poor." (The federal government's official definition of poverty is an annual income of $21,203 for a family of four, or $13,540 for a family of two.)

In early November 2009, the official U.S. unemployment rate hit 10.2 percent. But the actual number is substantially higher. Officials at the Federal Reserve Bank in Atlanta have estimated that the actual unemployment rate—when accounting for people who have dropped out of the job market or are underemployed—is likely closer to 16 percent.[2] That implies that the number of Americans living in poverty is closer to 50 million, or about 15 percent of the U.S. population.[3] Put another way, nearly 1 in 6 Americans are now living in poverty. The numbers are even worse if you are a child, and worse still if you are an African American child. Whereas 1 in 5 children in the United States are now living in poverty, among African American youths it's 1 in 3.[4]

In mid-November, the Mortgage Bankers Association provided yet more grim news, announcing that about 1 in 10 U.S. homeowners who had mortgages were at least one payment behind schedule, the highest level since the association began keeping records in 1972. But as the *New York Times* reported, that number does not include the homes that are already in foreclosure. "The combined percentage of those in foreclosure as well as delinquent homeowners is 14.41%, or about 1 in 7 mortgage holders."[5]

Given the high level of unemployment, the huge number of people living in poverty, and the ragged state of the economy, few Americans are willing to put carbon dioxide emissions and global warming at the top of their list of concerns. On July 1, 2009, Rasmussen Reports released a survey finding that 56 percent of Americans "say they are not willing to pay more in taxes and utility costs to generate cleaner energy and fight global warming." The same survey, which was conducted after the U.S. House of Representatives narrowly approved a bill that would create a cap-and-trade mechanism for carbon dioxide emissions, found that just 21 percent of Americans were willing to pay $100 more per year for cleaner energy and global warming efforts. Just 14 percent said they would pay more than $100 per year.[6]

Those results, and previous polls showing strong opposition to any form of gasoline or motor fuel tax, prove that most Americans are more interested in their economic well-being than they are in abstract issues such as climate change. Americans are practical. They understand their need for energy even if they don't understand the broad outlines of the energy business or its many complexities. Americans want gasoline, electricity, natural gas, and all the other forms of energy. When those energy sources are not available, or when Americans think they're too expensive, they get mad. And when Americans get mad, alas, they don't necessarily inspire politicians to create thoughtful policies. Instead, the political rhetoric gets degraded yet further, with predictable results.

The news throughout the decade of the 2000s was dominated by stories about cheats, liars, and bamboozlers. The decade began with Enron and ended with Tiger Woods. But Enron was just the beginning. George W. Bush and his cronies used trumped-up intelligence to justify the Second Iraq War, a ruinously expensive campaign that will haunt the United States for decades to come. We had the fraud perpetrated by Dennis Kozlowski of Tyco International, who felt entitled to a $6,000 shower curtain.[7] There were the two Bernies: Bernie Ebbers of WorldCom, who's now serving a twenty-five-year sentence for fraud and conspiracy, and, of course, Bernie Madoff, the gold-digging mastermind of a multibillion-dollar Ponzi scheme who's now serving 150 years in prison. The sports pages were full of news about cheaters, from Major League Baseball players such as Mark McGwire and Barry Bonds to the ongoing doping scandals at the Tour de France. And we saw the carnage created by the pirates on Wall Street who engineered a multitrillion-dollar mess of toxic derivatives—from collateralized debt obligations to credit default swaps—that would have made even a privateer such as Enron's Jeffrey Skilling blush in embarrassment.[8]

We cannot, must not, be Enroned when it comes to energy and energy policy. We must understand—as business author Jim Collins makes clear—that facts are better than dreams. Americans must reject the notion that energy should be scarce and expensive. And make no mistake, that's exactly what many of these "green" energy projects will achieve: scarce, expensive energy.

Cheap energy must be the priority. Cheap energy will help us adapt to changes in the climate—regardless of why those changes are occurring. Cheap energy will allow the production of more potable water. As the world's demand for fresh water continues to grow, the need for desalination and other water-treatment technologies becomes more acute. In the coming decades, the energy-water nexus will be ever more important as the need for safe drinking water, water distribution, and wastewater treatment grows. Cheap energy also means better mobility. As the global economy continues to shift, cheap energy will allow more people to travel farther to find jobs. And, as always, cheap energy will allow for greater increases in productivity. But the United States must not only aim to have cheap energy at home, it must pursue that goal globally. As Michael Shellenberger and Ted Nordhaus of the Breakthrough Institute have declared, "we need to make clean energy cheap worldwide."[9]

The pursuit of cheap energy means pursuing N2N. Natural gas and nuclear power offer the best no-regrets energy policy because they reduce the volumes of neurotoxins released into the environment, cut solid waste production, slash greenhouse gases, eliminate air pollution, and obviate the need for carbon capture and sequestration. The United States can, and should, lead the development of the technologies that will speed the global embrace of N2N. The historical and societal megatrends favoring N2N are indisputable. The United States need only adjust its energy policies so that they align with those trends.

In 1865, William Stanley Jevons called energy "the universal aid—the factor in everything we do." He went on to say that with sufficient energy resources, "almost any feat is possible or easy; without it, we are thrown back into the laborious poverty of early times."[10]

We must accept the essentiality of energy, and in particular, the essentiality of electricity. If we are going to avoid the "laborious poverty of early times," we must move beyond the puerile energy rhetoric that is poisoning our politics. We must depoliticize energy and move beyond the petty recriminations and small-minded approach to the world's most important commodity. If we can achieve that, the United States can get busy providing more energy to more people, which will mean more prosperity.

There is no silver bullet. We cannot rely solely on solar and wind power to fuel our economy. Nor can we solely focus on hydrocarbons and nuclear power. We will need all of those sources, because it is the nature of civilization to use energy.[11] More energy means more power. And we are, all of us, power hungry.

Epilogue to the Paperback Edition

FOR ABOUT A MONTH after the release of the hardback edition of *Power Hungry*, I was rather despondent. The book was released on April 27, 2010, exactly one week after the blowout of BP's Macondo well in the Gulf of Mexico.

The accident, which left eleven workers dead, led to an orgy of news coverage about the dangers of offshore drilling, the potential for other blowouts, BP's terrible safety history, and inevitably, the need for the United States and the rest of the world to quit using oil. The blowout was a made-for-TV disaster, complete with live video from the bottom of the ocean showing the oil gushing out of the well bore. Images of oil-soaked birds and miles-wide patches of soupy crude floating atop the ocean were a staple of the twenty-four-hour news cycle as the well spewed some 50,000 barrels of oil per day into the Gulf.

And yet here I was, an energy journalist from Texas, promoting a book that declares that hydrocarbons are here to stay and that, yes, "oil is green."

My despondency gradually lifted, thanks to a number of positive reviews of *Power Hungry*, as well as my realization that the spill was destined to be a short-term story. And that's exactly what it was. The Macondo well was effectively controlled in early August. By September, BP had killed the well for good with a relief well that pumped cement into the bore hole and permanently sealed it. More importantly, the Macondo well accident, which ended up costing BP tens of billions of dollars, turned out to be the ecological disaster that wasn't. Despite the myriad predictions of a catastrophe that would endure for years (about ten days after the accident, the state director of the Mississippi chapter of the Sierra Club called the accident "America's Chernobyl"), the Gulf of Mexico turned out to be far more resilient than anyone expected.[1]

Now don't get me wrong. I am not suggesting that BP is blameless. Nor am I suggesting that oil spills are a good thing. Far from it. The company's management of the Macondo well was beyond pathetic. If BP had invested a fraction of the talent, money, and focus that it invested to clean up the mess created by the blowout into actually drilling the well properly, the accident would never have happened. Further, the fact that the accident happened in the warm waters of the Gulf of Mexico rather than in the Arctic was a key reason the oil disappeared so quickly. A blowout in a region where waters are cold or icy could have devastating, long-term effects. And finally, it's readily apparent that the oil industry as a whole must take safety more seriously. That is certain to happen in the wake of the Macondo accident, but the industry should have been more diligent all along. Sure, the accident happened at BP, a company whose safety history is among the worst in the industry. But by being lax on safety, BP hurt the image of the entire oil and gas sector.

While it's difficult to image the general public ever changing its dim view of the oil business, the spill created by the Macondo blowout did not last long. In mid-August, researchers announced the discovery of a previously unknown microbe that just loves the taste of crude oil in the summer.[2] And by that time, there was almost no evidence that the blowout had even happened. No oil was floating on top of the water. No more oily birds. The main economic damage in the region from the Macondo well accident was not a result of any spilled crude. Instead, it was a result of the Obama administration's "permitorium" on deepwater drilling. Shortly after the blowout, the administration declared a moratorium on drilling the Gulf in order to conduct a safety review of the rigs in the region. While that ban was lifted in early October, by mid-November federal officials were still loathe to approve new permits for drilling in water depths greater than 500 feet.[3] The result of the slow permitting process: deepwater rigs were leaving the Gulf, and tens of thousands of jobs were left in a prolonged state of limbo.

The Obama administration has used the Macondo accident to hamper the continuation of offshore drilling in the United States. Never mind that about 36 percent of all U. S. oil production comes from offshore, and, of that 1.8 million barrels per day of offshore oil production, about 66 percent comes from federal leases.[4]

The accident in the Gulf of Mexico, awful as it was, given the loss of life and the tremendous waste of resources, will turn out to be a small blip in the global history of energy use. That assessment will not find favor among the ardent opponents of hydrocarbons. But the reality is that the modern world runs on oil, coal, and natural gas. And while those fuels take a toll on the environment, they are indispensible. In this epilogue, I'll do a quick rundown of the major developments that have occurred with respect to oil, coal, natural gas, nuclear, and renewables since *Power Hungry* came out in hardback. And given that oil is the world's single most important energy source, that's where I'll start.

Oil

For those who doubt the primacy of oil, the news headlines in the fourth quarter of 2010 provided plenty of reminders. For instance, in September, China's crude oil imports hit a record 5.7 million barrels per day, an increase of 24 percent over levels a year earlier. Yes, you read it right: an increase of 24 percent. But even with those huge imports, China was experiencing widespread shortages of oil products, and the country's oil reserves were being drawn down at alarming rates. In April 2010, China had stored enough oil and oil products to meet demand for about 36 days. By late October, those stores had fallen to just 16 days, and the *Wall Street Journal* reported that "that number is falling fast."[5]

In early November, the International Energy Agency (IEA) released its World Energy Outlook, which predicted that by 2035, global oil demand will reach 99 million barrels per day, an 18 percent increase over the 84 million barrels per day that was consumed in 2009.[6] And while the IEA offered three different scenarios in its outlook, in all but one of the scenarios, oil will still be the world's most important fuel in 2035.[7]

Coal

The importance of coal to the global economy can be best understood by looking at just one country: Pakistan. The average American uses about 18 times as much energy as the average Pakistani. And that lack of energy availability in Pakistan is a direct contributor to that country's ongoing poverty and instability.

Secretary of State Hillary Clinton recognized that fact in October 2009 on her visit to Lahore. During that visit, she told her Pakistani hosts that they should burn more coal. Yes, you read that right.

During her visit to the Governor's House in Lahore, she said "The more economic development, the greater the energy challenges." She went on:

> I asked Minister Qureshi whether there had been any prohibition nationally under developing your coal deposits. Now, obviously, that is not the best thing for the climate, but everybody knows that. But many of your neighbors are producing coal faster than they can even talk about it. It's unfortunate, but it's a fact that coal is going to remain a part of the energy load until we can transition to cleaner forms of energy.
>
> So getting the resources to exploit your coal as opposed to being dependent upon imported energy is a choice for you to make, but it is certainly a choice that your neighbors have made. And that's something that should attract foreign investment and should attract capital investment within your own country. And we don't know how we're going to proceed on the climate change issue. We're working hard to come to some framework before Copenhagen, but coal will be, for the foreseeable future, part of the energy mix. And if you have these kinds of reserves, you should seek the best and cleanest technology for their extraction and their use going forward.[8]

Clinton's endorsement of increased coal consumption in Pakistan came just a few days before Obama greeted European officials in Washington to talk about the need to do something about reducing global carbon dioxide emissions.

But for Pakistan, concerns about carbon dioxide emissions are of scant importance when compared to the dire need for more electricity. As Clinton told her hosts, "Pakistan has to have more internal investment in your public services and in your business opportunities." But if that's to happen, it will require more cheap, abundant, reliable electricity, a commodity that is all too scarce in Pakistan, a country that has about 180 million people but has just 20,000 megawatts of installed generating capacity.[9] By comparison, France, with about 61 million people, has 112,000 megawatts of electric generation capacity.[10]

About a year after Clinton visited Lahore, Pakistani news outlets reported that about 2 billion tons of coal had been discovered in Sindh province in southern Pakistan.[11] The wealthy countries of the world may want to talk about carbon dioxide and climate change, but developing countries like Pakistan are more interested in pulling their citizens out of dire energy poverty. And that means burning more coal.

Natural Gas

On November 10, 2010, the Energy Information Administration announced that the amount of gas in storage hit an all-time record of 3.8 trillion cubic feet. That volume of gas was a full 10 percent above the average amount in storage in the United States over the previous five years.[12] The reason for the surge in gas storage, according to the EIA: "robust domestic production."

Ah, now that's an understatement. In August 2010, U. S. natural gas production totaled 1.85 trillion cubic feet.[13] That's an increase of 21 percent over the level achieved in June 2005, the month that Lee Raymond, the famously combative CEO of Exxon Mobil, declared that "gas production has peaked in North America."

As I discussed in *Power Hungry*, neither Raymond nor anyone else foresaw the shale gas revolution. But by November 2010, the full impact of the new gas paradigm was starting to come into focus. The most obvious indicator of America's sudden gas riches could be seen in the price of the commodity itself. With record amounts of gas in storage, the spot market price for natural gas at Henry Hub was about $3.50 per thousand cubic feet, a level that many in the industry believed was simply too low to be sustainable.[14] (There is widespread belief that natural gas prices need to be in the $6 to $7 range for production levels to be sustainable over a long period of time.) And while the full-year production numbers weren't available, it was clear by late 2010 that the United States would likely set a new annual record for production, thereby eclipsing the record achieved in 1973 when U. S. gas production peaked at 21.7 trillion cubic feet.[15]

With the United States awash in gas, it has become apparent that the country will remain awash in methane until it does more to stimulate demand. But, given the lackluster economy, for natural gas demand to increase, gas will have to steal market share from coal or oil. And given the many problems facing coal in the electricity generation business, natural gas will be the fuel of choice. In August 2010, the Tennessee Valley Authority announced that it would shutter nine of its coal-fired power plants with a combined capacity of 1,000 megawatts and replace them with gas-fired units. In announcing the move, TVA President and CEO Tom Kilgore said that "replacing some coal with other, cleaner fuel sources allows a reduction in air emissions including carbon."[16]

The TVA move heralds a much broader trend in the United States away from coal and toward gas in the power generation sector. Better still, it appears that the Obama administration has finally begun to understand the merits of natural gas. During his press conference on November 3, 2010, a day after the Democrats took a brutal beating during the mid-term elections, President Barack Obama uttered two words that have been missing from nearly every energy-related discussion he has had since he began running for president.

Yes, Obama actually said the words "natural gas." And he didn't mention solar, or wind, or geothermal.

Obama's mention—"we've got terrific natural gas resources in this country"—should not be overly noteworthy. But throughout his first two years in the White House, the president has been almost completely silent about the potential for gas to make a difference in the U. S. energy mix.[17] While it would be foolish to assume immediate, significant changes in federal policy toward gas based solely on Obama's November 3 press conference, there is no doubt that a number of trends favor increased use of methane both in the United States and around the world. The most important trend favoring gas: price. And some forecasters are predicting an abundance of cheap gas for the next decade. On the day that the IEA released its World Energy Outlook, the agency's chief economist Fatih Birol said that the world is oversupplied with gas and that "the gas glut will be with us 10 more years."[18]

While that may be true, the gas industry must contend with significant opposition, particularly with regard to concerns about the possible dangers of hydraulic fracturing. Those concerns will not be alleviated quickly. And the industry will have to be far more open and transparent with regard to that technology. It may also face more regulation. But as I write this in late November 2010, it appears the industry will be able to deal with those concerns.

Electricity

The essentiality of electricity to modernity is incontrovertible. The corollary to that point is also obvious: given that hydrocarbons (and coal in particular) are essential to the production of cheap, reliable electricity, there is no chance—none—that the world's developing countries (China and India in particular) will ever agree to emission limits on carbon dioxide.

Those facts were underscored by the most recent data from the IEA that forecast that global electricity demand will soar by some 80 percent by 2035. Meanwhile, the IEA expects primary energy demand to grow by 36 percent over that time period. Of that total, oil demand will grow by 19 percent, coal by 20 percent, and natural gas by 44 percent.[19]

Electricity supplies are closely tied to the world's most serious problem: widespread energy poverty. The latest IEA data show that 1.4 billion people, about 20 percent of the world's population, do not have access to electricity. And about 85 percent of those people live in rural areas.[20] The IEA's latest outlook included this amazing paragraph:

Electricity consumption in sub-Saharan Africa, excluding South Africa, is roughly equivalent to consumption in New York. In other words, the

19.5 million inhabitants of New York consume in a year roughly the same quantity of electricity, 40 terawatt-hours (TWh), as the 791 million people of sub-Saharan Africa.[21]

Unless or until the countries of the world can provide near-universal access to cheap, abundant, reliable electricity—much of which will have to be generated by burning hydrocarbons—nearly all of the talk about limiting carbon dioxide emissions is simply a waste of time.

Wind Energy

Since the release of *Power Hungry*, two major developments have occurred in the wind energy sector. First, it is readily apparent that the wind sector faces a massive problem with regard to wind turbine noise. Second, the industry is being hammered by the persistence of low natural gas prices.

On page 85 in Chapter 8 of *Power Hungry*, there are two paragraphs on the issue of wind turbine noise. I added the paragraphs to the manuscript for the hardcover edition at the last minute and only did so after doing some hurried research on the matter. Since then, it's become obvious that the wind industry is running scared about the noise issue, and it's apparent that the problem is far more widespread than the wind boosters are willing to admit. Here's a fuller explanation of the issue.

On January 25, I got an e-mail from Charlie Porter, a Missouri-based horse trainer whose farm near King City had been surrounded by a phalanx of giant turbines. His message said "The overwhelming noise, sleep deprivation, constant headaches, anxiety, etc., etc., etc., forced us to abandon our home/horse farm of 15 years. We had to buy a house in town, away from the turbines and move!"

I called Porter immediately. What he told me was like a bolt from the blue. His twenty-acre farm was, he said, "surrounded by lots of acres that nobody lived on." He was training quarter horses and having good success with it. But the wind turbines, the closest of which was installed 1,800 feet from his home, changed the life his family had grown to love. The noise from the turbines "just ruined life out in the country like we knew it. . . . We never intended to sell that farm. Now we couldn't sell it if we wanted to."

I immediately began researching Porter and his background. I double-checked everything he told me. I talked to the Gentry County tax assessor's office to verify his property records, including his claim that he'd had to buy a house in town to escape the noise. Everything checked out. I also began looking at the health effects that Porter described, symptoms that are now known as "wind turbine syndrome"—a term created by Dr. Nina Pierpont, a Malone, New York, physician who has studied a number of people, like Porter, who are suffering ill health due to the noise from wind turbines.

Since then, I've talked to or corresponded with homeowners who have had wind turbines built near their homes in Wisconsin, Maine, New York, Nova Scotia, Ontario, the U.K., New Zealand, and Australia. All of them used almost identical language in describing the low-frequency noise and problems caused by turbines that were built near their homes. Janet Warren, who was raising sheep on her 500-acre family farm near Makara, New Zealand, told me via e-mail that the turbines put up near her home emit "continuous noise and vibration" which she said was resulting in "genuine sleep deprivation causing loss of concentration, irritability, and short-term memory effects." The turbines, installed about 2,900 feet from Warren's house, began generating electricity in July 2009, and she says "we started recording formal noise compliance complaints in August." In February 2010, she and her husband were forced to move out of her home and to another location.[22]

In mid-February, I interviewed Tony Moyer, a resident of Empire, Wisconsin, who was now living with the noise created by three turbines that were erected within 1,400 feet of his home in late 2008 "If you get up at night, you can't go back to sleep because you hear those things howling." Moyer and his wife have tried white noise machines, to no avail. Nor are they able to sell. "The option to sell your home isn't there. I have wind turbines east, west, and north of me. If you talk to realtors, they can't sell homes near a wind farm."[23]

The wind energy industry has tried to dismiss the many complaints about wind turbine noise, sleep disruption, and deleterious health effects caused by the turbines. But there is plenty of evidence that shows that the low-frequency noise emitted by the giant turbines can be problematic. For decades, scientists and audiologists have known that even though humans cannot hear "infrasound"—noise that is lower in frequency than 20 Hertz—that same noise can still cause physiological damage.[24]

Noise complaints are a central element of an emerging citizen backlash against the global wind industry. Lawsuits that focus on the noise issue have been filed in Maine, Pennsylvania, and New Zealand.[25] In New Zealand, more than 750 noise complaints were lodged against a large wind project near Makara in the first ten months of its operation.

Indeed, evidence of the growing backlash against wind can even be found in Denmark, a country that boosters of wind energy like to claim as their Valhalla. On September 1, the *Copenhagen Post* carried a story titled "Dong gives up on land-based turbines." The subhead said "Mass protests mean the energy firm will look offshore." Here's the lead sentence from the article: "State-owned energy firm Dong Energy has given up building more wind turbines on Danish land, following protests from residents complaining about the noise the turbines make." The article goes on to quote the Danish company's CEO Anders Eldrup who said "It is very difficult to get the public's acceptance if the turbines are built close to residential buildings, and therefore we are now looking at maritime options."[26]

The opposition to wind energy in Denmark is hardly unique. Europe now has about 400 anti-wind energy groups spread among twenty European countries.[27] Canada has more than two dozen such groups.[28] And the United States has about 100 anti-wind groups.[29] In October, I attended a symposium on wind turbine noise in Picton, Ontario. Presenters at the conference included numerous medical doctors and PhDs. The consensus among the various presenters: wind turbines should have setbacks of at least two kilometers from any residence to avoid adverse health effects due to the noise. Alec Salt, a PhD scientist who works in the Department of Otolaryngology at the Washington University School of Medicine in St. Louis, was among the experts who spoke at the conference. Salt, an expert on the workings of the inner ear, said that "A physiologic pathway exists for infrasound to affect the brain at levels that are not heard. The idea that infrasound effects can be dismissed because they are inaudible is absolutely incorrect." Salt continued, "We need to stop ignoring the infrasound component of wind turbine noise and find out why it bothers people."

The Canadian symposium occurred about the same time that a new documentary by Laura Israel about the wind industry was appearing at film festivals around the country. Watching the documentary, called *Windfall*, provided a bit of déjà vu, as it puts on the screen nearly all of the issues that I'd been hearing about in my own research since Porter had contacted me in January.[30]

Israel's documentary, which premiered at the Toronto Film Festival in September, focused on the wind industry's attempt to build a number of turbines in Meredith, New York. Israel, who owns a cabin in the town, interviewed local residents and let viewers see how the town became bitterly divided over the issue of permitting the turbines. Some large landowners favored the siting of the turbines, in part because they were going to get royalty payments from the wind industry. That faction was led by the town's long-time supervisor Frank Bachler, who is portrayed as a well-intentioned man who, in favoring the wind development, is only trying to help the area's struggling farmers.

But a majority of the townspeople opposed the turbines. The resulting battle for control of the town's board provided a textbook example of democracy in action. After the board voted to approve the siting of turbines, three wind opponents ran for election to the town board with the stated purpose of reversing the existing board's position on wind. In November 2007, the opponents won and quickly passed a measure that effectively banned industrial wind development in the town.[31]

Israel's film provides a much-needed view of the anger that rural residents are expressing toward the rapid expansion of the wind industry. One of the best examples of that backlash includes Israel's interview with Carol Spinelli, a resident of Bovina, a small town located a few miles east of Meredith that imposed a ban on industrial wind development. Spinelli led the fight against wind turbines in Bovina, and she declares that the controversy is about "big money, big companies, big politics." Discussing wind developers, she says "I refer to them as modern day carpetbaggers. And that's what they are."

Israel also talked to a few homeowners who live near large wind projects. One of them, Eve Kelley, used language much like what I'd heard in my own research into the infrasound problem caused by turbines. The noise from the turbines, says Kelley, led to "dizzy spells, sick to my stomach . . . Sounds like the noise is in the walls. The house is vibrating."

While the noise issue is bedeviling wind energy developers, the more immediate concern for the industry is low gas prices. Why? Wind competes primarily with natural gas–fired generation.[32] And when gas prices are low, wind energy is at a big disadvantage in the marketplace, even with huge federal subsidies such as the $0.022 per kilowatt-hour federal production tax credit.

In 2008, T. Boone Pickens, one of the wind industry's most reliable boosters, said that gas prices must be at least $9 per million Btu for wind energy to be competitive in the marketplace.[33] In March of this year, Pickens was once again talking up wind energy, and he declared that "the place where it works best is with natural gas at $7."[34] That same month, a reporter from Dow Jones summarized Pickens's position by writing "Wind power is profitable when natural gas prices are about $7 a million British thermal units, Pickens said."

The bad news for the wind industry is that, for much of 2010, natural gas on the spot market has been selling for less than $4.[35] In September 2010, Paul Sankey, an energy analyst at Deutsche Bank, wrote that gas is in "fundamental oversupply" and will continue to be in oversupply through 2015.[36] That fundamental oversupply is due to several factors including a surge in natural gas liquefaction capacity in places like Qatar as well as the enormous increases in U. S. gas supplies that are a direct result of the shale gas revolution.

Those low gas prices make offshore wind appear even more uneconomic. The cost of building offshore wind projects is about $5,000 per kilowatt, or about the same as building a new nuclear plant. For comparison, a new gas-fired generation plant costs about $850 per kilowatt.[37] Those high costs are reflected in the prices that the developers of Cape Wind, the controversial offshore wind project near Cape Cod, are seeking for the electricity that could be generated by the turbines to be located in the waters of one of America's most famous vacation spots. The likely cost for electricity from Cape Wind will be between $0.17 and $0.21 per kilowatt-hour. Another offshore project, off the coast of Rhode Island, Deepwater Wind, was recently rejected by that state's public utility commission because the cost of electricity from the project was expected to be $0.244 per kilowatt-hour with annual increases of 3.5 percent per year.[38] For reference, the average retail price of electricity in the United States is about $0.10.[39]

The result from the wind industry's high costs: devastation. In late October 2010, the American Wind Energy Association announced that during the first nine months of the year, just 1,600 megawatts of new wind capacity was installed in the United States," down 72 percent versus 2009, and the lowest level since 2006." In a press release, the lobby group said the solution for its woes were—wait for it—more subsidies and mandates. The group's CEO Denise Bode said that "the best way to galvanize the industry now will be continued tax credits and a federal benchmark [read: "mandate"] of 15 percent renewables in the national electricity mix by 2020." Bode continued, saying that those subsidies and mandates "will send a clear signal to investors that the U. S. is open for business."[40]

Sure. But the business sector has already signaled that it doesn't want what the wind industry is selling. The reason: even with huge subsidies available to wind-energy projects, natural gas–fired generation remains a cheaper, more reliable option.[41] And even worse for the wind industry is that some industry insiders are predicting that the number of new wind generation installations will fall again, by as much as 50 percent, in 2011.[42]

Unfortunately, officials in the Obama administration are apparently convinced that wind energy is the way of the future and that opposition to projects based on noise or other concerns is unwarranted. In late October, Energy Secretary Steven Chu dismissed the opponents of wind energy projects by saying, "There's always some group . . . that will really be against whatever."[43]

Perhaps that's true. But some hard-core environmentalists are starting to understand that more wind energy projects will mean yet more energy sprawl. On November 8, five people, several of them from Earth First!, were arrested near Lincoln, Maine, after they blocked a road leading to a construction site for a 60-megawatt wind project on Rollins Mountain. According to a story written by Tux Turkel of the *Portland Press Herald*, one of the protesters carried a sign which read "Stop the rape of rural Maine."[44]

Ethanol

I've been writing about the corn ethanol scam for more than five years, but I'm still not cynical enough. That was made obvious in October 2010, when the Environmental Protection Agency approved an increase in the amount of ethanol that can be blended into the U. S. gasoline supply from 10 percent to as much as 15 percent.[45]

The Obama administration, the same one that said it was going to follow the science rigorously, made the move even though the EPA's own data shows that adding more ethanol to gasoline makes air quality worse. By granting the bailout, the EPA will allow ethanol producers to blend more of their corrosive, hydrophilic, low-heat-content fuel into our gasoline. And while the agency's ruling limits the use of the higher-ethanol-content gasoline to model year 2007 and newer cars and trucks, the move further complicates the motor fuel market. America has the most balkanized motor fuel market on earth. Refiners are now producing about four dozen blends of gasoline and multiple blends of diesel fuel.

Furthermore, the EPA bailout of the ethanol sector allows the corn fuel scammers to continue gorging themselves at the public trough. In July, the Congressional Budget Office reported that corn ethanol subsidies cost U. S. taxpayers more than $7 billion per year. Those subsidies are larger than those given to any other form of renewable energy.

The increase in ethanol consumption was opposed by one of the strangest coalitions in modern American politics. In August, thirty-nine groups—ranging from the Alliance of Automobile Manufacturers and the American Petroleum Institute to the Natural Resources Defense Council and the Environmental Working Group—asked Congress to hold hearings about the proposed increase.[46] Congressional leaders ignored the request.

Of course, the ethanol lobby loved the EPA's decision. Growth Energy, an advocacy group issued a press release applauding the move but insisted that "much more must be done to reduce America's dependence on foreign oil." That statement implies that all of the subsidies and mandates for corn ethanol have helped cut U. S. foreign oil imports. Here's the reality: they haven't done anything.

Between 1999 and 2009, U. S. ethanol production increased seven-fold to more than 700,000 barrels per day, but during that same time period, U. S. oil imports actually *increased* by more than 800,000 bbl/d. Furthermore, and perhaps most surprising, is this: during that same time period, U. S. oil exports—yes, exports—more than doubled to some 2 million bbl/d. Data from the U. S. Energy Information Administration show that oil imports closely track U. S. oil consumption. Over the last decade, as U. S. oil demand grew, imports grew. When consumption fell, imports dropped. And ethanol production levels had no apparent effect on oil imports or consumption.[47]

Thus, despite more than three decades of subsidies that have cost taxpayers tens of billions of dollars, the ethanol industry has not, and cannot, show any decline in oil imports during the time period when it experienced its most rapid growth. Maddening as that is, the real outrage of the corn ethanol scam involves air quality. In 2007, the EPA admitted that increased use of ethanol in gasoline would increase emissions of key air pollutants like volatile organic compounds and nitrogen oxide by as much as 7 percent.[48] In the documents the EPA released on October 13, 2010, announcing the approval of the 15 percent ethanol blends, the agency again acknowledged that more ethanol consumption will mean higher emissions of key pollutants.[49]

That admission is driving environmental advocates like Frank O'Donnell, the president of Clean Air Watch, a Washington, DC-based group, to distraction. Right after the EPA decision, O'Donnell told me that the agency is saying that more ethanol will mean higher emissions of nitrogen oxide, and yet the ethanol bailout, is "coming at the same time that the EPA is setting tougher standards on smog." Indeed, the EPA is implementing new rules on ground-level ozone that could affect dozens of cities.[50] What contributes to the formation of ozone? You guessed it: nitrogen oxide.[51]

Donald Stedman, a professor emeritus of chemistry at the University of Denver, has been studying ethanol's impact on air quality for two decades. His assessment of the EPA's decision is nearly identical to O'Donnell's. "More ethanol means worse air quality, period," says Stedman, who adds that corn ethanol "doesn't do anything to reduce greenhouse gases."

Evidence that the Obama administration is more worried about the farm lobby than urban air quality came within minutes of the EPA's announcement. Agriculture Secretary Tom Vilsack quickly issued a statement praising the move, saying that the increased use of ethanol "is an important step toward making America more energy independent."

Here's a tip: whenever you hear the phrase "energy independent" or any of its variants, substitute the word "ripoff." The EPA's decision is yet another unfortunate win for the farm lobby and another loss for consumers and clean-air advocates.

Finally, just in case you need one more example of the egregiousness of the ethanol scam, here it is: U. S. ethanol producers and blenders are now exporting record amounts of ethanol. Through the first nine months of 2010, the U. S. exported about 251 million gallons of the alcohol fuel—that's more than double the export volume recorded in 2009. Among the countries getting U. S. ethanol exports: Saudi Arabia and the United Arab Emirates.

To summarize: In October, the Obama administration bailed out the ethanol industry because the industry had built too much capacity. Administration officials and the ethanol scammers justified the bailout by saying it will help the United States achieve energy independence and cut oil imports. But rather than reduce oil imports, the ethanol scammers are collecting about $7 billion per year in subsidies from U. S. taxpayers so that they can ship increasing amounts of American-made ethanol abroad.[52] And in doing so, the ethanol scammers are consuming nearly 40 percent of all the corn grown in the United States.[53]

I'm running out of adjectives that do justice to the stupidity of the ethanol madness.

APPENDIX A:
UNITS AND EQUIVALENTS

Electricity Units

1 watt (W) = 0.00134 horsepower, or 1 joule/second (J/s)
1 kilowatt (kW) = 1,000 watts, or 1.35 horsepower (hp)
1 kilowatt-hour (kWh) = 1,000 watts for 1 hour
1 megawatt-hour (MWh) = 1 megawatt for 1 hour
1 megawatt (MW) = 1,000 kilowatts, or 1 million watts
1 gigawatt (GW) = 1,000 megawatts, 1 million kilowatts, or 1 billion watts
1 terawatt (TW) = 1,000 gigawatts, 1 million megawatts, 1 billion kilowatts, or 1 trillion watts

Power Units and Equivalencies

1 electric lamp of 100 W = 0.1 kW
1 car engine with a 60 hp engine = 44 kW
1 turbine rated at 1 megawatt (MW) = 1,350 hp[1]
1 nuclear plant with 1,000 MW of capacity = 1,350,000 hp[2]
1 gallon of oil equivalent per day = 0.71 hp (529 W)
1 barrel of oil equivalent per day = 30 hp (22.1 kW)[3]
1,000 cubic feet of natural gas per day = 5 hp (3,819 W)[4]
1 day of Saudi Arabia's oil production = 250 million hp (186.5 billion W, or 186.5
gigawatts)[5]

Energy Units and Equivalencies

0.1 joule = energy used in average golf putt[6]
1 Btu = energy released by burning 1 wooden match = 1.055 kilojoules
1 cubic foot of natural gas = 1,031 Btu[7] = 1.09 megajoules
1 cubic foot = volume of a regulation basketball[8]
1 cubic meter of natural gas = 35.3 cubic feet of natural gas
1 kilowatt-hour (kWh) of electricity = 3,412 Btu = 3.6 megajoules[9]
1 gallon gasoline = 125,000 Btu = 125 megajoules
1 gallon gasoline = 36 kWh of electricity

1 ton of oil = 7.33 barrels (bbl) of oil
1 bbl of oil = 42 gallons, or 159 liters[10]
1 bbl of oil equivalent = 5,800,000 Btu = 5.8 gigajoules
1 bbl of oil equivalent = 1.64 megawatt-hours (MWh) of electricity[11]
1 bbl of oil equivalent = 5,487 cubic feet of natural gas[12]

Note: Equivalent units between oil and electricity are notoriously difficult. These equivalences only measure the Btu content of each and do not account for any heat lost during the conversion of oil to electricity, which normally results in a loss of about two-thirds of the heat content.

1. Bertrand Barre and Pierre-Rene Bauquis, *Understanding the Future: Nuclear Power* (Strasbourg: Editions Hirlé, 2007), 11.

2. Ibid.

3. This assumes continuous horsepower (twenty-four hours per day). The power metrics of oil were determined thusly:

   1 bbl of oil = 5,800,000 Btu.

   5,800 megajoules / 86,400 seconds = 67,129 watts (assumes 1 Btu = 1,000 joules).

   67,129 watts times 0.33 (to account for heat loss during conversion to
   electricity) = 22,152 W (22.1 kW); 22,152 W / 746 W = 29.7 hp. Call it
   30 hp per barrel.

4. Here's the math:

   1,000 cubic feet of gas = 1,000,000 Btu.

   1,000 megajoules / 86,400 seconds = 11,574 watts (assumes 1,000 Btu =
   1 megajoule).

   11,574 watts times 0.33 (to account for heat loss during conversion) = 3,819 W (3.8
   kW).

   3,819 / 746 = 5.1 hp. Call it 5 hp per 1 mcf of gas.

5. This assumes 1 barrel of oil = 30 hp.

6. Matthew Futterman, "The Terror of the 10-Foot Putt," *Wall Street Journal*, June 18, 2009, http://online.wsj.com/article/SB124528252062525413.html.

7. Energy Information Administration, "Natural Gas Basics," http://www.eia.doe.gov/kids/energyfacts/sources/non-renewable/naturalgas.html.

8. Alberta Government, "Energy Measurements," http://www.energy.gov.ab.ca/About_Us/1132.asp.

9. BP Statistical Review of World Energy 2009, http://www.bp.com/liveassets/bp_internet/globalbp/globalbp_uk_english/reports_and_publications/statistical_energy_review_2008/STAGING/local_assets/2009_downloads/renewables_section_2009.pdf.

10. Areva, *All About Nuclear Energy: From Atom to Zirconium* (Paris: Areva, 2008).

11. BP, Statistical Review of World Energy 2009.

12. "Oil Industry Conversions," http://www.eppo.go.th/ref/UNIT-OIL.html.

APPENDIX B:
SI NUMERICAL DESIGNATIONS

As discussed in Chapter 3, SI units are an essential part of modern life. We use many SI numerical designations—milli, mega, nano—on a regular basis without recognizing that they are part of a larger system. Given the fact that most Americans are only passingly familiar with these terms, it makes sense to examine all of the designations—starting with "yocto" and "yotta"—and understand what they mean.

The difference between yocto and yotta is the difference between a septillionth and a septillion. Between yocto, the SI prefix for 10^{-24}, and yotta (sometimes spelled yota), the SI prefix for 10^{24}, there are 48 zeroes. It's the difference between 0.000,000,000,000,000,000,000,001 and 1,000,000,000,000,000,000,000,000. But in SI, those numbers would be written without the commas, thus, yocto is: 0.000 000 000 000 000 000 000 001; and yotta is: 1 000 000 000 000 000 000 000 000.

Herewith, the SI numerical designations and their symbols.

TABLE B.1 SI Numerical Designations,
Prefixes, and Symbols

Number	Prefix	Symbol
10^{-24}	yocto-	y
10^{-21}	zepto-	z
10^{-18}	atto-	a
10^{-15}	femto-	f
10^{-12}	pico-	p
10^{-9}	nano-	n
10^{-6}	micro-	u (greek mu)
10^{-3}	milli-	m
10^{-2}	centi-	c
10^{-1}	deci-	d
10^{1}	deka-	da
10^{2}	hecto-	h
10^{3}	kilo-	k

10^6	mega-	M
10^9	giga-	G
10^{12}	tera-	T
10^{15}	peta-	P
10^{18}	exa-	E
10^{21}	zeta-	Z
10^{24}	yotta-	Y

Source: Math.com, "Number Notation: Hierarchy or Decimal Notation," http://www.math.com/tables/general/numnotation.htm.

APPENDIX C:
AMERICA'S CONVOLUTED
ENERGY REGULATORY STRUCTURE

In 1971, the *Christian Science Monitor* estimated that federal energy policy was determined by forty-eight federal agencies and fourteen congressional committees.* In 2009, the U.S. Chamber of Commerce's Institute for 21st Century Energy estimated that there were twenty-four federal agencies and twenty-five congressional committees playing roles in shaping energy policy:**

Federal Agencies

U.S. Government Departments

U.S. Department of Agriculture
U.S. Department of Commerce
U.S. Department of Defense
U.S. Department of Energy
U.S. Department of Health and Human Services
U.S. Department of Homeland Security
U.S. Department of Housing and Urban Development
U.S. Department of Justice
U.S. Department of State
U.S. Department of the Interior
U.S. Department of the Treasury
U.S. Department of Transportation

Other Government Agencies

Export Import Bank of the United States

---

*Monty Hoyt, "US Burning Its Way Toward Fuel Crisis," *Christian Science Monitor*, June 2, 1971, 1.
**Matt LeTourneau, Chamber communications official, U.S. Chamber of Commerce, personal communication with author, August 12, 2009.

Federal Energy Regulatory Commission
Northwest Power and Conservation Council
Nuclear Regulatory Commission
Overseas Private Investment Corporation
Tennessee Valley Authority
U.S. Agency for International Development
U.S. Commodity Futures Trading Commission
U.S. Environmental Protection Agency
U.S. International Trade Commission
U.S. Nuclear Waste Technical Review Board
U.S. Securities and Exchange Commission

Congressional Committees

House Agriculture
House Appropriations
House Armed Services
House Budget
House Energy and Commerce
House Foreign Affairs
House Homeland Security
House Natural Resources
House Oversight and Government Reform
House Rules Committee
House Science and Technology
House Select Committee on Energy Independence and Global Warming
House Transportation and Infrastructure
House Ways and Means
Senate Agriculture
Senate Appropriations
Senate Armed Services
Senate Budget
Senate Commerce, Science and Transportation
Senate Energy and Natural Resources
Senate Environment and Public Works
Senate Finance
Senate Foreign Relations
Senate Homeland Security and Government Affairs
Senate Select Committee on Indian Affairs

APPENDIX D:
COUNTRIES RANKED BY PRIMARY
ENERGY CONSUMPTION, 2007

TABLE D.1 World's Most Coal-Dependent Countries

Country	Coal Consumption (in million tons of oil equivalent)	Total Energy Use (in million tons of oil equivalent)	2007 Primary Energy: Coal
South Africa	97.7	127.8	76.5%
China	1311.4	1863.4	70.4%
Poland	57.1	94.4	60.5%
India	208.0	404.4	51.4%
Kazakhstan	29.9	60.2	49.8%
Czech Republic	18.9	43.3	43.6%
Australia	53.1	121.8	43.6%
Bulgaria	8.1	20.4	39.6%
Taiwan	41.1	115.1	35.7%
Turkey	31.0	101.7	30.5%
Ukraine	39.3	136.0	28.9%
WORLD	**3177.5**	**11099.3**	**28.6%**
Germany	86.0	311.0	27.7%
China Hong Kong SAR	7.0	26.5	26.3%
Denmark	4.7	18.2	26.0%
South Korea	59.7	234.0	25.5%
U.S.	**573.7**	**2361.4**	**24.3%**
Indonesia	27.8	114.6	24.3%
Japan	125.3	517.5	24.2%
Philippines	5.9	24.9	23.9%
Greece	8.1	34.1	23.7%
Romania	9.0	39.7	22.7%
Slovakia	4.0	17.5	22.7%
United Kingdom	39.2	215.9	18.1%
Finland	4.6	27.4	16.6%
Former Soviet Union	166.2	1035.2	16.1%
Portugal	3.3	24.0	13.8%
Russian Federation	94.5	692.0	13.7%
Spain	20.1	150.3	13.4%
Malaysia	6.9	57.4	12.1%
Hungary	2.9	24.5	11.8%
Chile	3.3	28.6	11.6%
Thailand	8.9	85.6	10.4%
Austria	3.2	32.6	9.8%
New Zealand	1.7	17.4	9.8%

Source: BP Statistical Review of World Energy 2008, http://www.bp.com/liveassets/
bp_internet/globalbp/globalbp_uk_english/reports_and_publications/statistical_energy
_review_2008/STAGING/local_assets/downloads/pdf/statistical_review_of_world_energy
_full_review_2008.pdf.

TABLE D.2 World's Most Oil-Dependent Countries

Country	Oil Consumption (in million tons of oil equivalent)	Total Energy Use (in million tons of oil equivalent)	2007 Primary Energy: Oil
Singapore	47.4	53.4	88.9%
Ecuador	8.1	10.6	76.6%
China Hong Kong SAR	16.9	26.5	63.7%
Greece	21.6	34.1	63.5%
Republic of Ireland	9.4	15.0	62.6%
Portugal	14.4	24.0	60.1%
Saudi Arabia	99.3	167.6	59.2%
Mexico	89.2	155.5	57.4%
Belgium & Luxembourg	41.2	73.6	56.1%
Philippines	13.9	24.9	55.9%
Chile	16.0	28.6	55.8%
Kuwait	14.0	25.4	55.3%
Netherlands	48.5	91.8	52.9%
Spain	78.7	150.3	52.4%
Denmark	9.3	18.2	51.4%
Thailand	43.0	85.6	50.3%
Egypt	30.6	63.2	48.4%
Peru	6.6	13.8	47.9%
Indonesia	54.4	114.6	47.5%
Italy	83.3	179.6	46.4%
South Korea	107.6	234.0	46.0%
Taiwan	52.5	115.1	45.6%
Brazil	96.5	216.8	44.5%
Japan	228.9	517.5	44.2%
Iran	77.0	182.9	42.1%
Austria	13.5	32.6	41.5%
Malaysia	23.6	57.4	41.1%
New Zealand	7.0	17.4	40.1%
U.S.	**943.1**	**2361.4**	**39.9%**
Switzerland	11.3	28.9	39.1%
Finland	10.6	27.4	38.7%
Venezuela	26.8	71.4	37.5%
United Kingdom	78.2	215.9	36.2%
Azerbaijan	4.5	12.5	36.2%
United Arab Emirates	22.0	60.9	36.2%
Germany	112.5	311.0	36.2%
France	91.3	255.1	35.8%
WORLD	**3952.8**	**11099.3**	**35.6%**

Source: BP Statistical Review of World Energy 2008, http://www.bp.com/liveassets/
bp_internet/globalbp/globalbp_uk_english/reports_and_publications/statistical_energy
_review_2008/STAGING/local_assets/downloads/pdf/statistical_review_of_world_energy
_full_review_2008.pdf.

TABLE D.3 World's Most Natural Gas–Dependent Countries

Country	Natural Gas Consumption (in million tons of oil equivalent)	Total Energy Use (in million tons of oil equivalent)	2007 Primary Energy: Natural Gas
Uzbekistan	41.1	49.6	82.8%
Qatar	18.5	22.6	81.7%
Turkmenistan	19.7	24.4	80.7%
Bangladesh	14.6	20.3	72.1%
Belarus	17.5	24.5	71.2%
United Arab Emirates	38.9	60.9	63.8%
Algeria	22.0	34.7	63.3%
Azerbaijan	7.4	12.5	59.5%
Russian Federation	394.9	692.0	57.1%
Iran	100.7	182.9	55.0%
Argentina	39.7	73.7	53.8%
Pakistan	27.7	58.3	47.5%
Egypt	28.8	63.2	45.6%
Kuwait	11.3	25.4	44.7%
Malaysia	25.4	57.4	44.3%
Hungary	10.6	24.5	43.3%
Ukraine	58.2	136.0	42.8%
Saudi Arabia	68.3	167.6	40.8%
Italy	70.0	179.6	39.0%
United Kingdom	82.3	215.9	38.1%
Lithuania	3.4	9.0	37.9%
Thailand	31.8	85.6	37.2%
Romania	14.7	39.7	37.1%
Netherlands	33.4	91.8	36.4%
Venezuela	25.6	71.4	35.9%
Mexico	48.7	155.5	31.3%
Turkey	31.6	101.7	31.0%
Slovakia	5.3	17.5	30.1%
Kazakhstan	17.8	60.2	29.6%
Republic of Ireland	4.3	15.0	28.5%
Indonesia	30.4	114.6	26.5%
Canada	84.6	321.7	26.3%
U.S.	**595.7**	**2361.4**	**25.2%**
Austria	8.0	32.6	24.6%
Germany	74.5	311.0	23.9%
WORLD	**2637.7**	**11099.3**	**23.8%**
Colombia	6.9	30.0	23.1%
Denmark	4.1	18.2	22.6%

Source: BP Statistical Review of World Energy 2008, http://www.bp.com/liveassets/
bp_internet/globalbp/globalbp_uk_english/reports_and_publications/statistical_energy
_review_2008/STAGING/local_assets/downloads/pdf/statistical_review_of_world_energy
_full_review_2008.pdf.

TABLE D.4 World's Most Nuclear-Dependent Countries

Country	Nuclear Energy Consumption (in million tons of oil equivalent)	Total Energy Use (in million tons of oil equivalent)	2007 Primary Energy: Nuclear
France	99.7	255.1	39.1%
Sweden	15.3	50.2	30.4%
Lithuania	2.2	9.0	24.7%
Switzerland	6.3	28.9	21.7%
Slovakia	3.5	17.5	19.8%
Finland	5.4	27.4	19.7%
Bulgaria	3.3	20.4	16.2%
Ukraine	20.9	136.0	15.4%
Belgium & Luxembourg	10.9	73.6	14.8%
South Korea	32.3	234.0	13.8%
Czech Republic	5.9	43.3	13.7%
Hungary	3.3	24.5	13.5%
Japan	63.1	517.5	12.2%
Germany	31.8	311.0	10.2%
Spain	12.5	150.3	8.3%
U.S.	**192.1**	**2361.4**	**8.1%**
Taiwan	9.2	115.1	8.0%
Canada	21.1	321.7	6.6%
United Kingdom	14.1	215.9	6.5%
Former Soviet Union	59.9	1035.2	5.8%
WORLD	**622.0**	**11099.3**	**5.6%**
Russian Federation	36.2	692.0	5.2%
Romania	1.6	39.7	4.0%
South Africa	3.0	127.8	2.3%
Other Europe & Eurasia	1.9	81.9	2.3%
Argentina	1.6	73.7	2.2%
Mexico	2.4	155.5	1.5%
Brazil	2.8	216.8	1.3%
Netherlands	1.0	91.8	1.0%
India	4.0	404.4	1.0%
Pakistan	0.5	58.3	0.9%
China	14.2	1863.4	0.8%

Source: BP Statistical Review of World Energy 2008, http://www.bp.com/liveassets/
bp_internet/globalbp/globalbp_uk_english/reports_and_publications/statistical_energy
_review_2008/STAGING/local_assets/downloads/pdf/statistical_review_of_world_energy
_full_review_2008.pdf.

TABLE D.5 World's Most Hydro-Dependent Countries

Country	Hydro Electric Consumption (in million tons of oil equivalent)	Total Energy Use (in million tons of oil equivalent)	2007 Primary Energy: Hydro
Norway	30.6	45.0	68.0%
Iceland	1.9	3.0	63.5%
Brazil	84.1	216.8	38.8%
Colombia	10.1	30.0	33.6%
Peru	4.4	13.8	31.9%
New Zealand	5.3	17.4	30.7%
Sweden	15.0	50.2	29.9%
Switzerland	8.3	28.9	28.7%
Venezuela	19.0	71.4	26.6%
Canada	83.3	321.7	25.9%
Austria	7.9	32.6	24.1%
Ecuador	2.2	10.6	21.1%
Chile	5.4	28.6	18.7%
Pakistan	7.5	58.3	12.9%
Finland	3.2	27.4	11.6%
Argentina	8.5	73.7	11.6%
Portugal	2.3	24.0	9.8%
Romania	3.6	39.7	9.1%
Turkey	8.0	101.7	7.9%
Philippines	1.9	24.9	7.8%
India	27.7	404.4	6.8%
WORLD	**709.2**	**11099.3**	**6.4%**
Slovakia	1.0	17.5	5.9%
China	109.3	1863.4	5.9%
Russian Federation	40.5	692.0	5.9%
France	14.4	255.1	5.6%
Spain	7.4	150.3	4.9%
Italy	8.8	179.6	4.9%
Egypt	2.9	63.2	4.6%
Azerbaijan	0.5	12.5	4.3%
Bulgaria	0.8	20.4	4.0%
Mexico	6.1	155.5	3.9%
Japan	18.9	517.5	3.7%
Australia	3.8	121.8	3.1%
Kazakhstan	1.8	60.2	3.0%
Uzbekistan	1.4	49.6	2.8%
Malaysia	1.4	57.4	2.5%
Lithuania	0.2	9.0	2.4%
U.S.	**56.8**	**2361.4**	**2.4%**
Iran	4.1	182.9	2.2%

Source: BP Statistical Review of World Energy 2008, http://www.bp.com/liveassets/
bp_internet/globalbp/globalbp_uk_english/reports_and_publications/statistical_energy
_review_2008/STAGING/local_assets/downloads/pdf/statistical_review_of_world_energy
_full_review_2008.pdf.

APPENDIX E:
U.S. AND WORLD PRIMARY ENERGY CONSUMPTION, BY SOURCE, 1973 AND 2008

TABLE E.1 U.S. and World Primary Energy Consumption, by Source, 1973 and 2008

US Primary Energy Consumption, 1973 & 2008	Source in bbls of oil equivalent/ day, 1973	Source in bbls of oil equivalent/ day, 2008	% Change 73–08	Source as % of primary energy, 1973	Source as % of primary energy, 2008
Source					
Oil	16,427,233	17,762,699	8%	46%	38%
Natural Gas	11,392,627	12,063,373	6%	32%	26%
Coal	6,564,868	11,346,438	73%	18%	25%
Nuclear	399,636	3,855,781	865%	1%	8%
Hydro	1,265,178	1,138,660	-10%	4%	2%
Corn ethanol	n/a	387,475	n/a		0.83%
Wind/solar	n/a	88,362	n/a		0.19%
Total Primary Energy	36,047,534	46,642,787	28%		
Hydrocarbons as % of US primary energy	95%	88%	-7%		
World Primary Energy Consumption, 1973 & 2008					
Source					
Oil	55,310,373	78,880,841	43%	48%	35%
Natural Gas	21,134,499	54,746,063	159%	18%	24%
Coal	31,727,855	66,345,537	109%	28%	29%
Nuclear	919,764	12,444,934	1253%	1%	5%
Hydro	5,964,411	14,408,973	142%	5%	6%
Ethanol	n/a	746,320	n/a		0.33%
Wind/solar*	n/a	441,810	n/a		0.19%
Total Primary Energy	115,265,756	228,014,478	97%		
Hydrocarbons as % of world primary energy	94%	88%	-6%		

*Estimate extrapolated from U.S. data. According to BP, global installed wind power capacity in 2008 was 122,158 megawatts, which was roughly five times the installed U.S. capacity of 25,237 megawatts. Thus, the global figure for wind and solar production is simply five times the estimated U.S. output.

Sources: BP Statistical Review of World Energy 2009, http://www.bp.com/liveassets/ bp_internet/globalbp/globalbp_uk_english/reports_and_publications/statistical_energy _review_2008/STAGING/local_assets/2009_downloads/renewables_section_2009.pdf. For ethanol data, see Renewable Fuels Association, "Statistics," http://www.ethanolrfa.org/ industry/statistics/. For U.S. wind and solar data, see Energy Information Administration, Table ES1.B, "Total Electric Power Industry Summary Statistics," http://www.eia.doe.gov/ cneaf/electricity/epm/tablees1b.html.

NOTES

Author's Note

1. Shannon Hale, "Write What You Don't Know," Squeetusblog, December 10, 2008, http://oinks.squeetus.com/2008/12/write-what-you-dont-know.html.

Introduction

1. For more information, see Joy Mining Machinery's website at http://www.joy.com.

2. Energy Information Administration, Table 9, "Major U.S. Coal Mines," http://www.eia.doe.gov/cneaf/coal/page/acr/table9.html; author communication with the mine's general manager, Eric Anderson, via e-mail, March 2, 2009. According to Anderson, the mine produced 5.75 tons of coal per employee per hour in 2008.

3. In 2007, the average for underground coal mines was 3.34 tons per employee per hour. Energy Information Administration, Table 21, "Coal Mining Productivity by State and Mine Type," http://www.eia.doe.gov/cneaf/coal/page/acr/table21.html.

4. Eric Anderson, interview with author, February 20, 2009. In 2008, the mine produced 5.6 million tons of coal. That's 15,342 tons per day.

5. A barrel of oil contains approximately 5.8 million Btu. Energy Information Administration, "Energy Calculators," http://www.eia.doe.gov/kids/energyfacts/science/energy_calculator.html.

6. Energy Information Administration, Table 1.1.A, "Net Generation by Other Renewables: Total (All Sectors)," http://www.eia.doe.gov/cneaf/electricity/epm/table1_1_a.html.

7. Converting electricity to oil terms is a straightforward calculation: One barrel of oil contains the energy equivalent of 1.64 megawatt-hours of electricity. Thus, 52,869,000 megawatt-hours, divided by 1.64 megawatt-hours (the amount per barrel of oil) = 32,237,195 barrels of oil equivalent from wind and solar for all of 2008. Dividing those 32,237,195 barrels of oil equivalent by 365 days (to determine daily energy production) gives us 88,321 barrels of oil equivalent per day from all solar and wind power generation in the United States.

8. Eric Anderson, interview with author, February 20, 2009. Wind-capacity data is from American Wind Energy Association, "Wind Energy Grows by Record 8,300 MW in 2008," January 27, 2009, http://www.awea.org/newsroom/releases/wind_energy_growth2008_27Jan09.html.

9. I would like to thank Tulsa-based professional land surveyor Chris Cauthon for his tutorial on Points of Beginning. See also Andro Linklater, *Measuring America: How the United States Was Shaped by the Greatest Land Sale in History* (New York: Plume, 2002), 2.

10. Central Intelligence Agency, World Factbook, "North America: United States," https://www.cia.gov/library/publications/the-world-factbook/geos/us.html#Econ.

11. See Van Jones, *The Green Collar Economy: How One Solution Can Fix Our Two Biggest Problems* (New York: HarperOne, 2008), and Greenjobs, http://www.greenjobs.com/public/index.aspx. For Obama's statement, see his energy platform, available at http://my.barackobama.com/issues/newenergy/index.php.

12. For more on the corn ethanol scam, see my book *Gusher of Lies: The Dangerous Delusions of "Energy Independence"* (New York: PublicAffairs, 2008).

13. The International Energy Agency puts recoverable global gas resources at 850 trillion cubic meters. See IEA, *World Energy Outlook 2009*, Executive Summary, 49. A cubic meter contains 35.3 cubic feet.

14. Global gas consumption is about 3 trillion cubic meters per year. See BP Statistical Review of World Energy 2009, http://www.bp.com/liveassets/bp_internet/globalbp/globalbp_uk_english/reports_and_publications/statistical_energy_review_2008/STAGING/local_assets/2009_downloads/renewables_section_2009.pdf.

15. White House, "Remarks by President Barack Obama," Prague, Czech Republic, April 5, 2009, http://www.whitehouse.gov/the_press_office/Remarks-By-President-Barack-Obama-In-Prague-As-Delivered/.

16. "Environmental Statement on Nuclear Energy and Global Warming," June 2005, http://www.citizen.org/documents/GroupNuclearStmt.pdf.

17. Jim Collins, *Good to Great: Why Some Companies Make the Leap . . . and Others Don't* (New York: HarperCollins, 2001), 65.

18. Ibid., 69.

Chapter 1

1. Areva, *All About Nuclear Energy: From Atom to Zirconium* (Paris: Areva, 2008), 8.

2. For Hannibal, see John Noble Wilford, "The Mystery of Hannibal's Elephants," *New York Times*, September 18, 1984, http://www.nytimes.com/1984/09/18/science/the-mystery-of-hannibal-s-elephants.html?sec=health. For Saturn V, see Boeing Integrated Defense Systems, "Apollo 11 Factoids," n.d., http://www.boeing.com/defense-space/space/apollo11/factoids.html.

3. BP Statistical Review of World Energy 2009, http://www.bp.com/liveassets/bp_internet/globalbp/globalbp_uk_english/reports_and_publications/statistical_energy_review_2008/STAGING/local_assets/2009_downloads/renewables_section_2009.pdf. Hydrocarbons provide about 89 percent of all the commercial energy consumed in the United States. Globally, it's 88 percent.

4. TED.com, "Stewart Brand Proclaims 4 Environmental Heresies," June 2009. Video available at http://www.youtube.com/watch?v=TUxwiVFgghE.

5. For Romm's rantings, see Climateprogress.org.

6. Set America Free, "Set America Free Update E-mail," January 29, 2007, www
.setamericafree.org/safupdate012907.htm.

7. Environment New Jersey, "Report Says Fossil Fuels Status Quo Will Cost New
Jersey Billions; Urges Clean Energy Solutions," June 30, 2009, http://www.environment
newjersey.org/newsroom/energy/energy-program-news/report-says-fossil-fuels-status
-quo-will-cost-new-jersey-billions-urges-clean-energy-solutions.

8. These issues are fully discussed in my last book, *Gusher of Lies: The Dangerous
Delusions of "Energy Independence"* (New York: PublicAffairs, 2008).

9. Energy Information Administration, "Petroleum Navigator," http://tonto.eia.doe
.gov/dnav/pet/hist/mttexus2M.htm.

10. Energy Information Administration, "Country Energy Profiles," http://tonto.eia
.doe.gov/country/index.cfm.

11. Keith Crane, Andreas Goldthau, Michael Toman, Thomas Light, Stuart E. Johnson, Alireza Nader, Angel Rabasa, and Harun Dogo, "Imported Oil and US National
Security," Rand Corporation, May 11, 2009, http://www.rand.org/pubs/monographs/
2009/RAND_MG838.pdf, xvi.

12. Edward Gismatullin, "BP Makes 'Giant' Oil Discovery in Gulf of Mexico,"
Bloomberg, September 2, 2009, http://www.bloomberg.com/apps/news?pid=news
archive&sid=adF31W9._rik.

13. BP press release, September 2, 2009, http://www.bp.com/genericarticle.do
?categoryId=2012968&contentId=7055818. The other participants in the well are
Petrobras, with a 20 percent stake, and ConocoPhillips, with 18 percent.

14. Steven Bodzin and Daniel Cancel, "Spain's Repsol Says It Makes Venezuela's
Biggest Gas Discovery," Bloomberg, September 12, 2009, http://www.bloomberg.com/
apps/news?pid=20601086&sid=aHpa4d9ORrz0. See also Repsol, "Venezuelan President Hugo Chavez Announces Repsol's Largest Ever Gas Find," September 22, 2009,
http://www.repsol.com/es_en/todo_sobre_repsol_ypf/sala_de_prensa/noticias/ultimas
_noticias/bloque-cardon.aspx.

15. Anadarko Petroleum, "Anadarko Announces Discovery Offshore Sierra Leone,"
September 16, 2009, http://www.anadarko.com/Investor/Pages/NewsReleases/News
Releases.aspx?release-id=1332358.

16. Petrobras, "New Discover [*sic*] in the Santos Basin Pre-Salt," September 14,
2009, http://www2.petrobras.com.br/portal/frame_ri.asp?pagina=/ri/ing/index.asp&
lang=en&area=ri.

17. BP Statistical Review of World Energy 2009.

18. Andrew C. Revkin, "Campaign Against Emissions Picks Number," *New York Times*,
October 24, 2009, http://www.nytimes.com/2009/10/25/science/earth/25threefifty
.html.

19. Climate Progress, "Rajendra Pachauri Endorses 350 ppm, Not as IPCC Chair
but 'as a Human Being,'" August 25, 2009, http://climateprogress.org/2009/08/25/
ipcc-chair-rajendra-pachauri-350-ppm-bill-mckibben/.

20. *Late Show with David Letterman*, November 3, 2009, http://www.cbs.com/
late_night/late_show/video/?pid=Pvb2AfVeGFHFfegRj7cbeENTolRNrecs&vs=Big%
20Show%20Highlights&play=true.

21. Nanet Poulsen, "192 Nations at UN Climate Conference in Copenhagen," Associated Press, December 7, 2009, http://en.cop15.dk/news/view+news?newsid=2855.

22. Andrew C. Revkin and John M. Broder, "A Grudging Accord in Climate Talks," *New York Times*, December 19, 2009, http://www.nytimes.com/2009/12/20/science/earth/20accord.html?scp=1&sq=copenhagen%20and%20broder&st=cse.

23. Vaclav Smil, "Moore's Curse and the Great Energy Delusion," *The American*, November 19, 2008, http://www.american.com/archive/2008/november-december-magazine/moore2019s-curse-and-the-great-energy-delusion.

24. Ibid.

25. According to the BP Statistical Review of World Energy 2009, global primary energy consumption in 2008 was about 11.3 billion tons of oil equivalent, or about 82.8 billion barrels of oil equivalent per year. Multiplying that figure by $60 per barrel gives us about $4.9 trillion.

Chapter 2

1. Amory Lovins, "Saving the Climate for Fun and Profit," Yahoo! Green, June 7, 2007, http://green.yahoo.com/blog/amorylovins/1/saving-the-climate-for-fun-and-profit.html.

2. Michael Mechanic, "Power Q&A: Amory Lovins," *Mother Jones*, May/June 2008, http://www.motherjones.com/politics/2008/05/power-qa-amory-lovins.

3. Stone Phillips, "A Simple Solution to Pain at the Pump?" *Dateline NBC*, May 7, 2006, http://www.msnbc.msn.com/id/12676374/.

4. Dana Childs, "Cellulosic Ethanol to Be Cost-Competitive by 2009, Says Khosla," Cleantech.com, March 23, 2007, http://cleantech.com/news/928/cellulosic-ethanol-to-be-cost-competiti.

5. National Public Radio, "Al Gore's Speech on Renewable Energy," July 17, 2008, http://www.npr.org/templates/story/story.php?storyId=92638501.

6. Al Gore, "The Climate for Change," *New York Times*, November 9, 2008, http://www.nytimes.com/2008/11/09/opinion/09gore.html.

7. Ted McKenna, "New Al Gore Campaign Applies Ads, Media for Grassroots Effort," *PR Week*, April 4, 2008.

8. Wecansolveit.org, "Grassroots Partners," http://www.wecansolveit.org/pages/partners/.

9. Jennifer Alsever, "Pickens' Natural Gas Idea Picking Up Steam," MSNBC, October 21, 2008, http://www.msnbc.msn.com/id/27052462/.

10. Pickens Plan, http://media.pickensplan.com/pdf/pickensplan.pdf.

11. WorldPublicOpinion.org, "World Publics Strongly Favor Requiring More Wind and Solar Energy, More Efficiency, Even If It Increases Costs," November 19, 2008, http://www.worldpublicopinion.org/pipa/articles/btenvironmentra/570.php.

12. League of Conservation Voters, "Board of Directors," http://www.lcv.org/about-lcv/board-of-directors/ (accessed December 29, 2008).

13. League of Conservation Voters, "Repower, Refuel, and Rebuild America," http://action.lcv.org/campaign/repower_d (accessed December 29, 2008).

14. NBC Universal, "Company Overview," http://www.nbcuni.com/About_NBC_Universal/Company_Overview/.

15. *The Daily Beast*, "Top 9 Moments from Miss USA," April 20, 2009, http://www.thedailybeast.com/blogs-and-stories/2009-04-20/the-best-and-worst-of-the-miss-usa-pageant/.

16. U.S. Department of Energy, "Dr. Steven Chu, Secretary of Energy," http://www.energy.gov/organization/dr_steven_chu.htm.

17. *The Daily Show with Jon Stewart*, "July 21, 2009: Steven Chu," video available at http://www.thedailyshow.com/watch/tue-july-21-2009/steven-chu.

18. CNN.com, "Then & Now: Heidi Fleiss," June 19, 2005, http://www.cnn.com/2005/US/02/28/cnn25.tan.fleiss/.

19. Henry Brean, "Heidi Fleiss Gives Up on Plan for Brothel for Women," *Las Vegas Review-Journal*, February 10, 2009, http://www.lvrj.com/news/39357657.html.

20. See the Center for Internet Addiction Recovery, http://www.netaddiction.com/.

21. Peter Tertzakian and Keith Hollihan, *The End of Energy Obesity: Breaking Today's Energy Addiction for a Prosperous and Secure Tomorrow* (New York: Wiley, 2009), 12.

22. CBS News, "Transcript of Barack Obama's Speech," February 10, 2007, http://www.cbsnews.com/stories/2007/02/10/politics/main2458099.shtml.

23. According to the BP Statistical Review of World Energy 2009 (http://www.bp.com/liveassets/bp_internet/globalbp/globalbp_uk_english/reports_and_publications/statistical_energy_review_2008/STAGING/local_assets/2009_downloads/renewables_section_2009.pdf), India's consumption is about 2.9 million barrels per day. The country has 1.1 billion people. See Central Intelligence Agency, World Factbook, https://www.cia.gov/library/publications/the-world-factbook/geos/IN.html.

24. Jerome Taylor, "Ethical Travel Company Drops Carbon Offsetting," *The Independent*, November 7, 2009, http://www.independent.co.uk/environment/green-living/ethical-travel-company-drops-carbon-offsetting-1816554.html.

25. Louise Story, "FTC Asks If Carbon-Offset Money Is Well Spent," *New York Times*, January 9, 2008, http://www.nytimes.com/2008/01/09/business/09offsets.html; Christopher Joyce, "Carbon Offsets: Government Warns of Fraud Risk," National Public Radio, January 3, 2008, http://www.npr.org/templates/story/story.php?storyId=17814838.

26. Taylor, "Ethical Travel Company."

27. See Ecorazzi website, http://www.ecorazzi.com/2007/02/16/live-earth-to-be-carbon-neutral-model-of-sustainable-entertainment/.

28. Jerd Smith, "Bet Made on Carbon Offsets," *Rocky Mountain News*, July 26, 2008, http://www.rockymountainnews.com/news/2008/jul/26/bet-made-on-carbon-offsets/.

29. Elisabeth Rosenthal, "Vatican Seeks to Be Carbon Neutral," *New York Times*, September 3, 2007, http://www.nytimes.com/2007/09/03/business/worldbusiness/03iht-carbon.4.7366547.html.

30. For specs on the Toyota Sequoia, see Cargurus.com, http://www.cargurus.com/Cars/Overview-c21323-2009-Sequoia.html.

31. Revenue data from Yahoo! Finance, "Income Statement," http://finance.yahoo.com/q/is?s=CVX&annual.

32. See http://www.willyoujoinus.com/commitment/mediagallery/.

33. For an example, see http://www.willyoujoinus.com/assets/downloads/media/Chevron_Iwill_use%20less%20energy.pdf.

34. Karlyn Bowman, "The Federal Government: Losing Public Support," *Roll Call*, October 5, 2006.

35. In 2008, Exxon Mobil paid about $116.2 billion in taxes. See Robert Bryce, "Exxon, Big Oil Profits Evil Only Until You Weigh Their Tax Bills," *US News & World Report*, February 11, 2009, http://www.usnews.com/articles/opinion/2009/02/11/exxon-big-oil-profits-evil-only-until-you-weigh-their-tax-bills.html.

36. Drew Thornley, "Energy & the Environment: Myths & Facts," Manhattan Institute, 2009, http://www.manhattan-institute.org/pdf/EnergyMyth_2ndEdition.pdf, 35.

37. James Howard Kunstler, *The Long Emergency: Surviving the Converging Catastrophes of the Twenty-First Century* (New York: Atlantic Monthly Press, 2005), jacket flap.

38. California Progress Report, "President Clinton: Why I Support Proposition 87 and Why the Oil Companies are Wrong—The Complete Speech Delivered at UCLA," October 14, 2006, http://www.californiaprogressreport.com/2006/10/first-_thank_y.html.

39. Michael Moore, "Goodbye, GM," *The Daily Beast*, June 1, 2009, http://www.thedailybeast.com/blogs-and-stories/2009-06-01/goodbye-gm/full/.

40. US Global Change Research Program, "Global Climate Change Impacts in the United States," June 16, 2009, http://www.globalchange.gov/publications/reports/scientific-assessments/us-impacts/download-the-report, 9.

41. Ibid., 12.

42. Ibid., 157.

43. For more on Smil, see his website, http://home.cc.umanitoba.ca/~vsmil/.

44. Robert Bryce, "An Interview with Vaclav Smil," *Energy Tribune*, July 2007, http://www.robertbryce.com/smil.

45. *ScienceDaily*, "Scientific Literacy: How Do Americans Stack Up?" February 27, 2007, http://www.sciencedaily.com/releases/2007/02/070218134322.htm.

46. California Academy of Sciences, "American Adults Flunk Basic Science," February 25, 2009, http://www.calacademy.org/newsroom/releases/2009/scientific_literacy.php.

47. The Pew Research Center for the People and the Press, "Public Praises Science; Scientists Fault Public, Media," July 9, 2009, http://people-press.org/report/528/.

48. C. P. Snow, *The Two Cultures* (Cambridge: Cambridge University Press, 1998), 4. Available at http://books.google.com/books?id=OyHm4sc6IPoC&dq=cp+snow+two+cultures&printsec=frontcover&source=bn&hl=en&ei=G7EySqSpE4G-MpvHiY8K&sa=X&oi=book_result&ct=result&resnum=4#PPA4,M1.

49. Ibid., 14.

50. Or, as Bill Spencer, the former director of Sematech, explained the first two laws of thermodynamics to me in the fall of 2009: "You can't get something for nothing; and you can't even break even."

51. Titu Andreescu, Joseph A. Gallian, Jonathan M. Kane, and Janet E. Mertz, "Cross-Cultural Analysis of Students with Exceptional Talent in Mathematical Problem Solving," *Notices of the AMS* 55, no. 10 (November 2008): 1256, available at http://graphics8.nytimes.com/packages/pdf/national/10math_report.pdf.

52. Ibid., 1254.

53. U.S. Department of Education, "Final Report of the National Mathematics Advisory Panel," 2008, http://www.ed.gov/about/bdscomm/list/mathpanel/report/final -report.pdf, xiii.

54. Ibid., 3.

55. Ibid.

56. The best analogy for electricity is to compare it to water flowing through a pipe. Voltage is akin to the pressure of the water flowing through the pipe. The current, which is akin to the flow of the water, is measured in amperes (amps). And the resistance to flow, which is measured in ohms, is similar to the resistance of a waterwheel that gets rotated by the flow of the water.

Chapter 3

1. Carl Lira, "Biography of James Watt," n.d, http://www.egr.msu.edu/~lira/supp/ steam/wattbio.html.

2. Peter W. Huber and Mark P. Mills, *The Bottomless Well: The Twilight of Fuel, the Virtue of Waste, and Why We Will Never Run Out of Energy* (New York: Basic Books, 2005), 27.

3. Marshall Brain, "How Horsepower Works," n.d., http://www.howstuffworks .com/horsepower.htm/printable.

4. Joule invented the British Thermal Unit (Btu).

5. One joule is the amount of energy needed to move an object with a force of 1 newton (N) over a distance of 1 meter (m). The newton is a unit of force named after Isaac Newton. One watt is equal to 1 joule per second (1 W = 1 J/s). Americans are well acquainted with the watt from buying lightbulbs, hair dryers, and various other appliances.

6. Wikipedia, "Joule," http://en.wikipedia.org/wiki/Joule.

7. Richard A. Muller, *Physics for Future Presidents: The Science Behind the Headlines* (New York: W. W. Norton, 2008), 72.

8. Renewableenergyworld.com, "US Geothermal Capacity Could Top 10 GW," October 2, 2009, http://www.renewableenergyworld.com/rea/news/article/2009/10/us -geothermal-capacity-could-top-10-gw.

9. Arnulf Grübler, "Transitions in Energy Use," Encyclopedia of Earth, 2008, http://www.eoearth.org/article/Energy_transitions, 163.

10. Energy-density metrics for area are uncommon.

11. John Pearley Huffman, "Generations," May 8, 2003, http://www.edmunds.com/ insideline/do/Features/articleId=93327#3.

12. "2010 Ford Fusion Review," n.d., http://www.edmunds.com/ford/fusion/2010/ review.html.

13. Here's the math. The Fusion produces 70 horsepower per liter, while the Model T produces 7.6, and 70 divided by 7.6 equals 9.2. Thus, the Fusion engine has 9.2 times greater power density than the engine from the Model T.

14. Vaclav Smil, *Energies: An Illustrated Guide to the Biosphere and Civilization* (Cambridge: MIT Press, 1999), 123.

15. Ibid., 120.

16. Jad Mouawad and Kate Galbraith, "Plugged-In Age Feeds a Hunger for Electricity," *New York Times*, September 20, 2009, http://www.nytimes.com/2009/09/20/business/energy-environment/20efficiency.html?scp=2&sq=mouawad%20and%20electricity&st=cse.

17. International Energy Agency, *World Energy Outlook 2008*, 180.

18. Ibid., 390.

19. Boeing Integrated Defense Systems, "Apollo 11 Factoids," n.d., http://www.boeing.com/defense-space/space/apollo11/factoids.html.

20. Announced by an unnamed American Airlines pilot during Dallas to Austin flight, July 11, 2009.

21. John Kiewicz, "Top Fuel by the Numbers," *Motor Trend*, February 2005, http://sciencececec.ep.profweb.qc.ca/physique/Documents/Mecanique/Top_Fuel_by_the_Numbers.pdf.

22. Federation of American Scientists, n.d., http://www.fas.org/man/dod-101/sys/land/m1.htm.

23. CarandDriver.com, "We Drive a Honda F1 Car—The Big Day," n.d., http://www.caranddriver.com/features/09q1/we_drive_a_honda_f1_car-sport/the_big_day_page_3.

24. MotorTrend.com, n.d., http://www.motortrend.com/new_cars/04/ferrari/f430/index.html.

25. InternetAutoguide, n.d., http://www.internetautoguide.com/car-specifications/09-int/1999/acura/tl/index.html.

26. "2010 Ford Fusion Review."

27. John Pearley Huffman, "Generations," May 8, 2003, http://www.edmunds.com/insideline/do/Features/articleId=93327#3.

28. Calculated by author from home A/C unit, which draws 19.2 amps at 220 volts, for 4,224 watts.

29. Wikipedia, "Honda Super Cub," http://en.wikipedia.org/wiki/Honda_Super_Cub.

30. Based on author's personal Yard Machines lawnmower.

31. Measured at author's home with a Kill A Watt, August 27, 2009.

32. "Home Wattage Calculator," n.d., http://www.poweredgenerators.com/wattage-calculator.html. This source puts a toaster at 1,250 watts.

33. Ben Hewitt, "Tour de Lance," *Wired*, July 2004, http://www.wired.com/wired/archive/12.07/armstrong.html.

34. "Home Wattage Calculator," n.d., http://www.poweredgenerators.com/wattage-calculator.html. This source puts a coffeemaker at 800 watts.

35. Measured at author's home with a Kill A Watt, August 27, 2009.

36. Stan Jakuba, "Power Consumption and Generation in Bicycling and in Walking," n.d. Jakuba estimates that walking at a speed of 2 miles per hour generates 106 watts.

37. Measured at author's home with a Kill A Watt, August 27, 2009.

38. Measured at author's home with a Kill A Watt, August 27, 2009.

39. Measured at author's home with a Kill A Watt, August 26, 2009.

40. "Home Wattage Calculator," n.d., http://www.poweredgenerators.com/wattage-calculator.html. This source puts a table fan at 25 watts.

41. Measured at author's home with a Kill A Watt, August 26, 2009.

Chapter 4

1. U.S. Census Bureau, *Historical Statistics of the United States*, "Series Q 148–162, Motor-Vehicle Factory Sales and Registrations, and Motor-Fuel Usage: 1900 to 1970," 716.

2. BP Statistical Review of World Energy 2009, http://www.bp.com/liveassets/bp_internet/globalbp/globalbp_uk_english/reports_and_publications/statistical_energy_review_2008/STAGING/local_assets/2009_downloads/renewables_section_2009.pdf.

3. Energy Information Administration, Table 1.3, "Primary Energy Consumption by Source, 1949–2008," http://www.eia.doe.gov/emeu/aer/txt/ptb0103.html.

4. Vaclav Smil, *Global Catastrophes and Trends: The Next Fifty Years* (Cambridge: MIT Press, 2008), 90.

5. Jeff Goodell, *Big Coal: The Dirty Secret Behind America's Energy Future* (New York: Houghton Mifflin, 2008), 75. Goodell reported that the first significant rail line was built in the British coal town of Darlington to carry coal to the port at Stockton.

6. Peter W. Huber and Mark P. Mills, *The Bottomless Well: The Twilight of Fuel, the Virtue of Waste, and Why We Will Never Run Out of Energy* (New York: Basic Books, 2005), 4–5.

7. American Coalition for Clean Coal Electricity, "Your State," n.d., http://www.cleancoalusa.org/docs/state/.

8. Jalal Torabzadeh, "A Message from the Chair," Los Angeles Basin SPE Section Newsletter, June 2009, http://www.laspe.org/newsletters/june09nltr.pdf.

9. Advanced Resources International, "Bringing Real Information on Energy Forward: Economic Considerations Associated with Regulating the American Oil and Natural Gas Industry," April 24, 2009, http://s3.amazonaws.com/propublica/assets/natural_gas/economic_consequences_report_april2009.pdf, 2.

10. That well was drilled about 43 miles south of Morgan City, Louisiana.

11. Halliburton, "Brown & Root and Kerr-McGee Celebrate 50th Anniversary of First Producing Offshore Oil Well Out-of-Sight-Of-Land," November 14, 1997, http://www.halliburton.com/news/archive/1997/bresnws_111497.jsp.

12. Anadarko CEO James Hackett, interview with author for *Energy Tribune*, June 12, 2009, http://www.energytribune.com/articles.cfm?aid=1906.

13. Transocean, "Transocean Inc. and ChevronTexaco Announce New World Water-Depth Drilling Record in 10,011 Feet of Water," November 17, 2003, http://www.deepwater.com/fw/main/Transocean_Inc_and_ChevronTexaco_Announce_New_World_Water_Depth_Drilling_Record_in_10_011_Feet_of_Water-20C4.html.

14. Note that "extended-reach" wells are known for having long lateral (horizontal) sections. Transocean, "Transocean GSF Rig 127 Drills Deepest Extended-Reach Well," May 21, 2008, http://www.deepwater.com/fw/main/Transocean-GSF-Rig-127-Drills-Deepest-Extended-Reach-Well-283C4.html.

15. Robert Bryce, *Cronies: Oil, the Bushes, and the Rise of Texas, America's Superstate* (New York: PublicAffairs, 2004), 26.

16. William Fisher, geology professor, University of Texas at Austin, personal communication with author, April 8, 2009.

17. Huber and Mills, *Bottomless Well*, 170.

Chapter 5

1. Maury Klein, *The Power Makers: Steam, Electricity, and the Men Who Invented Modern America* (New York: Bloomsbury, 2008), 168.

2. Ibid., 172.

3. Ibid., 159.

4. Vaclav Smil, *Creating the Twentieth Century: Technical Innovations of 1867–1914 and Their Lasting Impact* (New York: Oxford University Press, 2005), 89.

5. Ibid., 60.

6. "Yuhuan 1,000MW Ultra-Supercritical Pressure Boilers, China," n.d., http://www.power-technology.com/projects/yuhuancoal/. For more on ultra-supercritical technology, see "Supercritical Pressure Coal-Fired Thermal Power Plant," n.d., http://www.hitachi.com/environment/showcase/solution/energy/thermal_power.html.

7. Smil, *Creating the Twentieth Century*, 56. For weight and power rating, see American Society of Mechanical Engineers, "National Historic Mechanical Engineering Landmarks: Edison 'Jumbo' Engine-Driven Dynamo and Marine-Type Triple Expansion Engine-Driven Dynamo," May 29, 1980, http://files.asme.org/ASMEORG/Communities/History/Landmarks/5537.pdf.

8. "Ferrari 599 GTB Fiorano," n.d., http://www.autobytel.com/content/research/searchresults/index.cfm/action/selecttrim/make_vch/Ferrari/model_vch/599%20GTB%20Fiorano.

9. BP Statistical Review of World Energy 2008, http://www.bp.com/liveassets/bp_internet/globalbp/globalbp_uk_english/reports_and_publications/statistical_energy_review_2008/STAGING/local_assets/downloads/pdf/statistical_review_of_world_energy_full_review_2008.pdf.

10. Barbara Freese, *Coal: A Human History* (New York: Penguin, 2003), 97.

11. Vaclav Smil, *Energy: A Beginner's Guide* (Oxford: Oneworld, 2006), 120.

12. Keith Bradsher and David Barboza, "Pollution from Chinese Coal Casts a Global Shadow," *New York Times*, June 11, 2006, http://www.nytimes.com/2006/06/11/business/worldbusiness/11chinacoal.html?pagewanted=print.

13. Bryan Walsh, "Linfen China," *Time*, 2007, http://www.time.com/time/specials/2007/article/0,28804,1661031_1661028,00.html.

14. Matthew Knight, "Gore Calls for Coal Plant Protests," September 25, 2008, http://www.cnn.com/2008/TECH/science/09/25/gore.carbon/index.html.

15. Shaila Dewan, "At Plant in Coal Ash Spill, Toxic Deposits by the Ton," *New York Times*, December 29, 2008, http://www.nytimes.com/2008/12/30/us/30sludge.html?_r=1&em.

16. James Hansen, "Dear Michelle and Barack," December 29, 2008, http://www.columbia.edu/~jeh1/mailings/20081229_DearMichelleAndBarack.pdf.

17. James Hansen, "Coal-Fired Power Stations Are Death Factories. Close Them," *The Guardian*, February 15, 2009, http://www.guardian.co.uk/commentisfree/2009/feb/15/james-hansen-power-plants-coal.

18. Robert F. Kennedy Jr., "How to End America's Deadly Coal Addiction," *Financial Times*, July 19, 2009, http://www.ft.com/cms/s/0/58ec3258-748b-11de-8ad5-00144feabdc0.html.

19. Jeff Goodell, *Big Coal: The Dirty Secret Behind America's Energy Future* (New York: Houghton Mifflin, 2008), 134.

20. Ibid., 135.

21. Environmental News Service, "Mercury Found in Blood of One-Third of American Women," September 1, 2009, http://www.ens-newswire.com/ens/sep2009/2009-09-01-092.asp.

22. Goodell, *Big Coal*, 123.

23. BP Statistical Review of World Energy 2009, http://www.bp.com/liveassets/bp_internet/globalbp/globalbp_uk_english/reports_and_publications/statistical_energy_review_2008/STAGING/local_assets/2009_downloads/renewables_section_2009.pdf. In 2007 and 2008, consumption increased from 3,194.5 million tons of oil equivalent to 3,303.7 million tons of oil equivalent, an increase of 109.2 million tons. Converted to oil, that's an increase of 800.4 million barrels of oil equivalent per year, or about 2.2 million barrels of oil equivalent per day.

24. BP Statistical Review of World Energy 2009.

25. Ibid.

26. Keith Bradsher, "China's Unemployment Swells as Exports Falter," *New York Times*, February 5, 2009, http://www.nytimes.com/2009/02/06/business/worldbusiness/06yuan.html?_r=1&sq=china%20and%20electricity&st=cse&scp=2&pagewanted=all. Goldman Sachs analysts have indicated that they also use electricity data as a proxy for output in China.

27. "Energy Statistics: Electricity: Consumption (per capita) (most recent) by Country," http://www.nationmaster.com/graph/ene_ele_con_percap-energy-electricity-consumption-per-capita.

28. Alan D. Pasternak, "Global Energy Futures and Human Development: A Framework for Analysis," Lawrence Livermore National Laboratory, October 2000, https://e-reports-ext.llnl.gov/pdf/239193.pdf, 17.

29. BP Statistical Review of World Energy 2009.

30. Central Intelligence Agency, World Factbook, https://www.cia.gov/library/publications/the-world-factbook/print/ca.html.

31. For images of Africa at night, see "Satellite Photo of Earth at Night" and other NASA images, at http://geology.com/articles/satellite-photo-earth-at-night.shtml.

32. "Poverty Facts and Stats," n.d., http://www.globalissues.org/article/26/poverty-facts-and-stats.

33. Central Intelligence Agency, World Factbook, "Country Comparison: Death Rate," https://www.cia.gov/library/publications/the-world-factbook/rankorder/2066rank.html.

34. Central Intelligence Agency, World Factbook, "Country Comparison: Infant Mortality Rate," https://www.cia.gov/library/publications/the-world-factbook/rankorder/2091rank.html.

35. BP Statistical Review of World Energy 2008.

36. Ibid.

37. Xina Xie, "Capitalist Coal Versus Socialist Electricity," *Energy Tribune*, July 21, 2009, http://www.energytribune.com/articles.cfm?aid=2103.

38. Energy Information Administration, "Existing Capacity by Energy Source," http://www.eia.doe.gov/cneaf/electricity/epa/epat2p2.html.

39. "Graphic: The State of Nuclear Power," *National Post*, July 31, 2009, http://network.nationalpost.com/np/blogs/posted/archive/2009/07/31/graphic-the-state-of-nuclear-power.aspx.

40. BP Statistical Review of World Energy 2009.

41. Ibid.

42. "Graphic: The State of Nuclear Power."

43. World Bank, "World Bank Supports Modernization of Old, Polluting Coal-Fired Power Plants in India to Lower Carbon Emissions," June 18, 2009, http://web.world bank.org/WBSITE/EXTERNAL/COUNTRIES/SOUTHASIAEXT/0,,content MDK:22217371~menuPK:2246552~pagePK:2865106~piPK:2865128~the SitePK:223547,00.html?cid=ISG_E_WBWeeklyUpdate_NL.

44. "Pachauri Defends India's Climate Stand," July 22, 2009, http://www.hindu .com/thehindu/holnus/001200907220334.htm.

45. Robert Bryce, "From Lahore to Copenhagen," *Energy Tribune*, November 3, 2009, http://www.energytribune.com/articles.cfm?aid=2533. For the full text of Clinton's remarks, see U.S. Department of State, Diplomacy in Action, "Roundtable with Business Leaders Opening and Closing Remarks: Hillary Rodham Clinton, Secretary of State, Governor's House, Lahore, Pakistan, October 29, 2009," http://www.state .gov/secretary/rm/2009a/10/131073.htm.

46. Energy Information Administration, Table 8.2a, "Electricity Net Generation: Total (All Sectors) 1949–2008," http://www.eia.doe.gov/emeu/aer/txt/ptb0802a.html.

47. Energy Information Administration, Table 1.1, "Net Generation by Energy Source: Total (All Sectors)," http://www.eia.doe.gov/cneaf/electricity/epm/table1 _1.html.

48. Central Intelligence Agency, World Factbook, "Europe: Spain," https:// www.cia.gov/library/publications/the-world-factbook/geos/sp.html#Econ.

49. Energy Information Administration, Table 8.2a. Note that in 1994, coal provided 1,690 billion kWh. In 2008, coal provided 1,994 billion kWh, for an increase of about 300 billion kWh.

50. BP Statistical Review of World Energy 2009. In 2008, U.S. coal consumption equaled 565 million tons of oil equivalent. The total primary energy use for Central and South American countries was 579.6 million tons of oil equivalent.

51. Energy Information Administration, Table 1.1.A, "Net Generation by Other Renewables: Total (All Sectors)," http://www.eia.doe.gov/cneaf/electricity/epm/ table1_1_a.html.

52. Ibid.

53. American Society of Mechanical Engineers, "National Historic Mechanical Engineering Landmarks: Edison 'Jumbo' Engine-Driven Dynamo and Marine-Type Triple

Expansion Engine-Driven Dynamo," May 29, 1980, http://files.asme.org/ASMEORG/
Communities/History/Landmarks/5537.pdf.

54. Klein, *The Power Makers*, 202.

55. See EveryGenerator.com for the Briggs and Stratton 30207, http://www.every
generator.com/Briggs-and-Stratton-30207-BAS1008.html.

Chapter 6

1. U.S. Institute for Peace, "Iraqi Oil Revenues: 'Managing the Devil's Excrement,'"
n.d., http://www.usip.org/events/iraqi-oil-revenues-managing-devils-excrement.

2. For an analysis of various fuels and their energy densities, see Wikipedia, "Energy
Density," http://en.wikipedia.org/wiki/Energy_density.

3. Vaclav Smil, "The Two Prime Movers of Globalization: History and Impact of
Diesel Engines and Gas Turbines," *Journal of Global History* 2 (2007), 376–377,
http://home.cc.umanitoba.ca/~vsmil/pdf_pubs/jgh%202007.pdf.

4. Robert Bryce, *Gusher of Lies: The Dangerous Delusions of "Energy Independence"*
(New York: PublicAffairs, 2008), 63–64. Diesel consumption in 2008 was about 3.9
million barrels per day. See Energy Information Administration, "Product Supplied,"
http://tonto.eia.doe.gov/dnav/pet/pet_cons_psup_dc_nus_mbblpd_a.htm.

5. Ibid.

6. Energy Information Administration, "Diesel—A Petroleum Product," http://
www.scribd.com/doc/12860647/Encyclopedia-of-Energy-Basic.

7. Smil, "Two Prime Movers," 377–378.

8. Ibid., 391.

9. Ibid., 379.

10. Rachel Layne, "GE, Safran Plan Engine to Help Keep Lead into New Era,"
Bloomberg, July 13, 2008, http://www.bloomberg.com/apps/news?pid=20601087
&sid=arNVkr5BBKjI&refer=home.

Chapter 7

1. BP Statistical Review of World Energy 2009, http://www.bp.com/liveassets/
bp_internet/globalbp/globalbp_uk_english/reports_and_publications/statistical
_energy_review_2008/STAGING/local_assets/2009_downloads/renewables
_section_2009.pdf. See also Jad Mouawad, "OPEC Plans Further Output Cut," *New
York Times*, December 17, 2008, http://www.nytimes.com/2008/12/17/business/world
business/17opec.html?fta=y. Mouawad reports that the Saudis settled on a production
rate of 8.5 million barrels per day in November 2008.

2. Here's the math: 5,800,000,000 J / 86,400 s = 67,129 W. To account for heat lost
during the conversion of energy into electrical power, we must multiply the 67,129 W
by 0.33, which leaves us with 22,152 W.

3. BP Statistical Review of World Energy 2009.

4. Energy Information Administration, Table 1.1, "Primary Energy Overview, 1949–
2008," n.d., http://www.eia.doe.gov/emeu/aer/txt/ptb0101.html.

5. Gene Whitney, Carl E. Behrens, and Carol Glover, "US Fossil Fuel Resources:
Terminology, Reporting, and Summary," Congressional Research Service, October 28,

2009, http://epw.senate.gov/public/index.cfm?FuseAction=Files.View&FileStore_id =f7bd7b77-ba50-48c2-a635-220d7cf8c519, 17.

Chapter 8

1. Witold Rybczynski, "The Green Case for Cities," *Atlantic Monthly*, October 2009, http://www.theatlantic.com/doc/200910/solar-panels.

2. E. F. Schumacher, *Small Is Beautiful: Economics as if People Mattered* (New York: Harper and Row, 1973).

3. The calculations for the power densities of the renewable sources is based on work done by Jesse Ausubel. See Ausubel, "The Future Environment for the Energy Business," *APPEA Journal* (2007), http://phe.rockefeller.edu/docs/ausubelappea.pdf, 8. Ausubel's estimates for renewables are of the same orders of magnitude as those published by the Nature Conservancy. Furthermore, Ausubel's numbers are almost identical to estimates provided to the author by Stan Jakuba, an engineer who has collected power-density data from numerous sources.

4. Ibid.

5. Ibid.

6. Robert I. McDonald, Joseph Fargione, Joe Kiesecker, William M. Miller, and Jimmie Powell, "Energy Sprawl or Energy Efficiency: Climate Policy Impacts on Natural Habitat for the United States of America," August 26, 2009, http:// www.plosone.org/article/info:doi/10.1371/journal.pone.0006802#pone-0006802 -g001.

7. Author interviews with Porter by telephone, January 25 and 26, 2010.

8. See the work by Nina Pierpont at the website Windturbinesyndrome.com.

9. American Wind Energy Association, "Wind Turbine Sound and Health Effects, An Expert Review Panel," December 2009, ES-1, http://awea.org/newsroom/ releases/AWEA_CanWEA_SoundWhitePaper_12–11–09.pdf.

10. Richard Cockle, "Oregon wind farms whip up noise, health concerns," *The Oregonian*, March 26, 2009, http://www.oregonlive.com/news/index.ssf/2009/ 03/oregon_wind_farms_whip_up_nois.html. See also Dan Gunderson, "Wind turbine noise concerns prompt investigation," Minnesota Public Radio, August 4, 2009, http://minnesota.publicradio.org/display/web/2009/08/03/wind-turbine -noise.

11. Vaclav Smil, "A Reality Check on the Pickens Energy Plan," *environment360*, August 25, 2008, http://e360.yale.edu/content/feature.msp?id=2058.

12. The math here is simple: 40,000 miles times 5,280 feet/mile times 100 feet equals 21.1 billion square feet. Divide by 27.9 million square feet per square mile to get approximately 750 square miles. Data on rights-of-way from American Transmission Company, "Property Values," n.d., http://www.atcllc.com/PropertyValues.shtml. For Rhode Island info, see "U.S. States (plus Washington, D.C.): Area and Ranking," n.d., http://www.enchantedlearning.com/usa/states/area.shtml.

13. *Los Angeles Times*, "Regulators Approve $1.9 Billion Power Line," December 19, 2008, http://articles.latimes.com/2008/dec/19/business/fi-sunrise19.

14. Desert Protective Council, "Big Solar," February 26, 2009, http://www.dpcinc .org/_bigsolar.shtml.

15. Miriam Raftery, "Courts Likely to Overturn Sunrise Powerlink Approval, Consumer Attorney Predicts to Crowd of 600 Protestors in Alpine," *East County Magazine*, April 8, 2009, http://www.eastcountymagazine.org/?q=node/856.

16. Robert Bryce, *Pipe Dreams: Greed, Ego, and the Death of Enron* (New York: PublicAffairs, 2002), 241.

17. For Senate dates, see Wikipedia, "Phil Gramm," http://en.wikipedia.org/wiki/Phil_Gramm.

18. Lisa Sorg, "Power Play," *San Antonio Current*, October 30, 2003, http://www.sacurrent.com/columns/story.asp?id=57671. For info on Government Canyon, see "Government Canyon State Natural Area," n.d., http://www.tpwd.state.tx.us/spdest/findadest/parks/government_canyon/. Visitors to Government Canyon should request a chat with interpreter/peace officer/all-around good guy John Koepke.

19. See "StopNYRI.com: What You Can Do Today," n.d., http://www.stopnyri.com/towns.php.

20. Kate Galbraith, "Environmentalists Sue Over Energy Transmission Across Federal Lands," *New York Times*, July 8, 2009, http://greeninc.blogs.nytimes.com/2009/07/08/environmentalists-sue-over-energy-transmission-across-federal-lands/?hp.

21. Earthjustice.org has the entire July 7, 2009, complaint online at http://www.earthjustice.org/library/legal_docs/final-complaint-energy-corridors.pdf.

22. Map of project from Lower Colorado River Authority, available at http://www.statesman.com/news/content/news/stories/local/2009/07/22/WEB0722lcrapowerlinesfinal3.html.

23. Asher Price, "In the Line of Ire," *Austin American-Statesman*, July 22, 2009, http://www.statesman.com/news/content/news/stories/local/2009/07/22/0722wind.html.

24. Laura Hancock, "BLM OKs Milford Wind Project," *Deseret News*, October 21, 2008, http://www.deseretnews.com/article/1,5143,705256822,00.html.

25. All data provided by personnel who worked on the Milford project. Note that in addition to the massive amounts of steel and concrete needed for the project, it also consumed about 700 tons of copper, or about 2.3 tons per megawatt. When accounting for wind's intermittency, that means that each reliable megawatt of wind-power capacity requires about 6.9 tons of copper.

26. Per F. Peterson, "Issues for Nuclear Power Construction Costs and Waste Management," September 16, 2008, http://www.ostp.gov/galleries/PCAST/PCAST%20Sep.%202008%20Peterson%20slides.pdf, 4. Wind's resource intensity is also far greater than coal's. Peterson's report says that coal requires 98 tons of steel and 160 cubic meters of concrete per megawatt of capacity. That means that wind power's steel requirements are 4.7 times as great as those of a coal plant, and its concrete requirements are 5.4 times as great as those of a coal plant.

27. Rybczynski, "The Green Case for Cities."

28. "A Pragmatic Response to Climate Change," October 26, 2009, http://marketplace.publicradio.org/display/web/2009/10/26/pm-whole-earth-q/.

29. David Case, "Texas Oil Tycoon Tackles Renewable Energy," *Fast Company*, May 9, 2008, http://www.fastcompany.com/magazine/126/a-mighty-wind.html.

30. The entire STP facility covers 12,000 acres. STP Nuclear Operating Company, "About Us," n.d., http://www.stpnoc.com/About.htm. The math is straightforward:

12,000 acres is equal to 48 million square meters. The plant produces 2.7 billion watts. Thus 2,700,000,000 / 48,000,000 = 56.2.

31. Based on author calculations. Assumes that the average U.S. gas well produces 4.8 million Btu per hour. Converted to electricity (assuming a loss of two-thirds of the heat energy), that yields about 470 kilowatts. Assuming each well covers 2 acres, it works out to about 235,000 watts per acre.

32. Here's the math: 60,000 cubic feet = 60,000,000 Btu = 60,000 MJ, and 60,000 MJ / 86,400 = 694,444 W. So, 694,444 times 0.33 = 229,166 W, and 229,166 W / 746 W = 307 hp. Assuming a 2-acre well site: 307 / 2 = 153/5 hp per acre.

33. This definition of stripper well comes from ConocoPhillips, http://www.conoco phillips.com/newsroom/other_resources/energyglossary/glossary_s.htm. Here's the math: 10 bbls = 58,000,000 Btu, and 58,000 MJ / 86,400 seconds = 671, 296 W. So, 671,296 times 0.33 = 221,152 W (221 kW), and 221,152 / 746 = 297 hp. Again, assuming 2 acres: 297 / 2 = 148.5 hp per acre.

34. The calculations for the energy densities of the renewable sources is based on work done by Jesse Ausubel. See Ausubel, "The Future Environment for the Energy Business."

35. Ibid.

36. Ibid.

37. Ibid.

Chapter 9

1. Global Wind Energy Council, "Global Wind Energy Outlook 2008," http://www.gwec.net/index.php?id=92, 46.

2. When asked for data, officials at the American Wind Energy Association pointed to a report by the Electric Reliability Council of Texas (ERCOT) that estimates possible carbon reductions. The report, "Analysis of Transmission Alternatives for Competitive Renewable Energy Zones in Texas," dated December 2006, is available at http://www.ercot.com/news/presentations/2006/ATTCH_A_CREZ_Analysis _Report.pdf. The American Wind Energy Association officials also pointed to a report from the U.S. Department of Energy projecting that if wind power in the United States reached 20 percent of electricity generation, then some 825 million tons of carbon dioxide would be "saved." See U.S. Department of Energy, "20% Wind Energy by 2030: Increasing Wind Energy's Contribution to US Electricity Supply," May 2008, http://www.20percentwind.org/20percent_wind_energy_report_05-11-08_wk.pdf, 12.

3. Global Wind Energy Council, "Global Wind Energy Outlook 2008," 46.

4. Lawrence J. Makovich, Patricia DiOrio, and Douglas D. Giuffre, "Renewable Portfolio Standards: Getting Ahead of Themselves?" Cambridge Energy Research Associates, February 2008, summary page.

5. Ibid., 14.

6. Keith Johnson, "Wind Power: Everything's Bigger in Texas," WSJ blogs, April 13, 2009, http://blogs.wsj.com/environmentalcapital/2009/04/13/wind-power-everythings -bigger-in-texas/.

7. Office of the Governor (Texas), "Gov. Perry Dedicates Desert Sky Wind Farm in Pecos County," May 3, 2002, http://governor.state.tx.us/news/press-release/4304/.

8. Lone Star Sierra Club, "Cool Texas: A 12-Step Plan for Meeting Electricity Needs That Is Good for Texas' Economy . . . and Our Climate," n.d., http://lonestar .sierraclub.org/Conservation/coolTexas2page.pdf.

9. Richard S. Dunham, "President Points to a Benefit for Texas," Hearst Newspapers, June 25, 2009, http://www.mysanantonio.com/news/local_news/President _points_to_a_benefit_for_Texas.html.

10. Data from the Electric Reliability Council of Texas. See http://www.ercot .com/about/.

11. Electric Reliability Council of Texas, "ERCOT Response to US Rep. Joe Barton," March 29, 2007, http://www.ercot.com/news/press_releases/2007/ERCOT _Response_to_Rep._Barton.html.

12. Electric Reliability Council of Texas, "Report on the Capacity, Demand, and Reserves in the ERCOT Region," May 2009, http://www.ercot.com/content/news/ presentations/2009/2009%20ERCOT%20Capacity,%20Demand%20and%20 Reserves%20Report.pdf, 13.

13. National Renewable Energy Laboratory, "Wind Powering America," n.d., http://www.windpoweringamerica.gov/wind_installed_capacity.asp. Texas capacity numbers were current as of September 27, 2009.

14. Electric Reliability Council of Texas, "Report on the Capacity, Demand, and Reserves in the ERCOT Region," 13.

15. Rowena Mason, "Wind farms produced practically no electricity during Britain's cold snap," *Daily Telegraph*, January 11, 2010, http://www.telegraph.co.uk/finance/ newsbysector/energy/6957501/Wind-farms-produced-practically-no-electricity-during -Britains-col-snap.html.

16. Makovich et al., "Renewable Portfolio Standards," 15.

17. Peter Lang, "Cost and Quantity of Greenhouse Gas Emissions Avoided by Wind Generation," Carbon-sense.com, February 16, 2009, http://carbon-sense.com/wp -content/uploads/2009/02/wind-power.pdf, 5.

18. Renewable Energy Foundation, "Wind Power Study Reveals Hidden Cost and Reliability Issues," press release, June 7, 2008, http://www.ref.org.uk/Files/pr.07.07 .08.pdf. The full article, by James Oswald, Mike Raine, and Hezlin Ashraf-Ball, appeared in *Energy Policy* 36 (2008): 3212–3225, and is available at http://www.wind -watch.org/documents/wp-content/uploads/oswald-energy-policy-2008.pdf.

19. Jing Yang, "China's Wind Farms Come with a Catch: Coal Plants," *Wall Street Journal*, September 28, 2009, A17, http://online.wsj.com/article/SB1254097307 11245037.html.

20. Kent Hawkins, "Wind Integration: Incremental Emissions from Back-Up Generation Cycling, Part 1," MasterResource.org, November 13, 2009, http://www.master resource.org/2009/11/wind-integration-incremental-emissions-from-back-up-generation -cycling-part-i-a-framework-and-calculator/comment-page-1/#comment-3244.

21. Kent Hawkins, interview with author via phone, November 14, 2009. For more on Hawkins's analysis of wind, see his blog, available at http://whitherindustrialwind power.wordpress.com/.

22. Global Wind Energy Council, "Global Wind Energy Outlook 2008," 39.

23. Ibid, 46.

24. The math is straightforward: 731/18,708 = 3.9%.

Chapter 10

1. CBSNews.com, "Transcript: Obama's Earth Day Speech," April 22, 2009, http://www.cbsnews.com/blogs/2009/04/22/politics/politicalhotsheet/entry4962412.shtml.

2. BBC, "Denmark 'World's Happiest Nation,'" July 3, 2008, http://news.bbc.co.uk/2/hi/in_depth/7487143.stm.

3. BBC, "Denmark 'Happiest Place on Earth,'" July 28, 2006, http://news.bbc.co.uk/2/hi/5224306.stm.

4. Cal Fussman, "The Energizer," *Discover*, February 20, 2006, http://discovermagazine.com/2006/feb/energizer/article_print.

5. Hannah Sentenac, "Denmark Points Way in Alternative Energy Sources," Fox News, November 28, 2006, http://www.foxnews.com/story/0,2933,203293,00.html.

6. Thomas L. Friedman, "Flush with Energy," *New York Times*, August 9, 2008, http://www.nytimes.com/2008/08/10/opinion/10friedman1.html?_r=2&oref=slogin.

7. Joshua Green, "The Elusive Green Economy," *Atlantic Monthly*, July/August 2009, 79, http://www.theatlantic.com/doc/200907/carter-obama-energy.

8. Ibid., 86.

9. Energy Information Administration, "Denmark Energy Profile," http://tonto.eia.doe.gov/country/country_time_series.cfm?fips=DA#coal.

10. BP Statistical Review of World Energy 2009, http://www.bp.com/liveassets/bp_internet/globalbp/globalbp_uk_english/reports_and_publications/statistical_energy_review_2008/STAGING/local_assets/2009_downloads/renewables_section_2009.pdf. In 2007, Denmark got 26 percent of its primary energy from coal, whereas the United States got 24.3 percent of its primary energy from coal.

11. Energy Information Administration, "Denmark Energy Data," http://tonto.eia.doe.gov/country/excel.cfm?fips=DA. See also Erik Matzen, "Danish Oil Reserves 200 mln Cubic Metres on Jan. 1," Reuters, June 15, 2009, http://www.reuters.com/article/rbssEnergyNews/idUSLF39693020090615.

12. Danish Energy Agency, "Energy Statistics 2007," http://www.ens.dk/en-US/Info/FactsAndFigures/Energy_statistics_and_indicators/Annual%20Statistics/Sider/Forside.aspx.

13. International Energy Agency, *World Energy Outlook 2008*, 166.

14. BP Statistical Review of World Energy 2009. One ton of coal is equal to 7.33 barrels of oil.

15. BP Statistical Review of World Energy 2009.

16. Energy Information Administration, "Denmark Energy Data."

17. Tony Lodge, "Wind Chill: Why Wind Energy Will Not Fill the UK's Energy Gap," Centre for Policy Studies, June 2008, http://www.cps.org.uk/cpsfile.asp?id=1026, 7.

18. David Pimentel, ed., *Biofuels, Solar and Wind as Renewable Energy Systems* (Ithaca, NY: Springer, 2008), 147.

19. Lodge, "Wind Chill," 7.

20. Hugh Sharman, e-mail to author, August 11, 2008.

21. BP Statistical Review of World Energy 2009, data from 2007.

22. Danish Center for Political Studies (CEPOS), "Wind Energy: The Case of Denmark," September 2009, http://www.cepos.dk/fileadmin/user_upload/Arkiv/PDF/Wind_energy_-_the_case_of_Denmark.pdf, 2.

23. According to BP, Denmark's total primary energy consumption in 1981 was 18.2 million tons of oil equivalent per year. That's about 365,000 barrels of oil equivalent per day. By 2007, the figure was 18.1 million tons of oil equivalent per year. Source: BP Statistical Review of World Energy 2009.

24. Energy Information Administration, "Electricity Prices for Households," http://www.eia.doe.gov/emeu/international/elecprih.html.

25. John Goerten and Daniel Cristian Ganea for Eurostat, "Environment and Energy," 2008, http://epp.eurostat.ec.europa.eu/cache/ITY_OFFPUB/KS-QA-08-045/EN/KS-QA-08-045-EN.PDF; "Dutch Gas and Electricity Prices Among the Highest in Europe," May 3, 2007, http://www.cbs.nl/en-GB/menu/themas/industrie-energie/publicaties/artikelen/archief/2007/2007-2187-wm.htm.

26. International Energy Agency, "Key World Energy Statistics, 2008," http://www.iea.org/textbase/nppdf/free/2008/key_stats_2008.pdf, 43.

27. Ibid.

28. German Technical Cooperation (GTZ), "GTZ International Fuel Price Survey-Data: Super Gasoline and Diesel Retail Prices as of Mid-November 2008," n.d., http://www.gtz.de/de/dokumente/en-international-fuel-prices-data-preview-2009.pdf.

29. BP Statistical Review of World Energy 2008.

30. European Environment Agency, "Greenhouse Gas Emission Trends and Projections in Europe 2008: Tracking Progress Towards Kyoto Targets," http://www.eea.europa.eu/publications/eea_report_2008_5/at_download/file, 13.

31. Ibid., 19.

32. Ibid., 23.

33. For more on the company, see Energinet.dk.

34. Energinet.dk, "Wind Power to Combat Climate Change: How to Integrate Wind Energy into the Power System," n.d., http://www.e-pages.dk/energinet/126/fullpdf/full.pdf. (Note: In April 2009, the company's website said the brochure on wind was "new.")

35. Ibid., 38–39.

36. Ibid., 34.

37. Ibid.

38. Nelson D. Schwartz, "In Denmark, Ambitious Plan for Electric Cars," *New York Times*, December 2, 2009, http://dealbook.blogs.nytimes.com/2009/12/02/in-denmark-ambitious-plan-for-electric-cars/.

39. The annual reports are available online at http://www.energinet.dk/en/servicemenu/Library/Library.htm#.

40. Energinet.dk, "Environmental Report 2008," n.d., http://www.energinet.dk/NR/rdonlyres/EC3E484D-08D5-4179-9D85-7B9A9DBD3E08/0/Environmentalreport2008.pdf, 13, 17.

41. Ibid., 27.

42. International Energy Agency, "CO_2 Emissions from Fuel Combustion 2009," http://www.iea.org/co2highlights/co2highlights.pdf, 56.

43. The 2007 number is an estimate based on demand in 2006, which was 34.7 billion kilowatt-hours. See Energy Information Administration, "Denmark Energy Data."

44. Central Intelligence Agency, World Factbook, "Europe: Denmark," https://www.cia.gov/library/publications/the-world-factbook/geos/da.html.

45. Based on CIA data for 2008 and an estimate of 5.29 million in 1998 from Populstat, "Denmark: General Data of the Country," http://www.populstat.info/Europe/denmarkg.htm.

46. Based on a 1998 population estimate of 270 million from CIA available at http://www.populstat.info/. The CIA puts the U.S. population in 2008 at 303.8 million.

47. Energinet.dk, "Environmental Report 2008," n.d., http://www.energinet.dk/NR/rdonlyres/EC3E484D-08D5-4179-9D85-7B9A9DBD3E08/0/Environmental report2008.pdf, 5.

48. Ibid., 12.

49. Energy Information Administration, "World Carbon Intensity: World Carbon Dioxide Emissions from the Consumption and Flaring of Fossil Fuels Per Thousand Dollars of Gross Domestic Product Using Market Exchange Rates, 1980–2006," http://www.eia.doe.gov/pub/international/iealf/tableh1gco2.xls.

50. Danish Energy Agency, "Large Drop in Energy Consumption and CO_2 Emissions in 2008," March 18, 2009, http://www.ens.dk/EN-US/INFO/NEWS/NEWS_ARCHIVES/2009/Sider/LargedropinenergyconsumptionandCO2emissionsin2008.aspx.

51. International Energy Agency, "Key World Energy Statistics 2008," 51, 57.

52. Energy Information Administration, "World Carbon Intensity."

53. Energy Information Administration, "Denmark Energy Data."

54. Ibid.

55. Ibid.

56. Danish Energy Agency, "Close to DKK 35.9 Billion State Revenue from Oil and Gas Activities in 2008," June 15, 2009, http://www.ens.dk/en-us/info/news/news_archives/2009/sider/20090615_oilgasandsubsoilusereport2008.aspx.

57. In 2008, wind power production was 6.9 billion kilowatt-hours. See the Danish Energy Agency website, http://www.ens.dk/da-DK/Info/TalOgKort/Statistik_og_noegletal/Maanedsstatistik/Documents/El-MonthlyStatistics.xls, to download the latest data in Excel.

58. Energy Information Administration, "Denmark Energy Data."

59. Friedman, "Flush with Energy."

60. Danish Center for Political Studies (CEPOS), "Wind Energy," 2.

61. Ibid., 37.

Chapter 11

1. TexasMonthly.com, "The Last Pickens Show," September 2008, http://www.texasmonthly.com/multimedia/slideshow/13087.

2. Slate.com, "80 over 80," September 11, 2008, http://www.slate.com/id/2199926/.

3. Matthew McDermott, "T. Boone Pickens Talks Natural Gas, Energy Independence, Peak Oil and Swift Boating with Katie Couric," July 15, 2008, http://www.treehugger.com/files/2008/07/t-boone-pickens-talks-with-katie-couric.php.

4. Steve Hargreaves, "Pickens' Wind Plan Hits a Snag," CNNMoney.com, November 12, 2008, http://money.cnn.com/2008/11/12/news/economy/pickens/index.htm.

5. PickensPlan.com, http://www.pickensplan.com/media/.

6. Facebook, http://www.facebook.com/Pickensplan?ref=s.

7. Pickens Plan e-mail, August 16, 2009.

8. Forbes.com, "The 400 Richest Americans," September 17, 2008, http://www.forbes.com/lists/2008/54/400list08_The-400-Richest-Americans_NameProper_12.html.

9. Gregory Zuckerman, "Pickens Funds Down About $1 Billion," *Wall Street Journal*, September 24, 2008, http://online.wsj.com/article/SB122221505732769415.html.

10. Elizabeth Souder, "Pickens Paring Down Wind Farm Project," *Dallas Morning News*, July 6, 2009, http://www.dallasnews.com/sharedcontent/dws/bus/industries/energy/stories/DN-pickenswind_05bus.State.Edition1.19e1daf.html.

11. Ibid.

12. Energy Information Administration, "Natural Gas Navigator," http://tonto.eia.doe.gov/dnav/ng/hist/n3045us2a.htm.

13. International Energy Agency, "Natural Gas Market Review 2009," 109.

14. PickensPlan.com, n.d., http://www.pickensplan.com/theplan/.

15. Energy Information Administration, "Imports by Area of Entry," http://tonto.eia.doe.gov/dnav/pet/pet_move_imp_dc_NUS-Z00_mbblpd_a.htm.

16. NGVAmerica.org, http://www.ngvamerica.org/media_ctr/fact_ngv.html.

17. NGVAmerica.org, "Fact Sheet: Potential Contribution of NGVs to Displacing 35 Billion Gallons of Non-Petroleum Fuels by 2017," n.d., http://www.ngvamerica.org/pdfs/PotentialNGVs.pdf. While the 1,500-gallon figure is used here, keep in mind that the average personal car in the United States only uses about 600 gallons of fuel per year. Thus, any move toward NGVs in the passenger car fleet will have an even smaller impact on overall oil consumption.

18. To convert gallons per year to barrels per day, divide by 15,330.

19. International Association of Natural Gas Vehicles, "Natural Gas Vehicle Statistics," n.d., http://www.iangv.org/tools-resources/statistics.html.

20. FWS.gov, "Utility Giant to Pay Millions for Eagle Protection," July 10, 2009, http://www.fws.gov/news/newsreleases/showNews.cfm?newsId=750629CF-E286-4B51-379292C1D9377C41.

21. Amy Littlefield, "ExxonMobil Pleads Guilty to Killing Protected Birds," *Los Angeles Times*, August 14, 2009, http://www.latimes.com/news/nationworld/nation/la-na-exxon-birds14-2009aug14,0,626783.story.

22. "Migratory Bird Treaty Act," http://alaska.fws.gov/ambcc/ambcc/treaty_act.htm.

23. U.S. Fish and Wildlife Service, "Utility Giant to Pay Millions for Eagle Protection," July 10, 2009, http://www.fws.gov/news/newsreleases/showNews.cfm?newsId=750629CF-E286-4B51-379292C1D9377C411.

24. U.S. Fish and Wildlife Service, "Bald Eagle Management Guidelines and Conservation Measures: The Bald and Golden Eagle Protection Act," May 27, 2008, http://www.fws.gov/midwest/Eagle/guidelines/bgepa.html.

25. Alameda County Community Development Agency, "Altamont Pass Wind Resource Area Bird Fatality Study," July 2008, http://www.altamontsrc.org/alt_doc/m30_apwra_monitoring_report_exec_sum.pdf, 1–3.

26. The Encyclopedia of the Earth points to a study done in 1994 at Altamont that discussed the raptor deaths but doesn't give specific numbers. The encyclopedia also mentions that a 2003 study estimated the annual bird kill at 1,000, including 24 gold eagles. See Encyclopedia of the Earth, "Altamont Pass, California," n.d., http://www.eoearth.org/article/Altamont_Pass,_California. A 2004 study estimated that as many as 4,700 birds per year were being killed at Altamont. See Golden Gate Audubon Society, "Reducing Bird Kills at Altamont Pass," *The Gull*, May 2005, http://www.goldengateaudubon.org/html/thegull/archive/2005/2005-05-thegull.pdf, 1.

27. Carl G. Thelander, "Bird Fatalities at Wind Energy Facilities: An Overview," BioResource Consultants, June 27, 2006, http://www.fws.gov/midwest/GreatLakes/windpowerpresentations/Thelander.pdf.

28. Michael Fry, interview with author by phone, August 21, 2009.

29. AWEA.org, "Wind Energy and Wildlife," n.d., http://www.awea.org/pubs/factsheets/Wind_Energy_and_Wildlife_Mar09.pdf, 1.

30. U.S. Department of Energy, "20% Wind Energy by 2030: Increasing Wind Energy's Contribution to US Electricity Supply," July 2008, http://www1.eere.energy.gov/windandhydro/pdfs/41869.pdf, 7.

31. AWEA.org, "Wind Energy and Wildlife."

32. Justin Blum, "Researchers Alarmed by Bat Deaths from Wind Turbines," *Washington Post*, January 1, 2005, http://www.washingtonpost.com/wp-dyn/articles/A39941-2004Dec31.html.

33. Brian Handwerk, "Wind Turbines Give Bats the 'Bends,' Study Finds," National Geographic News, August 25, 2008, http://news.nationalgeographic.com/news/2008/08/080825-bat-bends.html.

34. Todd Woody, "Judge Halts Wind Farm over Bats," *New York Times*, December 10, 2009, http://greeninc.blogs.nytimes.com/2009/12/10/judge-halts-wind-farm-over-bats/.

Chapter 12

1. Rebuttal Testimony and Exhibits of Thomas A. Imbler, Before the Public Utilities Commission of the State of Colorado, June 9, 2009, http://www.xcelenergy.com/SiteCollectionDocuments/docs/CRPImblerRebuttal.pdf.

2. Bonneville Power Administration data. For current wind generation information, see: bpa.gov, especially http://www.transmission.bpa.gov/business/operations/wind/baltwg.aspx.

3. Cameron Walker, "Blowout," *Outside*, August 2007, http://outside.away.com/outside/destinations/200708/dams.html.

4. Robert F. Kennedy Jr., "How to End America's Deadly Coal Addiction," *Financial Times*, July 19, 2009, http://www.ft.com/cms/s/0/58ec3258-748b-11de-8ad5-00144feabdc0.html.

5. Pöyry, "Impact of Intermittency: How Wind Variability Could Change the Shape of the British and Irish Electricity Markets," July 2009, http://www.ilexenergy.com/pages/documents/reports/renewables/Intermittency%20Public%20Report%202_0.pdf, 23.

6. International Energy Agency, "Natural Gas Market Review 2009," 114.

7. Ibid.

8. "Colorado: Incentives/Policies for Renewable Energy," n.d., http://www.dsireusa .org/incentives/incentive.cfm?Incentive_Code=CO24R&re=1&ee=0.

9. Xcel Energy, http://www.xcelenergy.com/Company/AboutUs/Pages/Temp.aspx.

10. Xcel Energy, "2008 Wind Integration Team, Final Report," December 1, 2008, http://www.xcelenergy.com/SiteCollectionDocuments/docs/CRPWindIntegration StudyFinalReport.pdf, 2.

Chapter 13

1. This group is also sometimes referred to as lanthanoids. Among the best interactive periodic table of the elements is available at http://www.dayah.com/periodic/.

2. Mark P. Mills, "Go Long on Lithium," Forbes.com, May 5, 2008, http://www.forbes .com/2008/05/05/lithium-batteries-electricity-pf-ii-in_mm_0505energyintelligence _inl_print.html.

3. Peter Day, "Prosperity Promise of Bolivia's Salt Flats," BBC.co.uk, August 15, 2009, http://news.bbc.co.uk/2/hi/programmes/from_our_own_correspondent/8201058.stm.

4. Robert Farago, "Editorial: The Truth About Rare Earths and Hybrids," *International Business Times*, July 23, 2009, http://www.ibtimes.com/contents/20090723/ editorial-truth-about-rare-earths-and-hybrids.htm.

5. Energy Information Administration, "U.S. Imports by Country of Origin," http:// tonto.eia.doe.gov/dnav/pet/pet_move_impcus_a2_nus_ep00_im0_mbblpd_a.htm.

6. BP Statistical Review of World Energy 2009, http://www.bp.com/liveassets/bp _internet/globalbp/globalbp_uk_english/reports_and_publications/statistical _energy_review_2008/STAGING/local_assets/2009_downloads/renewables_section _2009.pdf.

7. For the first ten months of 2009, OPEC production was about 28 million barrels per day. Global consumption is about 84 million barrels per day. See MEES.com, "OPEC Crude Oil Production," n.d., http://www.mees.com/Energy_Tables/crude-oil .htm.

8. Wikipedia, "Deng Xiaoping," http://en.wikipedia.org/wiki/Deng_Xiaoping. For Deng quote, see Farago, "Editorial: The Truth About Rare Earths and Hybrids."

9. "Molycorp Minerals: Global Outlook," n.d., http://www.molycorp.com/global outlook.asp.

10. Farago, "Editorial: The Truth About Rare Earths and Hybrids."

11. Leo Lewis, "Crunch Looms for Green Technology as China Tightens Grip on Rare-Earth Metals," *Times* (London), May 28, 2009, http://business.timesonline.co .uk/tol/business/industry_sectors/natural_resources/article6374603.ece.

12. Steve Gorman, "As Hybrid Cars Gobble Rare Metals, Shortage Looms," Reuters, September 2, 2009, http://www.reuters.com/article/GCA-BusinessofGreen/ idUSTRE57U02B20090902; Keith Bradsher, "China Tightens Grip on Rare Minerals," *New York Times*, September 1, 2009, http://www.nytimes.com/2009/09/01/business/ global/01minerals.html?_r=1&sq=neodymium&st=Search&scp=1&pagewanted=all.

13. Gorman, "As Hybrid Cars Gobble Rare Metals."

14. David Trueman, interview with author by phone, August 31, 2009.

15. National Science Foundation, "Science and Engineering Indicators 2008: Presentation Slides," January 2008, http://www.nsf.gov/statistics/seind08/slides.htm.

16. Ibid. See slide entitled "World Share of High-Technology Manufacturing by Region/Country: 1985–2005."

17. First Solar, 2009 10-K report, http://investor.firstsolar.com/phoenix.zhtml?c=201491&p=irol-sec&seccat01.1_rs=11&seccat01.1_rc=10.

18. Jack Lifton, interview with author, August 28, 2009; David Trueman, interview with author, August 30, 2009.

19. Kate Galbraith, "More Sun for Less: Solar Panels Drop in Price," *New York Times*, August 26, 2009, http://www.nytimes.com/2009/08/27/business/energy-environment/27solar.html.

20. "Molycorp Minerals: Global Outlook."

21. Keith Bradsher, "China Racing Ahead of US in the Drive to Go Solar," *New York Times*, August 24, 2009, http://www.nytimes.com/2009/08/25/business/energy-environment/25solar.html.

22. Christopher Calnan, "City Council Gives Austin Energy the Go-Ahead for Major Solar Project," *Austin Business Journal*, March 5, 2009, http://austin.bizjournals.com/austin/stories/2009/03/02/daily49.html.

23. Tom Zeller Jr. and Keith Bradsher, "Jobs Question Jeopardizes Wind Farm's Stimulus Deal," *New York Times*, November 5, 2009, http://www.nytimes.com/2009/11/06/business/energy-environment/06wind.html?_r=1.

Chapter 14

1. Robert Bryce, "Not by Energy Efficiency Alone," *Energy Tribune*, March 18, 2008, http://www.energytribune.com/articles.cfm?aid=820.

2. House Energy and Commerce Committee, American Clean Energy and Security Act, 2009, http://energycommerce.house.gov/Press_111/20090515/hr2454.pdf.

3. Energy Information Administration, "World Carbon Intensity—World Carbon Dioxide Emissions from the Consumption and Flaring of Fossil Fuels Per Thousand Dollars of Gross Domestic Product Using Market Exchange Rates, 1980–2006," http://www.eia.doe.gov/pub/international/iealf/tableh1gco2.xls.

4. Energy Information Administration, "World Energy Intensity—Total Primary Energy Consumption Per Dollar of Gross Domestic Product Using Purchasing Power Parities, 1980–2006," http://www.eia.doe.gov/pub/international/iealf/tablee1p.xls. Note that energy intensity fell faster in about seventeen countries not listed here, including Chad, Guam, Laos, and Afghanistan.

5. Energy Information Administration, Table 1.7, "Primary Energy Consumption Per Real Dollar of Gross Domestic Product," Monthly Energy Review, May 2009, http://www.eia.doe.gov/emeu/mer/pdf/pages/sec1_16.pdf.

6. Energy Information Administration, Table 1.7, "Primary Energy Consumption Per Real Dollar of Gross Domestic Product," http://www.eia.doe.gov/emeu/mer/pdf/pages/sec1_16.pdf.

7. The U.S. population in 1980 was 227.7 million. See U.S. Census Bureau, "No. HS-1. Population: 1900 to 2002," in *Statistical Abstract of the United States*, 2003, http://www.census.gov/statab/hist/HS-01.pdf. In 2006, the U.S. population was about 300 million. See Associated Press, "US Population Hits 300 Million Mark," October 17, 2006, http://www.msnbc.msn.com/id/15298443/.

8. It's worth noting that more than forty countries saw bigger per-capita declines during that time period, including Afghanistan, North Korea, and nearly two dozen African countries. That data is omitted here for the sake of brevity. The five countries with the biggest per-capita declines in energy use include, in order, Liberia, Guam, Afghanistan, Chad, and Somalia. All saw declines of more than 50 percent.

9. Ibid.

10. EnchantedLearning.com, "U.S. States (plus Washington, D.C.)," n.d., http://www.enchantedlearning.com/usa/states/area.shtml.

11. *New York Times*, "France," http://topics.nytimes.com/top/news/international/countriesandterritories/france/index.html.

12. *New York Times*, "Switzerland," http://topics.nytimes.com/top/news/international/countriesandterritories/switzerland/index.html.

13. According to EnchantedLearning.com, Oklahoma covers 69,900 miles (http://www.enchantedlearning.com/usa/states/area.shtml). According to the CIA World Factbook, Puerto Rico covers 13,790 square kilometers, or about 5,300 square miles (https://www.cia.gov/library/publications/the-world-factbook/rankorder/2147rank.html).

14. Bureau of Transportation Statistics, Table 1-11, "Number of U.S. Aircraft, Vehicles, Vessels, and Other Conveyances," n.d., http://www.bts.gov/publications/national_transportation_statistics/html/table_01_11.html. The 1973 number is my estimate based on 1970 and 1975 data.

15. Energy Information Administration, "Petroleum Navigator," http://tonto.eia.doe.gov/dnav/pet/hist/mttupus2a.htm.

16. U.S. Department of Transportation, "Historical Monthly VMT Report," n.d., http://www.fhwa.dot.gov/policyinformation/travel/tvt/history/.

17. Energy Information Administration, "Petroleum Navigator."

18. Nuclear Energy Institute, "Resources and Stats: Generation Statistics," n.d., http://www.nei.org/resourcesandstats/graphicsandcharts/generationstatistics/.

19. In July 2009, sales of large SUVs had fallen by 45 percent compared to July 2008. Pickup truck sales fell by 20 percent. During the same time period, sales of small and medium-sized cars were also down, but their declines were much smaller, 5 percent and 9 percent, respectively. There were only two vehicles to show positive sales growth in July 2009, and they were both midsized cars: the Ford Fusion and the Volkswagen Jetta, which saw their sales increase by 4.3 percent and 5.6 percent, respectively. Data from Wall Street Journal online, "Market Data Center: Auto Sales. Overview Charts," http://online.wsj.com/mdc/public/page/2_3022-autosales.html, accessed August 7, 2009.

20. Whitehouse.gov, "President Obama Announces National Fuel Efficiency Policy," May 19, 2009, http://www.whitehouse.gov/the_press_office/President-Obama-Announces-National-Fuel-Efficiency-Policy/.

21. *New York Times*, "Culling the Gas Hogs," August 6, 2009, http://www.nytimes.com/2009/08/07/opinion/07fri1.html?_r=1&scp=1&sq=fuel%20efficiency%20and%20ford%20f-series&st=cse.

22. Leslie Guevarra, "Empire State Building to Become a Model of Energy Efficiency," GreenerBuildings.com, April 6, 2009, http://www.GreenerBuildings.com/news/2009/04/06/empire-state-building-become-model-energy-efficiency.

23. David Heacock, Dominion Resources, interview with author via phone, May 26, 2009.

24. McKinsey & Company, "Unlocking Energy Efficiency in the US Economy," July 2009, http://www.mckinsey.com/clientservice/electricpowernaturalgas/downloads/US_energy_efficiency_full_report.pdf, iv.

25. Ibid., 109.

26. Ibid., viii.

27. William Stanley Jevons, *The Coal Question: An Inquiry Concerning the Progress of the Nation, and the Probable Exhaustion of Our Coal Mines* (London: Macmillan, 1866).

28. John Polimeni, interview with author by e-mail, October 20, 2008.

29. For example, see Lorna A. Greening, David L. Greene, and Carmen Difiglio, "Energy Efficiency and Consumption—the Rebound Effect—A Survey," *Energy Policy* (June 2000): 389–401. See also Kenneth A. Small and Kurt Van Dender, "Fuel Efficiency and Motor Vehicle Travel: The Declining Rebound Effect," *The Energy Journal*, January 1, 2007.

30. Wikipedia, "James Watt," http://en.wikipedia.org/wiki/James_Watt.

31. William Tucker, *Terrestrial Energy: How Nuclear Power Will Lead the Green Revolution and End America's Energy Odyssey* (Savage, MD: Bartleby Press, 2008), 133.

Chapter 15

1. My interviews with climate scientists include talks with Roger Pielke Sr. and Roger Pielke Jr. Both are concerned about the issue of carbon dioxide emissions, but both agree that carbon dioxide may not be the most important issue when talking about potential climate change. The Pielke Sr. interview was published May 16, 2007. See Robert Bryce, "Energy Tribune Speaks with Roger Pielke, Sr.," http://www.energytribune.com/articles.cfm?aid=487&idli=3. The Pielke Jr. interview was published March 16, 2009. See Robert Bryce, "An Interview with Roger Pielke, Jr., Center for Science and Technology Policy Research," http://www.energytribune.com/articles.cfm?aid=1441.

2. White House, "Remarks by the President at the National Academy of Sciences Annual Meeting," April 27, 2009, http://www.whitehouse.gov/the_press_office/Remarks-by-the-President-at-the-National-Academy-of-Sciences-Annual-Meeting/.

3. Vaclav Smil, *Global Catastrophes and Trends: The Next Fifty Years* (Cambridge: MIT Press, 2008), 180.

4. John M. Broder, "With Something for Everyone, Climate Bill Passed," *New York Times*, June 30, 2009, http://www.nytimes.com/2009/07/01/us/politics/01climate.html.

5. For an example of this, see Climateprogress.org.

6. Mark Twain, *Adventures of Huckleberry Finn*, Chapter 21. Available at http://ebooks.adelaide.edu.au/t/twain/mark/finn/chapter21.html.

7. BP Statistical Review of World Energy 2009, http://www.bp.com/liveassets/bp_internet/globalbp/globalbp_uk_english/reports_and_publications/statistical_energy_review_2008/STAGING/local_assets/2009_downloads/renewables_section_2009.pdf.

8. Population figures via Central Intelligence Agency, World Factbook, https://www.cia.gov/library/publications/the-world-factbook/geos/xx.html.

9. International Energy Agency, *World Energy Outlook 2008*, Table 16.2, "Energy-Related CO$_2$ Emissions by Region in the Reference Scenario."

10. BP Statistical Review of World Energy 2009.

11. Bibhudatta Pradhan, "India Rejects Any Greenhouse-Gas Cuts Under New Climate Treaty," Bloomberg, June 30, 2009, http://www.bloomberg.com/apps/news ?pid=20601091&sid=aWs0Pts2Kxes.

12. For an interview with Dyson, see the May 6, 1999, edition of Charlie Rose, http://www.charlierose.com/view/interview/4308.

13. Freeman Dyson, "Heretical Thoughts About Science and Society," Edge.org, August 8, 2007, http://www.edge.org/3rd_culture/dysonf07/dysonf07_index.html.

14. Robert Bryce, "Gore's Zero Emissions = Zero Sense," Energy Tribune, August 2007, http://www.robertbryce.com/01.

15. United Nations Framework Convention on Climate Change, http://unfccc .int/kyoto_protocol/items/2830.php.

16. William Tucker, *Terrestrial Energy: How Nuclear Power Will Lead the Green Revolution and End America's Energy Odyssey* (Savage, MD: Bartleby Press, 2008), 11.

17. David Kestenbaum, "Japan Wrestles with Kyoto Accord Promises," National Public Radio, October 1, 2007, http://www.npr.org/templates/story/story.php?storyId= 14087783.

18. Isabel Reynolds, "Japan Should Set Mid-Term Emissions Targets—Envoy," Reuters, February 13, 2009, http://www.reuters.com/article/latestCrisis/idUST53159.

19. Energy Information Administration, "Carbon Dioxide Emissions from the Consumption and Flaring of Fossil Fuels, 1980–2006," December 8, 2008, http://www.eia .doe.gov/pub/international/iealf/tableh1co2.xls.

20. Central Intelligence Agency, World Factbook, "Field Listing: Population," n.d., https://www.cia.gov/library/publications/the-world-factbook/fields/2119.html.

21. Steven F. Hayward, "The Real Cost of Tackling Climate Change," *Wall Street Journal*, April 28, 2008, http://online.wsj.com/article/SB120934459094348617.html.

22. U.S. Bureau of Economic Analysis, "State Personal Income 2008," March 24, 2009, http://www.bea.gov/newsreleases/regional/spi/2009/xls/spi0309.xls.

23. U.S. Census Bureau, "Table 1, Projections of the Population Components of Change for the United States: 2010 to 2050," http://www.census.gov/population/ www/projections/files/nation/summary/np2008-t1.xls.

24. International Energy Agency, "Key World Energy Statistics, 2008," http:// www.iea.org/textbase/nppdf/free/2008/key_stats_2008.pdf, 49, 51, 57.

25. Ibid., 51.

26. Hayward, "The Real Cost of Tackling Climate Change."

27. Ibid., 53.

28. For more on Pielke, see his homepage at the University of Colorado at Boulder, http://sciencepolicy.colorado.edu/about_us/meet_us/roger_pielke/.

29. Roger A. Pielke Jr., "Statement to the Committee on Government Reform of the United States House of Representatives," July 20, 2006, http://sciencepolicy.colorado .edu/admin/publication_files/resource-2466-2006.09.pdf.

30. Institution of Mechanical Engineers, "Climate Change: Adapting to the Inevitable?" February 2009, http://www.imeche.org/NR/rdonlyres/D72D38FF-FECF

-480F-BBDB-6720130C1AAF/0/Adaptation_Report.PDF, 2 (hereafter cited as "full report").

31. Institution of Mechanical Engineers, news release, "Climate Change: Adapting to the Inevitable?: New Research Paints True Picture of Our Planet's Future" (hereafter cited as "news release"), n.d., http://www.imeche.org/NR/rdonlyres/D961AE6B -645C-403A-A0D4-19BA0D3E9B95/0/AdaptationRelease.pdf.

32. Institution of Mechanical Engineers, full report, 2.

33. Ibid., 25.

34. Institution of Mechanical Engineers, news release.

35. National Academy of Sciences, "Restructuring Federal Climate Research to Meet the Challenges of Climate Change," 2009, http://cart.nap.edu/cart/deliver .cgi?record_id=12595, 14.

36. Energy Information Administration, "U.S. Annual Energy Expenditures as Percent of Gross Domestic Product," http://www.eia.doe.gov/emeu/steo/pub/gifs/Fig23.gif.

37. Jeff Goodell, "Coal's New Technology: Panacea or Risky Gamble?" *Yale Environment 360*, July 14, 2008, http://e360.yale.edu/content/feature.msp?id=2036.

38. Fred Krupp, speech at CERAWeek conference, hosted by Cambridge Energy Research Associates, Houston, February 12, 2009.

39. Howard Herzog, speech at CERAWeek conference, Houston, February 13, 2009.

40. Massachusetts Institute of Technology, "The Future of Coal: Options for a Carbon-Constrained World," 2007, http://web.mit.edu/coal/The_Future_of_Coal.pdf, ix.

41. Ibid., x.

42. Kate Galbraith, "Salazar Talks Guns, Parks and Solar Power," *New York Times*, March 24, 2009, http://greeninc.blogs.nytimes.com/2009/03/24/salazar-talks-guns -parks-and-solar-power/.

43. Nobuo Tanaka, speech in Bergen, Norway, May 27, 2009, http://www.iea.org/ Textbase/speech/2009/Tanaka/Bergen_CCS_notes.pdf, 2.

44. Associated Press, "US Energy Sec: World Wants Stable Oil Prices, Spikes Could Hurt Economic Recovery," May 23, 2009, http://abcnews.go.com/International/ wireStory?id=7662471.

45. David Sandalow, "Statement of David Sandalow, Assistant Secretary of Energy for Policy and International Affairs, Before the Committee on Environment and Public Works, United States Senate," August 6, 2009, http://epw.senate.gov/public/index .cfm?FuseAction=Files.View&FileStore_id=e2ec9004-e94d-41d6-81c6-a3f8c9bd9975.

46. Paul W. Parfomak, "Carbon Control in the US Electricity Sector: Key Implementation Uncertainties," Congressional Research Service, December 23, 2008, https://secure.wikileaks.org/leak/crs/R40103.pdf, 15.

47. U.S. Department of Energy, "Secretary Chu Announces $2.4 Billion in Funding for Carbon Capture and Storage Projects," May 15, 2009, http://www.energy.gov/ news2009/7405.htm.

48. Andres Cala, "Europe Bets $1.4B on Carbon Capture," *Energy Tribune*, May 29, 2009, http://www.energytribune.com/articles.cfm?aid=1844.

49. Goodell, "Coal's New Technology."

50. Ibid.

51. Greenpeace, "False Hope: Why Carbon Capture and Storage Won't Save the Climate," n.d. http://www.greenpeace.org/raw/content/usa/press-center/reports4/false-hope-why-carbon-capture/executive-summary-false-hope.pdf.

52. Energy Information Administration, "Carbon Dioxide Emissions from the Consumption and Flaring of Fossil Fuels, 1980–2006."

53. That 3 billion tons is also approximately equal to emissions from the U.S. electric utility sector. (That sector accounts for about 40 percent of total U.S. emissions, or about 2.4 billion tons.) Carbon dioxide emissions data is from Energy Information Administration, "Carbon Dioxide Emissions from the Consumption and Flaring of Fossil Fuels, 1980–2006."

54. Vaclav Smil, "Energy at the Crossroads: Background Notes for a Presentation at the Global Science Forum Conference on Scientific Challenges for Energy Research, Paris," May 17–18, 2006, http://home.cc.umanitoba.ca/~vsmil/pdf_pubs/oecd.pdf, 21. Also, carbon dioxide reaches a critical point at 74 atmospheres, which is about 1,087 psi. For more on this, see Economy-point.org, "Critical Point (thermodynamics)," n.d., http://www.economy-point.org/c/critical-point-thermodynamics.html.

55. BP Statistical Review of World Energy 2009. In 2008, global production was 81.8 million barrels per day.

56. Apache Corporation, "Topic Report: Tanker Market Review," July 14, 2008, http://www.apachecorp.com/explore/explore_features/browse_archives/View_Article/?docdoc=742.

57. Smil, "Energy at the Crossroads."

58. Paul Freund and Olav Kaarstad, *Keeping the Lights On: Fossil Fuels in the Century of Climate Change* (Oslo: Universitetsflorlaget, 2007), 156. The authors estimate that the parasitic load would be about 18 percent on a combined-cycle natural gas–fired power plant and about 28 percent on a coal-fired plant.

59. Guy Chazan, "Locals Try Sinking Plan to Store CO_2 Underground," *Wall Street Journal*, October 6, 2009, http://online.wsj.com/article/SB125476964655765445.html.

60. Peabody Energy, "Message from Chairman and CEO," n.d., http://www.peabodyenergy.com/Profile/ceomessage.asp. For Svec quote, see Clifford Krauss, "Natural Gas Hits a Roadblock in New Energy Bill," *New York Times*, September 7, 2009, http://www.nytimes.com/2009/09/07/business/07gas.html?_r=1&ref=business.

Chapter 16

1. Environmental Protection Agency, "Mercury: Human Health," http://www.epa.gov/mercury/health.htm.

2. U.S. Geological Survey, "Major Findings from USGS Mercury in Stream Ecosystem Studies," n.d., http://water.usgs.gov/nawqa/mercury/majorfindings.html. Barbara C. Scudder, Lia C. Chasar, Dennis A. Wentz, Nancy J. Bauch, Mark E. Brigham, Patrick W. Moran, and David P. Krabbenhoft, "Mercury in Fish, Bed Sediment, and Water from Streams Across the United States, 1998–2005," n.d. (mid-August 2009), http://pubs.usgs.gov/sir/2009/5109/pdf/sir20095109.pdf, 10.

3. U.S. Geological Survey, "Major Findings from USGS Mercury in Stream Ecosystem Studies."

4. Scudder et al., "Mercury in Fish," 2.

5. Ibid., 1.

6. Environmental Protection Agency, "Mercury: Basic Information," n.d., http://www.epa.gov/mercury/about.htm.

7. Matt Pottinger, Steve Stecklow, and John J. Fialka, "Invisible Export—A Hidden Cost of China's Growth: Mercury Migration," *Wall Street Journal*, December 20, 2004, http://www.aug.edu/~sbajmb/Clippings/2004-12-17-ChinaMercury.pdf.

8. *The Oregonian*, "China's Mercury Flushes into Oregon's Rivers," November 24, 2006, http://research.uwb.edu/jaffegroup/publications/116400a.pdf.

9. WorldWideWords.org, "Mad as a Hatter," http://www.worldwidewords.org/qa/qa-mad2.htm.

10. Newsrx.com, Women's Health Weekly: Environmental Health, "Maternal Exposure to Neurotoxins May Not Be Risk Factor for Fetal Neural Tube Defects," June 29, 2006, http://www.newsrx.com/newsletters/Womens-Health-Weekly/2006-06-29/0629200633367WW.html.

11. Jesse H. Ausubel, "Reasons to Worry About the Human Environment," *COSMOS* (Journal of the Cosmos Club of Washington, D.C.) 8 (1998): 1–12, http://phe.rockefeller.edu/reasons-to-worry/.

12. Environmental News Service, "Mercury Found in Blood of One-Third of American Women," September 1, 2009, http://www.ens-newswire.com/ens/sep2009/2009-09-01-092.asp.

Chapter 17

1. "Viewpoints: An Interview with Professor Kirk R. Smith," *Boiling Point*, no. 56 (2009): 10, http://ehs.sph.berkeley.edu/krsmith/publications/2009%20pubs/BP56_Smith.pdf.

2. Juan Forero, "Texaco Goes on Trial in Ecuador Pollution case," *New York Times*, October 23, 2003, http://www.nytimes.com/2003/10/23/business/texaco-goes-on-trial-in-ecuador-pollution-case.html.

3. Twenty years ago, I published numerous articles about the pollution problems and bird kills caused by the oil and gas industry in Oklahoma and Texas. In the late 1980s and early 1990s, I wrote articles for the *Tulsa Tribune* and other papers about the thousands of migratory birds that were being killed in open oil waste pits. I also wrote about careless industry practices that were polluting aquifers. Three of my articles from that time period are: "Crackdown Due on Foul Water Holes: Birds Fall Victim to Slime in 3 States," *Tulsa Tribune*, November 17, 1989, http://www.robertbryce.com/node/298; "Oil Wastes Taint Water Supply," *Christian Science Monitor*, April 18, 1991, http://www.csmonitor.com/1991/0418/18061.html; and "Caustic Pits," *Field and Stream*, March 1991. In 1991 I also wrote a long article for *Texas Monthly* that exposed the lax pollution enforcement of the Texas Railroad Commission, the agency charged with regulating the oil and gas industry in the Lone Star State. See Robert Bryce, "More Precious Than Oil," *Texas Monthly*, February 1991.

4. Transparency International, "Corruption Perceptions Index 2009," n.d., http://www.transparency.org/policy_research/surveys_indices/cpi/2009/cpi_2009_table.

5. Ian Urbina, "Taint of Corruption Is No Barrier to U.S. Visa," *New York Times*, November 16, 2009, http://www.nytimes.com/2009/11/17/us/17visa.html.

6. CBS News, *60 Minutes*, "Gorillas: Kings of Congo," December 9, 2007, http://www.cbsnews.com/stories/2007/12/07/60minutes/main3591264.shtml?source=related_story.

7. Ibid.

8. Wildlife Conservation Society, "What Is the Albertine Rift," n.d., http://programs.wcs.org/albertine/WhatistheAlbertineRift/tabid/1998/Default.aspx.

9. Gorilla.cd, "Who We Are," n.d., http://gorilla.cd/who-we-are/.

10. Emmanuel de Merode, e-mail correspondence with author, April 17, 2009.

11. Gorilla.cd, "Briquettes, Gorillas & Virunga Need Your Help," April 16, 2009, http://gorillacd.org/2009/04/16/briquettes-gorillas-virunga-need-your-help/.

12. Ibid.

13. Emmanuel de Merode, e-mail correspondence with author, April 17, 2009. For more information, see Gorillacd.org.

14. Tom Knudson, "The Cost of the Biofuel Boom on Indonesia's Forests," *The Guardian*, January 21, 2009, http://www.guardian.co.uk/environment/2009/jan/21/network-biofuels.

15. Ibid.

16. Ibid.

17. Kounteya Sinha, "'Indoor' Air Pollution Is the Biggest Killer," *Times of India*, March 22, 2007, http://timesofindia.indiatimes.com/articleshow/1790711.cms.

18. Fatih Birol, "Energy Economics: A Place for Energy Poverty in the Agenda?" *The Energy Journal* 28, no. 3 (2007): 3, 4.

19. Kirk R. Smith, "Wood: The Fuel That Warms You Thrice," *Human Health and Forests* (2008): 99, http://ehs.sph.berkeley.edu/krsmith/publications/2008%20pubs/Colfer%20book%20chapter.pdf.

20. Sinha, "'Indoor' Air Pollution."

21. "Viewpoints: An Interview with Professor Kirk R. Smith."

22. Kirk R. Smith, "Editorial: In Praise of Petroleum?" *Science*, December 6, 2002, http://ehs.sph.berkeley.edu/krsmith/publications/02_smith_3.pdf, 1847.

23. Robert Bryce, "An Interview with Kirk R. Smith on Indoor Air Pollution and Why the Rural Poor Need Propane and Butane," July 23, 2009, http://www.energytribune.com/articles.cfm?aid=2110.

24. Zeke Hausfather, "Black Carbon and Global Warming: A Promising Short-Term Approach?" Yale Forum on Climate Change and the Media, July 9, 2009, http://www.yaleclimatemediaforum.org/2009/07/black-carbon-and-global-warming/.

25. Ibid.

26. Bryce, "Interview with Kirk R. Smith."

Chapter 18

1. Bill Kovarik, "Henry Ford, Charles F. Kettering and the Fuel of the Future," *Automotive History Review*, no. 32 (Spring 1998): 7–27, http://www.runet.edu/~wkovarik/papers/fuel.html.

2. Eric Kroh, "Food Prices and Shortages Boost Support for Next Generation Ethanol," *Medill Reports*, May 5, 2008, http://news.medill.northwestern.edu/chicago/news.aspx?id=88157.

3. Joel Wendland, "Obama Pushes Cellulosic Ethanol, Wind Energy, and Solar Power," *Political Affairs*, May 14, 2008, http://www.politicalaffairs.net/article/view/6852/1/334/.

4. Amanda Little, "Who's Afraid of the Big Bad Woolsey?" Grist.org, June 7, 2005, http://www.grist.org/news/maindish/2005/06/07/little-woolsey/.

5. Amory Lovins, testimony before the U.S. Senate Committee on Energy and Natural Resources, March 7, 2006, http://energy.senate.gov/public/index.cfm?FuseAction =Hearings.Testimony&Hearing_ID=1534&Witness_ID=4345.

6. T. Boone Pickens and Ted Turner, "New Priorities for Our Energy Future," *Wall Street Journal*, August 16, 2009, http://online.wsj.com/article/SB10001424052970 203863204574348432504983734.html.

7. Amory Lovins, "Energy Strategy: The Road Not Taken?" *Foreign Affairs*, October 1976, 82.

8. Commercial paper makers have reportedly been producing ethanol from cellulose for years. But the quantities of ethanol being produced by the paper makers are nowhere close to the volumes envisioned by Congress for consumption as motor fuel.

9. Amory Lovins, E. Kyle Datta, Odd-Even Bustnes, Jonathan G. Koomey, and Nathan J. Glasgow, *Winning the Oil Endgame: Innovation for Profits, Jobs, and Security* (Snowmass, CO: Rocky Mountain Institute, 2004), executive summary, 3, http://www.oilendgame.com/ExecutiveSummary.html.

10. Lovins testimony, March 7, 2006.

11. S. David Freeman, *Winning Our Energy Independence: An Energy Insider Shows How* (Layton, UT: Gibbs Smith, 2007), 176.

12. Ibid., 59.

13. Ibid.

14. Robert Bryce, "The Senator from Big Corn," *Guardian*, September 4, 2008, http://www.guardian.co.uk/commentisfree/2008/sep/04/uselections2008.biofuels; Nancy Pelosi, "Statement on Passage of Farm Bill," May 15, 2008, http://www.house .gov/pelosi/press/releases/May08/farm-bill.html.

15. American Coalition for Ethanol website, http://www.ethanol.org/index.php ?id=78&parentid=26.

16. Energy Information Administration, "U.S. Imports by Country of Origin," http:// tonto.eia.doe.gov/dnav/pet/pet_move_impcus_a2_nus_ep00_im0_mbblpd_a.htm.

17. Barack Obama and Joe Biden, "New Energy for America," n.d. (approximately August 3, 2008), http://www.barackobama.com/pdf/factsheet_energy_speech_080308 .pdf.

18. PBS, "In Iowa, Questions Arise on Impact of Ethanol Production," January 28, 2009, http://www.pbs.org/newshour/bb/environment/jan-june09/mixedyield_01-28.html.

19. Steven Chu, "Pulling the Plug on Oil," *Newsweek*, April 4, 2009, http://www .newsweek.com/id/192481.

20. See transcripts of two Obama speeches from October 23 and October 27, 2009, http://www.whitehouse.gov/the-press-office/remarks-president-challenging -americans-lead-global-economy-clean-energy and http://www.whitehouse.gov/the -press-office/remarks-president-recovery-act-funding-smart-grid-technology.

21. Jan F. Kreider and Peter S. Curtiss, Kreider and Associates, "Comprehensive Life Cycle Analysis of Future, Liquid Fuels for Light Vehicles," September 2008, http://www.fuelsandenergy.com/presentations/Kreider_LCA.pdf, 36.

22. Jesse Ausubel, in his article in "The Future Environment for the Energy Business," *APPEA Journal* (2007), http://phe.rockefeller.edu/docs/ausubelappea.pdf, 8, uses 0.4 W/m². Note that Vaclav Smil, in *Energy: A Beginner's Guide* (Oxford: Oneworld Publications, 2006), 73, puts the sustainable power density of wood harvests at 0.2 W/m².

23. Vaclav Smil, *Energy at the Crossroads: Global Perspectives and Uncertainties* (Cambridge: MIT Press, 2003), 265. Smil puts corn ethanol at a maximum of 0.7 W/m².

24. In 2007, the United States got an average of 2.16 million barrels per day from the Persian Gulf. That includes these countries: Iraq, Kuwait, Qatar, Saudi Arabia, and the United Arab Emirates. Energy Information Administration, "U.S. Imports by Country of Origin," http://tonto.eia.doe.gov/dnav/pet/pet_move_impcus_a2_nus_ep00_im0_mbblpd_a.htm.

25. From 2004 to 2007, daily consumption averaged 20.7 million barrels per day. Energy Information Administration, "Product Supplied," http://tonto.eia.doe.gov/dnav/pet/pet_cons_psup_dc_nus_mbblpd_a.htm.

26. This assumes that the average distance to the Moon is 240,000 miles.

27. Oak Ridge National Laboratory, "Biofuels from Switchgrass: Greener Energy Pastures," n.d., http://bioenergy.ornl.gov/papers/misc/switgrs.html.

28. The math: 42.1 million acres / 640 acres per square mile = 65,781 square miles. Oklahoma covers 68,679 square miles.

29. There are about 440 million acres now planted in the United States. See Dennis Avery, "Biofuels, Food, or Wildlife? The Massive Land Costs of US Ethanol," Competitive Enterprise Institute, September 21, 2006, Available: http://cei.org/pdf/5532.pdf, 6.

30. QuoteOil.com, "What's in a Barrel of Oil: Understanding the Demand," n.d., http://www.quoteoil.com/oil-barrel.html.

31. International Energy Agency, "Medium-Term Oil Market Report," July 1, 2008, 20.

32. Ibid., 21.

33. Energy Information Administration, Appendix Table A2, "Annual Energy Outlook 2009," http://www.eia.doe.gov/oiaf/aeo/pdf/appendixes.pdf, 112.

34. Energy Information Administration, "Petroleum Navigator," http://tonto.eia.doe.gov/dnav/pet/hist/mapupus1A.htm.

Chapter 19

1. Peter Tertzakian, *The End of Energy Obesity*, May 6, 2009, http://media.wiley.com/product_data/excerpt/45/04704354/0470435445.pdf, 5.

2. Sebastian Blanco, "Audi of America President de Nysschen Calls Chevy Volt 'a Car for Idiots,' Slams Electric Vehicles," autoblog.com, September 3, 2009, http://www.autoblog.com/2009/09/03/report-audi-of-america-president-de-nysschen-calls-chevy-volt/.

3. Thomas Friedman, "Texas to Tel Aviv," *New York Times*, July 27, 2008, http://www.nytimes.com/2008/07/27/opinion/27friedman.html.

4. Daniel Roth, "Driven: Shai Agassi's Audacious Plan to Put Electric Cars on the Road," *Wired*, August 18, 2008, http://www.wired.com/cars/futuretransport/magazine/16-09/ff_agassi?currentPage=all.

5. Alan Salzman, "Shai Agassi; the 2009 Time 100," *Time*, May 4, 2009, http://www.time.com/time/specials/packages/article/0,28804,1894410_1893209_1893476,00.html.

6. Daniel Roth, "Driven."

7. Daniel Lyons, "Time for a Trade-In," *Newsweek*, May 2, 2009, http://www.newsweek.com/id/195662.

8. Ibid.

9. Menahem Anderman, testimony before the U.S. Senate Energy Committee, January 30, 2007, http://evworld.com/article.cfm?storyid=1184&first=13911&end=13910.

10. Eric Williams, "Plug-in and Regular Hybrids: A National and Regional Comparison of Costs and CO_2 Emissions," Climate Change Policy Partnership, November 2008, http://www.nicholas.duke.edu/ccpp/ccpp_pdfs/plug_in_hybrid.pdf, 4.

11. Ching-Shin Norman Shiau, Constantine Samaras, Richard Hauffe, and Jeremy J. Michalek, "Impact of Battery Weight and Charging Patterns on the Economic and Environmental Benefits of Plug-in Hybrid Vehicles," *Energy Policy*, February 2009, http://www.cmu.edu/me/ddl/publications/2009-EP-Shiau-Samaras-Hauffe-Michalek-PHEV-Weight-Charging.pdf, 2653.

12. Tomoko A. Hosaka, "Report: Honda to Sell Electric Cars in US," Associated Press, August 22, 2009, http://abcnews.go.com/Business/wireStory?id=8388341.

13. Economist.com, "Sticker Shock," August 21, 2009, http://www.economist.com/sciencetechnology/displaystory.cfm?story_id=14292008.

14. Sam Fletcher, "Mitsubishi iMiEV All-Electric Car Goes On Sale Next Month," popsci.com, June 5, 2009, http://www.popsci.com/cars/article/2009-06/mitsubishi's-electric-car-goes-production. See also Autozone.com, "Mitsubishi i-Miev Specifications and Sample Pics," http://autogang.blogspot.com/2009/07/mitsubishi-i-miev-specifications-and.html.

15. Robert Bryce, "'Galactically Stupid' and Other Random (Tardy) Notes from CERAWeek," *Energy Tribune*, February 27, 2009, http://www.energytribune.com/articles.cfm?aid=1377.

16. Chuck Squatriglia, "Nissan Turns Over an Electric Leaf," Wired.com, August 2, 2009, http://www.wired.com/autopia/2009/08/nissan-electric-leaf/.

17. Tomoko A. Hosaka, "Report."

18. John Reed and Jim Pickard, "Electric Car Buyers to Get £5,000 Grant," *Financial Times*, April 17, 2009, http://www.ft.com/cms/s/0/fb1643a6-2ae5-11de-8415-00144feabdc0.html.

19. Ibid.

20. Geir Moulson and Matt Moore, "Germany Invests Millions in Race for Battery-Powered Car," Associated Press, August 27, 2009, http://www.detnews.com/article/20090827/AUTO01/908270354/1148/rss25.

21. This is known as the "gravimetric" energy density. The other common measure is "volumetric" energy density.

22. For curb weight, see Tesla Motors, "2010 Tesla Roadster: Technical Specifications," http://www.teslamotors.com/performance/tech_specs.php. For battery weight,

see Tesla Motors, "2010 Tesla Roadster: Performance Specifications," http://www .teslamotors.com/performance/perf_specs.php.

23. Robert Bryce, "Energy Tribune Speaks with Bill Reinert, Designer of the Toyota Prius," *Energy Tribune*, February 2, 2009, http://www.energytribune.com/articles .cfm?aid=1270.

24. David Howell, "Progress Report for Energy Storage Research and Development," U.S. Department of Energy, Office of Vehicle Technologies, January 2009, http://www1.eere.energy.gov/vehiclesandfuels/pdfs/program/2008_energy_storage .pdf, 4.

25. *Los Angeles Times*, "Edison's New Storage Battery," May 19, 1901, 8.

26. *New York Times*, "Foreign Trade in Electric Vehicles," November 12, 1911, C8.

27. *Washington Post*, "Prophecies Come True," October 31, 1915, E18.

28. Joseph C. Ingraham, "Old Electric Car May Be the Car of Tomorrow," *New York Times*, July 26, 1959, X19.

29. Bob Thomas, "AMC Does a Turnabout: Starts Running in Black," *Los Angeles Times*, December 17, 1967, K10.

30. Jerry Knight, "GM Unveils Electric Car, New Battery," *Washington Post*, September 26, 1979, D7.

31. *Washington Post*, "Plug 'Er In?" June 7, 1980, A10.

32. Kim Reynolds, "First Drive: 2008 Tesla Roadster," *Motor Trend*, March 1, 2008, http://www.motortrend.com/roadtests/alternative/112_0803_2008_tesla_roadster.

33. Anthony ffrench-Constant, "Tesla Roadster Review," Telegraph.co.uk, June 4, 2009, http://www.telegraph.co.uk/motoring/carreviews/5446199/Tesla-Roadster -review.html.

34. Reynolds, "First Drive."

35. Ibid.

36. Phelps's record is about 1:51. See http://www.youtube.com/watch?v=bh9F7 DMMdVc; Wikipedia, "World Record Progression 200 Metres Butterfly," http://en .wikipedia.org/wiki/World_record_progression_200_metres_butterfly.

37. U.S. Government Accountability Office, "Federal Energy and Fleet Management: Plug-in Vehicles Offer Potential Benefits, but High Costs and Limited Information Could Hinder Integration into the Federal Fleet," June 2009, http://www .gao.gov/new.items/d09493.pdf, 21.

38. Josh Mitchell and Stephen Power, "Gore-Backed Firm Gets Large US Loan," *Wall Street Journal*, September 25, 2009, http://online.wsj.com/article/SB125383 160812639013.html#mod=most_viewed_month.

39. Steven Mufson, "A Jump-Start for New Battery Plants," *Washington Post*, July 25, 2009, http://www.washingtonpost.com/wp-dyn/content/article/2009/07/24/AR 2009072403163.html?hpid=sec-tech.

40. *Phoenix Business Journal*, "Phoenix Companies Get $99.8 Million in Stimulus Money for Electric Vehicle Network," August 5, 2009, http://www.bizjournals.com/ phoenix/stories/2009/08/03/daily39.html?jst=b_ln_hl.

41. U.S. Department of Energy, "Recovery Act Awards for Electric Drive Battery and Component Manufacturing Initiative," n.d., http://www1.eere.energy.gov/recovery/ pdfs/battery_awardee_list.pdf.

42. Katherine Brandon, "Spurring Innovation, Creating Jobs," Whitehouse.gov, August 5, 2009, http://www.whitehouse.gov/blog/Spurring-Innovation-Creating-Jobs/.

43. Associated Press, "China Drives Electric Bike, Scooter Boom," July 27, 2009, http://www.msnbc.msn.com/id/32172301/ns/world_news-world_environment/.

44. For example, see Urbanscooters.com.

45. Ultracapacitors.org, "How an Ultracapacitor Works," n.d., http://www.ultra capacitors.org/ultracapacitors.org-articles/how-an-ultra-capacitor-works.html.

46. Kevin Bullis, "Ultracaps Could Boost Hybrid Efficiency," *Technology Review*, August 20, 2009, http://www.technologyreview.com/energy/23289/.

47. Lawrence Ulrich, "A High-Mileage Masterpiece," *New York Times*, November 20, 2008, http://www.nytimes.com/2008/11/23/automobiles/autoreviews/23-vw-jetta.html.

48. BMWUSA.com, http://www.bmwusa.com/Standard/Content/Uniquely/BMW EfficientDynamics/ExploreAdvancedDiesel.aspx#Diesel_Vehicles/335d. The car sells for about $44,000.

49. MBUSA.com, http://www.mbusa.com/mercedes/#/greenBluetec/. See also Jerry Garrett, "Mercedes-Benz to Bring 4-Cylinder Diesels to the US?" *New York Times*, April 8, 2009, http://wheels.blogs.nytimes.com/2009/04/08/mercedes-benz-to-bring -4-cylinder-diesels-to-the-us/?apage=3.

50. Audi of America, http://www.audiusa.com/us/brand/en/exp/innovation/audi_tdi/ audi_a3_tdi_2_0.html.

51. Neil Dowling, "Test Drive: Hyundai i30 Diesel," November 26, 2008, http:// www.carsguide.com.au/site/news-and-reviews/story/test_drive_hyundai_i30_diesel/ Carsguide.com.au.

52. Stuart Birch, "Mercedes-Benz Plans Hybrid Super Saloon," Telegraph.co.uk, September 9, 2009, http://www.telegraph.co.uk/motoring/news/6157048/Mercedes -Benz-plans-hybrid-super-saloon.html.

53. Greenercars.org, http://www.greenercars.org/highlights.htm.

54. Lawrence Ulrich, "More with Less," *Popular Science*, October 2009, 24. See also Martin LaMonica, "Ford's EcoBoost Tech Busts into Showrooms," Cnet.com, July 17, 2009, http://news.cnet.com/8301-11128_3-10288882-54.html.

55. For more on french-fry grease to motor fuel, see Aimee Levitt, "Wash. U. Turns French Fry Grease into Biodiesel," *Riverfront Times*, December 1, 2009, http://blogs .riverfronttimes.com/dailyrft/2009/12/wash_u_turns_french_fry_grease_into_biodiesel .php.

56. Chuck Squatriglia, "Prius Sales Top 1 Million. Want One? Better Move Fast," *Wired*, May 15, 2008, http://www.wired.com/autopia/2008/05/prius-sales-top/.

Chapter 20

1. ClimateProgress.org, "Southern Company Embraces the Only Practical and Affordable Way to 'Capture' Emissions at a Coal Plant Today—Run It on Biomass," March 18, 2009, http://climateprogress.org/2009/03/18/southern-company-biomass -georgia-power-coal-cofiring/.

2. Matthew Wald, "A Bid to Cut Emissions Looks Away from Coal," *New York Times*, October 31, 2009, http://www.nytimes.com/2009/11/01/science/earth/01carbon .html?_r=1&scp=1&sq=wald%20and%20a%20bid%20to%20cut%20emissions&st=cse.

3. ClimateProgress.org, "Energy and Global Warming News for November 3: Yet Another Coal Plant to Be Replaced by a 'Plant' Plant! And South Dakota's Big Stone 2 Coal Plant Is Dead," November 3, 2009, http://climateprogress.org/2009/11/03/energy-and-global-warming-news-for-november-3-coal-plant-to-biomass-power/.

4. Katy Cummins, "Austin Energy Is Key in Race for Mayor," KXAN.com, April 17, 2009, http://www.kxan.com/dpp/news/politics/local_politics/austin_energy_key_in_race_for_mayor.

5. Daniel Mattola, "Biomass: A Question of Wood, Not Could," *Austin Chronicle*, August 22, 2008, http://www.austinchronicle.com/gyrobase/Issue/story?oid=oid:663310.

6. American Renewables data, "Nacogdoches Power," n.d., http://www.amrenewables.com/our-projects/nacogdoches-power.php.

7. Energy Information Administration, "Electricity: U.S. Data," http://www.eia.doe.gov/fuelelectric.html.

8. Here's the math: 33,629 megawatts x 10,000 tons/year = 336.29 million tons/year.

9. This estimate is from the author based on United Nations Environmental Program data. In 2004, North American per-capita wood consumption was about 700 kilograms. That's equal to about 1,540 pounds per person for 307 million Americans, and 1,540 x 307 million = 472.78 billion pounds, or about 236.4 macro metric (MM) tons. United Nations Environmental Program, "Global Wood Consumption," http://maps.grida.no/go/graphic/global-wood-consumption.

10. ClimateProgress, "Southern Company Embraces."

11. Vaclav Smil, *Energies: An Illustrated Guide to the Biosphere and Civilization* (Cambridge: MIT Press, 1999), 118. Note that Smil in this book gives a range for forest output of 0.4 to 0.9 watts per square meter. See also, Smil, *Global Catastrophes and Trends: The Next Fifty Years* (Cambridge: MIT Press, 2008), 85; Smil, *Energy at the Crossroads: Global Perspectives and Uncertainties* (Cambridge: MIT Press, 2003), 265.

Chapter 21

1. Steve Connor, "Warning: Oil Supplies Are Running Out Fast," *Independent*, August 3, 2009, http://www.independent.co.uk/news/science/warning-oil-supplies-are-running-out-fast-1766585.html.

2. BP Statistical Review of World Energy 2009, http://www.bp.com/liveassets/bp_internet/globalbp/globalbp_uk_english/reports_and_publications/statistical_energy_review_2008/STAGING/local_assets/2009_downloads/renewables_section_2009.pdf.

3. Selena Williams and Liam Moloney, "Enel, EDF to Build Nuclear Plants in Italy," *Wall Street Journal*, August 4, 2009, http://online.wsj.com/article/SB124932829249202353.html.

4. CNN, "China, Australia Ink $41 Billion Gas Deal," August 19, 2009, http://edition.cnn.com/2009/BUSINESS/08/18/china.aus.gas/.

5. BP Statistical Review of World Energy 2009.

6. Masumi Suga and Shunichi Ozasa, "China to Build More Nuclear Plants, Japan Steel Says," Bloomberg, September 7, 2009, http://www.bloomberg.com/apps/news?pid=20601101&sid=a2lUkzmYNGWI.

7. World Nuclear Association, "World Nuclear Power Reactors and Uranium Requirements," http://www.world-nuclear.org/info/reactors.html.

8. Energy Information Administration, "Mexico Energy Profile," http://tonto .eia.doe.gov/country/country_energy_data.cfm?fips=MX. In 2008, Mexico had about 54,000 megawatts of installed capacity.

9. World Nuclear Association, "World Nuclear Power Reactors and Uranium Requirements."

10. BP Statistical Review of World Energy 2009.

11. World Nuclear Association, "World Nuclear Power Reactors and Uranium Requirements."

12. Ibid.

13. Ibid., BP Statistical Review of World Energy 2009.

14. Hewitt Crane, Edwin Kinderman, and Ripudaman Malhotra, *A Cubic Mile of Oil: Realities and Options for Averting the Looming Global Energy Crisis* (New York: Oxford University Press), prepublication copy, 188.

15. See, for example, Cesare Marchetti, "On Decarbonization: Historically and Perspectively," International Institute for Applied Systems Analysis, January 2005, http://www.iiasa.ac.at/Admin/PUB/Documents/IR-05-005.pdf.

16. Ibid.

17. International Energy Agency, "IEA Executive Director: The Climate Challenge Is Immense but We Have the Clean Technology," November 25, 2008, http://www.iea .org/Textbase/press/pressdetail.asp?PRESS_REL_ID=277.

18. Pew Center on Global Climate Change, "History of Kyoto Protocol," n.d., http://www.pewclimate.org/history_of_kyoto.cfm.

19. Nicole Winfield and Alessandra Rizzo, "UN Chief Rebukes G8 over Climate Failures," Associated Press, July 9, 2009, http://www.google.com/hostednews/ap/ article/ALeqM5jd7k_fE2R4LuS8q9w0fglmjvKlBgD99B03085.

20. Environmental Protection Agency, "Clean Air Interstate Rule," http://www.epa .gov/interstateairquality/.

21. Roberto F. Aguilera, "The Past, Present, and Future of the Global Energy Market," EU Energy Policy Blog, March 5, 2009, http://www.energypolicyblog.com/?p=549.

22. Robert Bryce, "Energy Tribune Speaks with Duncan MacLeod," *Energy Tribune*, September 9, 2008, http://www.energytribune.com/articles.cfm?aid=978.

23. Reuters, "Exxon Says N. America Gas Production Has Peaked," June 21, 2005, http://www.reuters.com/article/Utilities/idUSN2163310420050621.

24. Ibid. For Raymond's retirement, see ABCNews.com, "Oil: Exxon Chairman's $400 Million Parachute," April 14, 2006, http://abcnews.go.com/GMA/PainAtThe Pump/story?id=1841989.

25. Julian Darley, *High Noon for Natural Gas: The New Energy Crisis* (White River Junction, VT: Chelsea Green, 2004), 2.

26. Mike Ruppert, "Interview with Matthew Simmons," Oilcrash.com, August 18, 2003, http://www.oilcrash.com/articles/blackout.htm.

27. Richard Heinberg, *The Party's Over: Oil, War, and the Fate of Industrial Societies* (Gabriola Island, British Columbia: New Society Publishers, 2003), 129.

28. BP Statistical Review of World Energy 2009.

29. H. deForest Ralph, "Shale Gas: The Black Swan in the Gas Patch," *Energy Tribune*, March 4, 2009, http://www.energytribune.com/articles.cfm?aid=1400.

30. Leta Smith, "Shale Gas Outside of North America: High Potential but Difficult to Realize," Cambridge Energy Research Associates, April 2009, summary page.

31. One barrel of oil contains the energy equivalent of 5,487 cubic feet of natural gas.

32. Navigant Consulting, "North American Natural Gas Supply Assessment," July 4, 2008, http://www.cleanskies.org/upload/MediaFiles/Files/Downloads2/finalncippt2.pdf, 14.

33. ICF International, Table 7, "Availability, Economics, and Production Potential of North American Unconventional Natural Gas Supplies," November 2008, http://www.ingaa.org/cms/31/7306/7628/7833.aspx, 51.

34. Potential Gas Committee, "Potential Gas Committee Reports Unprecedented Increase in Magnitude of US Natural Gas Resource Base," June 18, 2009, http://www.mines.edu/Potential-Gas-Committee-reports-unprecedented-increase-in-magnitude-of-US-natural-gas-resource-base.

35. BP Statistical Review of World Energy 2009.

36. Potential Gas Committee, "Potential Gas Committee Reports Unprecedented Increase."

37. BP Statistical Review of World Energy 2009.

38. Ibid.

39. Energy Information Administration, "International Natural Gas and Liquefied Natural Gas (LNG) Imports and Exports," http://www.eia.doe.gov/emeu/international/gastrade.html.

40. International Energy Agency, "Natural Gas Market Review 2009," 99.

41. Guy Chazan, "Shell Plans to Build Floating LNG Plant," *Wall Street Journal*, October 9, 2009, B2.

42. Petrobras, press release, "Natural Gas Liquefaction Project in the Pre-Salt," November 17, 2009.

43. Robert Campbell, "Natgas Giants Still Reeling from US Shale Shock," Reuters, October 9, 2009, http://uk.reuters.com/article/idUKTRE5983V020091009.

44. Oilshalegas.com, "The Montney Shale Play Formation," n.d., http://oilshalegas.com/montneyshale.html.

45. For more, see Kitimat LNG website, http://www.kitimatlng.com.

46. Mark Williams, "US Drilling Methods Hot Item," Associated Press, February 22, 2009, http://www.timesleader.com/news/hottopics/shale/U_S__drilling_methods_hot_item_02-22-2009.html.

47. Edward Klump, "StatoilHydro, Chesapeake Go Shale-Gas Shopping in Europe, Asia," Bloomberg, May 21, 2009, http://www.bloomberg.com/apps/news?pid=20601081&sid=a6u2iGw6EJ7U.

48. Doris Leblond, "European Shale Gas Prospects Heat Up," *Oil & Gas Journal*, May 29, 2009, http://www.petronaire.com/news/Exploration-Development/200905/1243662576971.html.

49. Clifford Krauss, "Gas Extraction Method Could Greatly Increase Global Supplies," *New York Times*, October 10, 2009, http://www.nytimes.com/2009/10/10/business/energy-environment/10gas.html.

50. BP Statistical Review of World Energy 2009.

51. Ibid.

52. Ibid.

53. Ibid.

54. Ibid.

55. Ibid.

56. Kenneth Deffeyes, "Join Us as We Watch the Crisis Unfolding," January 16, 2004, http://www.princeton.edu/hubbert/current-events-04-01.html.

57. Henry Groppe, interview with author, March 2, 2009.

58. Steve Andrews, "Interview with Marshall Adkins," *Peak Oil Review*, July 20, 2009, http://www.aspousa.org/index.php/2009/07/interview-with-marshall-adkins/, 4.

59. Peter Wells, correspondence with author, Neftex Petroleum Consultants, July 17, 2009.

60. For more on Groppe, see the website for his firm, Groppe, Long, and Littell, http://www.groppelong.com/About/Biographies/biographies.htm.

61. Energy Information Administration, "Petroleum Navigator," http://tonto.eia.doe.gov/dnav/pet/hist/LeafHandler.ashx?n=PET&s=MTTUPUS2&f=A.

62. Exxon Mobil, "Energy Outlook 2008," http://www.exxonmobil.com/corporate/files/news_pub_2008_energyoutlook.pdf, 10.

63. Mark Long, "BP CEO Says US Gasoline Demand Peaked in 2007," Dow Jones Newswires, November 19, 2009, http://online.wsj.com/article/BT-CO-20091119-718834.html.

64. Liam Denning, "Refining Business Enters the Twilight Zone," *Wall Street Journal*, October 19, 2009, C10.

65. Matt Daily and Rebekah Kebede, "Valero Shuts Delaware Refinery, Takes Big Charge," Reuters, November 20, 2009, http://www.reuters.com/article/reutersComService_3_MOLT/idUSTRE5AJ2RU20091120.

66. Energy Information Administration, "Petroleum Navigator," http://tonto.eia.doe.gov/dnav/pet/hist/mttupus2a.htm.

67. Energy Information Administration, "Short-Term Energy Outlook: Real Petroleum Prices," http://www.eia.doe.gov/emeu/steo/pub/fsheets/real_prices.html.

68. The standard model for this type of work uses of the "Hubbert curve," a model developed by M. King Hubbert, one of the originators of the peak-oil theory.

69. For more on Rutledge, see his website at http://www.its.caltech.edu/~mmic/people2/Rutledge.html.

70. Rutledge's PowerPoint slides and supporting documents can be viewed on his website, http://rutledge.caltech.edu/.

71. David Rutledge, interview with author by phone, August 5, 2009.

72. Tad W. Patzek and Gregory D. Croft, "A Global Coal Production Forecast with Multi-Hubbert Cycle Analysis," unpublished paper, accepted for publication by *Energy Journal* in late 2009, 14.

73. Ibid.

74. David Rutledge, "Hubbert's Peak, the Coal Question, and Climate Change," revised June 2009, http://www.its.caltech.edu/~rutledge/Hubbert's%20Peak,%20The%20Coal%20Question,%20and%20Climate%20Change.ppt, 63.

75. Stewart Brand, *Whole Earth Discipline: An Ecopragmatist Manifesto* (New York: Viking, 2009), 26.

Chapter 22

1. *Los Angeles Times*, "Natural Gas: What to Do," May 26, 1975.

2. See Robert Bryce, *Pipe Dreams: Greed, Ego, and the Death of Enron* (New York: PublicAffairs, 2002), 21–22.

3. David Prindle, *Petroleum Politics and the Texas Railroad Commission* (Austin: University of Texas Press, 1981), 55.

4. Ibid.

5. Ibid, 57.

6. Texas Railroad Commission, "History of the Railroad Commission," http://www.rrc.state.tx.us/about/history/chronological/chronhistory02.php.

7. Prindle, *Petroleum Politics*, 58.

8. Energy Information Administration, Table 1.3, "Primary Energy Consumption by Source, 1949–2008," http://www.eia.doe.gov/emeu/aer/txt/ptb0103.html.

9. Naturalgas.org, "The History of Regulation," http://www.naturalgas.org/regulation/history.asp.

10. Monty Hoyt, "US Burning Its Way Toward Fuel Crisis," *Christian Science Monitor*, June 2, 1971, 1.

11. Thomas O'Toole, "Price Curbs Helped Bring on Fuel Crisis," *Washington Post*, January 13, 1973, A2.

12. S. David Freeman, "Weak on Energy . . . " *Los Angeles Times*, October 13, 1974, G1.

13. *Los Angeles Times*, "Natural Gas: What to Do," May 26, 1975.

14. Steven Rattner, "Cold Wave Causes Energy Shortages and Plant Closings," *New York Times*, January 18, 1977, 1.

15. Richard T. Cooper, "Carter Seeks Emergency Natural Gas Deregulation," *Los Angeles Times*, January 26, 1977, B1.

16. Robert A. Hefner III, *The GET: Grand Energy Transition* (Oklahoma City: Hefner Foundation, 2008), 35.

17. Lawrence Goodwyn, *Texas Oil, American Dreams: A Study of the Texas Independent Producers and Royalty Owners Association* (Austin: Texas State Historical Association, 1996), 141.

18. J. P. Smith, "Glut of Natural Gas in Texas Leads to Red Faces on the Hill," *Washington Post*, March 3, 1978, A2.

19. Ruth Sheldon Knowles, "Price Controls on Natural Gas—Are They Helpful or Harmful?" *Los Angeles Times*, July 27, 1977.

20. Energy Information Administration, "Natural Gas Policy Act of 1978," http://www.eia.doe.gov/oil_gas/natural_gas/analysis_publications/ngmajorleg/ngact1978.html.

21. Energy Information Administration, Table 2, "Supply and Disposition of Natural Gas in the United States, 1930–2000," http://www.eia.doe.gov/pub/oil_gas/natural_gas/data_publications/historical_natural_gas_annual/current/pdf/table_02.pdf. For 1949 to 2008, see Energy Information Administration, "U.S. Natural Gas Total Consumption," http://tonto.eia.doe.gov/dnav/ng/hist_xls/N9140US2a.xls.

22. Energy Information Administration, "Repeal of the Powerplant and Industrial Fuel Use Act (1987)," http://www.eia.doe.gov/oil_gas/natural_gas/analysis_publications/ngmajorleg/repeal.html.

23. BP Statistical Review of World Energy 2008, http://www.bp.com/liveassets/bp_internet/globalbp/globalbp_uk_english/reports_and_publications/statistical_energy_review_2008/STAGING/local_assets/downloads/pdf/statistical_review_of_world_energy_full_review_2008.pdf.

24. Nuclear Energy Institute, "Resources and Stats: Generation Statistics," http://www.nei.org/resourcesandstats/graphicsandcharts/generationstatistics/.

25. BP Statistical Review of World Energy 2008.

26. Robert L. Bradley Jr., *Oil, Gas & Government: The US Experience*, vol. 2 (Lanham, MD: Rowman and Littlefield, 1996), 1267.

27. U.S. Congress, Office of Technology Assessment, "US Natural Gas Availability: Conventional Gas Supply Through the Year 2000—A Technical Memorandum," Report OTA-TM-E-12, September 1983, http://www.fas.org/ota/reports/8326.pdf, 4.

28. Energy Information Administration, "Natural Gas Navigator," http://tonto.eia.doe.gov/dnav/ng/hist/n9010us2a.htm.

29. Office of Technology Assessment, "US Natural Gas Availability," 3.

30. Hart Seely, "The Poetry of D. H. Rumsfeld," *Slate*, April 2, 2003, www.slate.com/id/2081042/.

31. Hefner, *The GET*, 39–40.

32. Energy Information Administration, "World Proved Natural Gas Reserves, January 1, 1980–January 1, 2009 Estimates," http://www.eia.doe.gov/pub/international/iealf/naturalgasreserves.xls.

33. Ibid.

34. Energy Information Administration, "U.S. Natural Gas Marketed Production," n.d., http://tonto.eia.doe.gov/dnav/ng/hist/n9050us2a.htm.

35. U.S. Department of Energy, "Marginal & Stripper Well Revitalization," n.d., http://fossil.energy.gov/programs/oilgas/marginalwells/index.html; Interstate Oil & Gas Compact Commission, "Marginal Wells: Fuel for Economic Growth," 2008, http://iogcc.publishpath.com/Websites/iogcc/pdfs/2008-Marginal-Well-Report.pdf, 5.

36. Here's the math: 60,000 cubic feet = 60,000,000 Btu = 60,000 MJ, and 60,000 MJ / 86,400 = 694,444 W. Thus, 694,444 times 0.33 = 229,166 W, and 229,166 W / 746 W = 307 hp. Assuming a 2-acre well site: 307 / 2 = 153.5 hp per acre. For watts per square meter: 229,166 W / 2 = 114,583. Divide by 4,000 meters/acre, which equals 28.6 watts per square meter.

37. Here's the math: 10 bbls = 58,000,000 Btu, and 58,000 MJ / 86,400 seconds = 671, 296 W. So, 671,296 times 0.33 = 221,152 W (221 kW), and 221,152 / 746 = 297 hp. Finally, 297 / 2 = 148.5 hp per acre. For watts per square meter, the math would be 221,152 / 2 = 110,576. Divide that by 4,000 meters per acre, which equals 27.6 watts per square meter.

38. Interstate Oil & Gas Compact Commission, "Marginal Wells: Fuel for Economic Growth," 3, 7.

39. Ibid. The IOGCC report shows that 396,000 marginal U.S. oil wells produced 291 million barrels in 2007. Total U.S. oil output that year was 1.8 billion barrels.

See Energy Information Administration, "Crude Oil Production," http://tonto.eia.doe
.gov/dnav/pet/pet_crd_crpdn_adc_mbbl_a.htm. Regarding gas, the IOGCC report (p.
10), shows that 322,000 marginal U.S. gas wells produced 1.7 trillion cubic feet of gas
in 2007. Total U.S. gas output that year reached 19 trillion cubic feet. See Energy
Information Administration, "Natural Gas Gross Withdrawals and Production: U.S.,"
http://tonto.eia.doe.gov/dnav/ng/ng_prod_sum_dcu_NUS_a.htm.

40. National Stripper Well Association, "Facts," http://nswa.us/dyn/showpage
.php?id=25. Total stripper well production in 2007 amounted to 291 million barrels,
which works out to just under 800,000 barrels per day.

41. BP Statistical Review of World Energy 2009, http://www.bp.com/liveassets/
bp_internet/globalbp/globalbp_uk_english/reports_and_publications/statistical_energy
_review_2008/STAGING/local_assets/2009_downloads/renewables_section_2009
.pdf.

Chapter 23

1. Lawrence Goodwyn, *Texas Oil, American Dreams: A Study of the Texas Independent Producers and Royalty Owners Association* (Austin: Texas State Historical Association, 1996), 142.

2. Michael E. (Gene) Powell Jr., "Recent Developments in the Barnett Shale,"
March 2009, http://www.barnettshalenews.com/documents/2009TAEPExpo/Recent%
20Developments%20in%20the%20Barnett%20Shale.pdf, 7.

3. International Energy Agency, *World Energy Outlook 2009*, 403–404.

4. Robert Bryce, *Cronies: Oil, The Bushes, and the Rise of Texas, America's Superstate*
(New York: PublicAffairs, 2004), 26–34.

5. Ibid., 133.

6. For more on this, see Bryce, *Cronies*.

7. Central Intelligence Agency, World Factbook, "Country Comparisons—Oil—
Proved Reserves," n.d., https://www.cia.gov/library/publications/the-world-factbook/
rankorder/2178rank.html.

8. Lisa Sumi, "Shale Gas: Focus on the Marcellus Shale," Oil & Gas Accountability Project/Earthworks, May 2008, http://www.earthworksaction.org/pubs/OGAP
MarcellusShaleReport-6-12-08.pdf, 2.

9. Joshua Schneyer, "Brazil, the New Oil Superpower," *Business Week*, November
19, 2007, http://www.businessweek.com/bwdaily/dnflash/content/nov2007/db2007
1115_045316.htm.

10. Ibid.

11. EnCana, "2009 Key Play Conference Call Series Haynesville & Deep Bossier,"
May 27, 2009, http://www.encana.com/investors/presentationsevents/pdfs/20090527
-haynesville-deep-bossier-key-play-call.pdf, 14.

12. Richard Heinberg, *The Party's Over: Oil, War, and the Fate of Industrial Societies*
(Gabriola Island, British Columbia: New Society, 2003), 105.

13. Robert A. Hefner III, *The GET: Grand Energy Transition* (Oklahoma City:
Hefner Foundation, 2008), 37–40.

14. Goodwyn, *Texas Oil, American Dreams*, 36.

15. Hunt Oil, "Hunt Oil History Window," n.d., http://www.huntoil.com/history.asp.

16. Barnett Shale Energy Education Council, http://www.bseec.org/index.php/content/facts/about_barnett_shale/. Connecticut covers about 5,500 square miles (Enchanted Learning, "US States [plus Washington, D.C.]: Area and Ranking," http://www.enchantedlearning.com/usa/states/area.shtml).

17. Bryce, *Cronies*, 24–36.

18. Barnett Shale News, "Barnett Shale: Drilling Statistics by RigData," n.d., http://www.barnettshalenews.com/documents/RigData%20Barnett%20Shale%20Rig%20Count%20Booklet%20Barnett%20Shale%20EXPO%203-11-2009.pdf.

19. Baker Hughes, "North American Rotary Rig Count," http://investor.shareholder.com/bhi/rig_counts/rc_index.cfm. According to Baker Hughes, an average of 1,880 drilling rigs were active during 2008.

20. Powell, "Recent Developments in the Barnett Shale," 7.

21. East Texas Oil Museum, "A Brief History of the East Texas Oil Field," 2000, http://www.easttexasoilmuseum.com/Pages/history.html.

22. Energy Information Administration, "Petroleum Navigator," http://tonto.eia.doe.gov/dnav/pet/hist/LeafHandler.ashx?n=PET&s=f000000__3&f=A.

23. Goodwyn, *Texas Oil, American Dreams*, 47.

24. In mid-2008, the Barnett was producing about 4.7 billion cubic feet of gas per day. See "Gene Powell: Barnett Guru," *Oil & Gas Investor*, January 2009, http://www.barnettshalenews.com/documents/Gene%20Powell%20Barnett%20Guru%20Article%20in%20Oil%20&%20Gas%20Investor%20Magazine%20January%202009.pdf, 18. For U.S. gas production data, see Energy Information Administration, "Natural Gas Navigator," http://tonto.eia.doe.gov/dnav/ng/hist/n9070us2m.htm. In July 2008, total production was 1.734 trillion cubic feet, or about 57.6 billion cubic feet per day.

25. Energy Information Administration, "Natural Gas Navigator," http://tonto.eia.doe.gov/dnav/ng/hist/rngc1d.htm.

26. Ibid.

Chapter 24

1. For more on mineral rights, see Robert Bryce, "The Meek Need Mineral Rights," *Energy Tribune*, December 26, 2007, http://www.energytribune.com/articles.cfm?aid=737.

2. For more on de Soto, see Dario Fernandez-Morera, "Hernando de Soto Interview," Reason.com, November 30, 1999, http://www.reason.com/news/show/32213.html.

3. This is my estimate. In 1992, the Environmental Protection Agency estimated that 1.2 million wells had been abandoned. See Roberto Suro, "Abandoned Oil and Gas Wells Become Pollution Portals," *New York Times*, May 3, 1992, http://www.nytimes.com/1992/05/03/us/abandoned-oil-and-gas-wells-become-pollution-portals.html. Since 1992, more than 500,000 wells have been drilled in the United States. See Energy Information Administration, "Petroleum Navigator," http://tonto.eia.doe.gov/dnav/pet/hist/e_ertw0_xwc0_nus_ca.htm.

4. Yahoo! Finance, "Google, Inc.: Income Statement," http://finance.yahoo.com/q/is?s=GOOG&annual.

5. The easiest way to estimate U.S. royalty payments is to calculate the value of all the oil and gas produced in the country and then multiply it by one-eighth. (A 12.5 percent

royalty rate is often used as a benchmark amount for landowners, but many mineral rights agreements call for royalties of 18.75 percent or more.) In 2007, the total value of all U.S. oil and gas production was about $243.9 billion, and domestic oil production totaled 1.848 billion barrels. (See Energy Information Administration, "Crude Oil Production," http://tonto.eia.doe.gov/dnav/pet/pet_crd_crpdn_adc_mbbl_a.htm.) The average well-head price was $66.52 per barrel. (See Energy Information Administration, "Petroleum Navigator," http://tonto.eia.doe.gov/dnav/pet/hist/f000000__3a.htm.) Therefore, the total value of U.S. oil production in 2007 was about $122.9 billion. Meanwhile, U.S. natural gas production was 19 trillion cubic feet. (See Energy Information Administration, "Natural Gas Navigator," http://tonto.eia.doe.gov/dnav/ng/hist/n9070us2a.htm.) And the average wellhead price was $6.37 per thousand cubic feet. (See Energy Information Administration, "Natural Gas Navigator," http://tonto.eia.doe.gov/dnav/ng/hist/n9190us3A .htm.) Thus, the value of all U.S. gas production that year was about $121 billion. At a royalty rate of 12.5 percent, that means that some $30.5 billion in mineral royalties were paid out in 2007. Of that, about $9 billion was collected by the federal government for oil and gas produced from federal onshore and offshore leases. (See U.S. Government Accountability Office, "Oil and Gas Royalties: The Federal System for Collecting Oil and Gas Revenues Needs Comprehensive Reassessment," September 2008, http://www.gao .gov/new.items/d08691.pdf.) Subtracting the feds' take means that private mineral owners were likely paid about $21.5 billion in 2007.

This estimate appears reasonable, given findings by other researchers. Energy In Depth, a group that represents independent oil and gas producers, estimates 2007 royalty payments to public and private landowners at $30 billion. See Energy In Depth, "Quick Facts: Jobs, Economic and Energy Benefits of Domestic Production," http://www.energyindepth.org/about/quick-facts/. In addition, a 2009 report by Advanced Resources International for the Independent Petroleum Association of America put Oklahoma's 2007 royalty payments at $1.85 billion. That sum is about 8.6 percent of my estimate of $21.5 billion for the whole United States. That percentage ties out nicely when you consider that in 2007, Oklahoma gas production was 1.65 trillion cubic feet, or about 8.6 percent of total U.S. gas production. See Advanced Resources International, "Bringing Real Information on Energy Forward: Economic Considerations Associated with Regulating the American Oil and Natural Gas Industry," April 24, 2009, http://s3.amazonaws.com/propublica/assets/natural_gas/economic _consequences_report_april2009.pdf, 3. For more information on Oklahoma production, see Energy Information Administration, "Natural Gas Gross Withdrawals and Production: Oklahoma," http://tonto.eia.doe.gov/dnav/ng/ng_prod_sum_dcu_sok _a.htm. Note that the calculation of $21.5 billion is based on 2007 data. Given the oil and gas price spikes of 2008, the royalty payments that year were likely far higher.

6. Bryce, "The Meek Need Mineral Rights."

7. PegasusNews.com, "Arlington Neighborhood Gets Record Deal for Barnett Shale Rights," March 18, 2008, http://www.pegasusnews.com/news/2008/mar/18/ arlington-neighborhood-gets-record-deal-barnett-sh/.

8. Business Images Northwest Louisiana, "Haynesville Shale Fuels Boom in Northwest Louisiana," May 18, 2009, http://imagesnwlouisiana.com/index.php/site/articles/ energy/haynesville_shale_fuels_boom_in_northwest_louisiana.

9. *Haynesville*, directed by Gregory Kallenberg, produced by Mark Bullard and Gregory Kallenberg (Threepenny Productions, 2009). For more information, see the official site for the movie at http://www.haynesvillemovie.com/. Also, Gregory Kallenberg, interview with author, approximately October 15, 2009.

10. Colorado Energy Research Institute, "Oil and Gas Economic Impact Analysis," June 2007, http://www.energyindepth.org/PDF/CERIOil&Gas.pdf, ix.

11. Oklahoma Energy Resources Board, "The Local Impact of Oil and Gas Production and Drilling in Oklahoma," October 2008, http://www.energyindepth.org/PDF/Local%20Oil%20Gas%20Impact%20Draft%2020080916.pdf, 24, 33.

12. Independent Petroleum Association of America, "The Oil & Gas Producing Industry in Your State," 2007–2008, http://www.energyindepth.org/PDF/IPAA07_ADB.indd.pdf, 118.

Chapter 25

1. Energy Information Administration, "World Carbon Dioxide Emissions from the Consumption and Flaring of Fossil Fuels, 1980–2006," posted December 8, 2008, http://www.eia.doe.gov/pub/international/iealf/tableh1co2.xls.

2. New York State Department of Environmental Conservation, "Gas Well Drilling in the Marcellus Shale," n.d., http://www.dec.ny.gov/energy/46288.html.

3. David O. Williams, "DeGette, Polis Introduce FRAC Act Aimed at Closing Hydraulic Fracturing 'Loophole,'" *The Colorado Independent*, June 9, 2009, http://coloradoindependent.com/30784/degette-polis-introduce-frac-act-aimed-at-closing-hydraulic-fracturing-loophole.

4. *Oil & Gas Journal*, "API Opposes Efforts to Federally Regulate Hydraulic Fracturing," June 9, 2009, http://www.laserfocusworld.com/display_article/364231/7/none/none/Gener/API-opposes-efforts-to-federally-regulate-hydraulic-fracturin.

5. Jeremy Miller, "Of Hydraulic Fracturing and Drinking Water," Green Inc., June 30, 2009, http://greeninc.blogs.nytimes.com/2009/06/30/of-hydraulic-fracturing-and-drinking-water/?pagemode=print.

6. Abrahm Lustgarten, "Buried Secrets: Is Natural Gas Drilling Endangering US Water Supplies?" ProPublica, November 13, 2008, http://www.propublica.org/feature/buried-secrets-is-natural-gas-drilling-endangering-us-water-supplies-1113.

7. Miller, "Of Hydraulic Fracturing."

8. Associated Press, "Wyo. Community Blames Fracking for Water Woes," September 6, 2009, http://www.lasvegassun.com/news/2009/sep/06/wyo-community-blames-fracking-for-water-woes/.

9. Jad Mouawad and Clifford Krauss, "Gas Company Won't Drill in New York Watershed," *New York Times*, October 27, 2009, http://www.nytimes.com/2009/10/28/business/energy-environment/28drill.html?_r=1&adxnnl=1&adxnnlx=1256732225-obw5RJbKNjJ9k+kgTB5w0Q.

10. Energy Information Administration, Table 6.4, "Natural Gas Gross Withdrawals and Natural Gas Well Productivity, 1960–2008," http://www.eia.doe.gov/emeu/aer/txt/ptb0604.html.

11. Energy Information Administration, "Crude Oil and Natural Gas Exploratory and Development Wells," http://tonto.eia.doe.gov/dnav/pet/pet_crd_wellend_s1_a.htm.

12. *Barnett Shale Newsletter*, November 16, 2009, 8.

13. Ibid., 13.

Chapter 26

1. James Lovelock, "Nuclear Power Is the Only Green Solution," *Independent*, May 24, 2004, http://www.independent.co.uk/opinion/commentators/james-lovelock-nuclear -power-is-the-only-green-solution-564446.html.

2. Stewart Brand, *Whole Earth Discipline: An Ecopragmatist Manifesto* (New York: Viking, 2009), 98.

3. Robert Bryce, "Bryce Interviews Amory Lovins," *Energy Tribune*, November 2007, http://www.robertbryce.com/node/161.

4. International Atomic Energy Agency, "Nuclear Power Plants Information: Operational Reactors by Age," http://www.iaea.org/cgi-bin/db.page.pl/pris.reaopag.htm.

5. "World Nuclear Power Reactors and Uranium Requirements," December 1, 2009, http://www.world-nuclear.org/info/reactors.html.

6. BP Statistical Review of World Energy 2009, http://www.bp.com/liveassets /bp_internet/globalbp/globalbp_uk_english/reports_and_publications/statistical _energy_review_2008/STAGING/local_assets/2009_downloads/renewables _section_2009.pdf.

7. Brand, *Whole Earth Discipline*, 99.

8. International Energy Agency, *World Energy Outlook 2009*, 266.

9. Ibid., 267.

10. Ibid., 381.

11. Brand, *Whole Earth Discipline*, 99. For more on Amory Lovins, see Democracy Now, "Amory Lovins: Expanding Nuclear Power Makes Climate Change Worse," video, July 16, 2008, http://www.democracynow.org/2008/7/16/amory_lovins_expanding _nuclear_power_makes.

12. Greenpeace International, "End the Nuclear Age," n.d., http://www.green peace.org/international/campaigns/nuclear. Also see Greenpeace International, "Nuclear Power Belongs in the Past," April 25, 2008, http://www.greenpeace.org/usa/ news/nuclear-power-belongs-in-past.

13. Sierra Club, "Nuclear Power," n.d., http://www.sierraclub.org/policy/conservation/ nuc-power.aspx.

14. For 15 percent of electricity, see International Energy Agency, "Key World Energy Statistics, 2008," http://www.iea.org/Textbase/nppdf/free/2008/Key_Stats_2008 .pdf, 24, and *World Energy Outlook 2008*, 509. For 5 percent of primary energy, see BP Statistical Review of World Energy 2009.

15. Richard Rhodes and Denis Beller, "The Need for Nuclear Power," *Foreign Affairs*, February 2000, http://www.iaea.org/Publications/Magazines/Bulletin/Bull422/ 42204784350.pdf, 47.

16. Nuclear Energy Institute, "US State-by-State Commercial Nuclear Used Fuel and Payments to the Nuclear Waste Fund," current as of March 31, 2009, http://www.nei.org/filefolder/US_State_by_State_Used_Fuel_and_Payments_to _NWF_1.xls.

17. Gwyneth Cravens, *Power to Save the World: The Truth About Nuclear Energy* (New York: Knopf, 2008), 269.

18. Halimah Abdullah and Renee Schoof, "Boxer: EPA Should Regulate Coal-Fired Power Plant Waste," McClatchy Newspapers, June 26, 2009, http://www.thesunnews.com/1014/story/956439.html.

19. Cravens, *Power to Save the World*, 269. For carbon data, see Energy Information Administration, "World Carbon Dioxide Emissions from the Consumption and Flaring of Fossil Fuels, 1980–2006," posted December 8, 2008, http://www.eia.doe.gov/pub/international/iealf/tableh1co2.xls.

20. James Kanter, "In Finland, Nuclear Renaissance Runs Into Trouble," *New York Times*, May 28, 2009, http://www.nytimes.com/2009/05/29/business/energy-environment/29nuke.html.

21. Ibid.

22. Jim Forsyth, "San Antonio Sees Cost for New Tex Reactors Rising," Reuters, October 27, 2009, http://www.reuters.com/article/rbssConsumerGoodsAndRetailNews/idUSN2726904520091027.

23. NewEnergyFocus.com, "Onshore Construction Begins for Sheringham Shoal Wind farm," September 7, 2009, http://www.newenergyfocus.com/do/ecco.py/view_item?listid=1&listcatid=32&listitemid=2978&section=Wind. See also Statoil, "Pioneering Wind Power at Sheringham Shoal," http://www.statoilhydro.com/en/TechnologyInnovation/NewEnergy/RenewablePowerProduction/Offshore/SheringhamShoal/Pages/default.aspx.

24. Alex Kuffner, "National Grid Objects to Proposed Cost of Wind Power," *Providence Journal*, November 19, 2009, http://www.projo.com/business/content/GRID_DEEPWATER_DEAL_11-19-09_ROGGH5M_v38.3c1d2cb.html.

25. Katherine Gregor, "Cool City: Solar Subtleties," *Austin Chronicle*, March 6, 2009, http://www.austinchronicle.com/gyrobase/Issue/story?oid=oid:751802.

26. Christopher Calnan, "City Council Gives Austin Energy the Go-Ahead for Major Solar Project," *Austin Business Journal*, March 5, 2009, http://austin.bizjournals.com/austin/stories/2009/03/02/daily49.html.

27. Zac Anderson, "Solar Plant Set to Open, Even as Shadows Loom," Herald Tribune.com, October 14, 2009, http://www.heraldtribune.com/article/20091014/ARTICLE/910141033/2055/NEWS?Title=Solar-plant-set-to-open-even-as-shadows-loom.

28. Energy Information Administration, "Federal Financial Interventions and Subsidies in Energy Markets 2007," April 2008, http://www.eia.doe.gov/oiaf/servicerpt/subsidy2/pdf/subsidy08.pdf, xvi.

29. Ibid.

30. Simon Lomax, "Nuclear Industry 'Restart' Means More Loan Guarantees, Chu Says," Bloomberg, October 27, 2009, http://www.bloomberg.com/apps/news?pid=20601072&sid=aR1MVERYEgAs.

31. This is commonly called the Price-Anderson Act. For more, see Wikipedia, "Price Anderson Nuclear Industries Indemnity Act," http://en.wikipedia.org/wiki/Price-Anderson_Nuclear_Industries_Indemnity_Act.

32. David Bradish, "Amory Lovins vs. Stewart Brand—Part Four," NEI Nuclear Notes, November 16, 2009, http://neinuclearnotes.blogspot.com/2009/11/amory-lovins-vs-stewart-brand-part-four.html#links.

33. Note that the total in this report includes petroleum liquids. The EIA's official statistics on electricity generation show that in 2007, natural gas total generation was 896.5 billion kilowatt-hours. See Energy Information Administration, "Net Generation by Energy Source by Type of Producer," http://www.eia.doe.gov/cneaf/electricity/epa/epat1p1.html.

34. Recall that wind gets $23.37 per megawatt-hour, and gas gets $0.25. Gas generated 900 billion kilowatt-hours, and wind generated 31 billion kilowatt-hours.

35. Energy Information Administration, "Federal Financial Interventions and Subsidies in Energy Markets 2007," xviii.

Chapter 27

1. Environmental Defense Fund, "Fact Sheet," n.d., http://www.edf.org/documents/2978_FactSheet_aboutus.pdf.

2. Fred Krupp speech at CERAWeek, Houston, Texas, February 12, 2009.

3. H. Josef Hebert, "Nuclear Waste Won't Be Going to Nevada's Yucca Mountain, Obama Official Says," Associated Press, March 6, 2009, http://www.chicagotribune.com/news/nationworld/chi-nuke-yucca_frimar06,0,2557502.story.

4. *The Hill*, "Reid Vows to Block New Push for Yucca Mountain Nuke Site," April 6, 2006, http://hill6.thehill.com/index2.php?option=com_content&do_pdf=1&id=59718.

5. Hebert, "Nuclear Waste Won't Be Going to Nevada's Yucca Mountain."

6. Republican Party, "2008 Republican Platform," http://platform.gop.com/2008Platform.pdf, 32.

7. Democratic Party, "Renewing America's Promise: Democratic Party Platform, 2008," http://www.democrats.org/a/party/platform.html, 47.

8. William Tucker, *Terrestrial Energy: How Nuclear Power Will Lead the Green Revolution and End America's Energy Odyssey* (Savage, MD: Bartleby Press, 2008), 345, 352.

9. World Nuclear Association, "Waste Management in the Nuclear Fuel Cycle," n.d., www.world-nuclear.org/info/info04.html.

10. Amory B. Lovins and Imran Sheikh, "The Nuclear Illusion," Rocky Mountain Institute, May 27, 2008, http://www.rmi.org/rmi/Library/E08=01_NuclearIllusion.

11. Vaclav Smil, *Energies: An Illustrated Guide to the Biosphere and Civilization* (Cambridge: MIT Press, 1999), 172.

12. Institute for 21st Century Energy, "Revisiting America's Nuclear Waste Policy," May 2009, http://www.energyxxi.org/reports/Nuclear_Waste_Policy.pdf, 2.

13. TED.com, "Stewart Brand Proclaims 4 Environmental Heresies," June 2009, video, http://www.youtube.com/watch?v=TUxwiVFgghE.

14. USEC.com, "Metagons to Megawatts," http://www.usec.com/megatonstomegawatts.htm.

15. See CEA.fr, "What Is Transmutation?" n.d., http://www.cea.fr/var/cea/storage/static/gb/library/Clefs53/pdf-gb/ence-abf_53gb.pdf.

16. Gwyneth Cravens, *Power to Save the World: The Truth About Nuclear Energy* (New York: Knopf, 2008), 265.

17. Swadesh Mahajan, interview with author, October 14, 2009.

18. Nuclear Regulatory Commission, "Transuranic Element," n.d., http://www.nrc.gov/reading-rm/basic-ref/glossary/transuranic-element.html. For neptunium info, see

"Neptunium Facts: Chemical and Physical Properties," http://chemistry.about.com/od/elementfacts/a/neptunium.htm.

19. Malcolm Browne, "Modern Alchemists Transmute Nuclear Waste," *New York Times*, October 29, 1991, http://www.nytimes.com/1991/10/29/science/modern-alchemists-transmute-nuclear-waste.html.

20. Martin Avery Snyder and Harold Weitzner, "The Road to Endless Energy," Courant Institute of Mathematical Sciences, New York University, December 12, 2008.

21. Hans A. Bethe, "The Fusion Hybrid," *Physics Today*, May 1979, 50.

22. Ed Gerstner, "The Hybrid Returns," *Nature*, July 2, 2009, 27.

23. U.S. Department of Energy, "Waste Isolation Pilot Plant," http://www.wipp.energy.gov/.

Chapter 28

1. Energy Information Administration, "Annual Energy Outlook 2009 with Projections to 2030," http://www.eia.doe.gov/oiaf/aeo/electricity.html.

2. Energy Information Administration, "Total Electric Power Industry Summary Statistics," http://www.eia.doe.gov/cneaf/electricity/epm/tablees1b.html.

3. Energy Information Administration, "Annual Energy Outlook 2009," Figure 55, http://www.eia.doe.gov/oiaf/aeo/excel/figure55_data.xls.

4. CBS News, "Al Gore: Energy Crisis Can Be Fixed," July 17, 2008, http://www.cbsnews.com/stories/2008/07/17/eveningnews/main4270123.shtml.

5. The U.S. Navy doesn't publish actual output data. John Pike of Globalsecurity.org estimates output at about 50,000 horsepower, which equals 37.3 megawatts.

6. Robert Bryce, "Nukes Get Small," *Energy Tribune*, July 16, 2008, http://www.energytribune.com/articles.cfm?aid=948.

7. Ibid.

8. NuScale, "NuScale Fact Sheet," http://www.nuscalepower.com/nr-Press-Fact-Sheet-NuScale.php.

9. Babcock.com, "B&W Power Generation Group: Historic Milestones," n.d., http://www.babcock.com/business_units/power_gen_group/facts_historic_milestones.html.

10. Yahoo! Finance, "McDermott International, Inc.: Income Statement," http://finance.yahoo.com/q/is?s=MDR&annual.

11. U.S. Nuclear Regulatory Commission, "Design Certification Applications for New Reactors," http://www.nrc.gov/reactors/new-reactors/design-cert.html.

12. Richard A. Muller, *Physics for Future Presidents: The Science Behind the Headlines* (New York: W. W. Norton, 2008), 334.

13. Tim Dean, "New Age Nuclear," *Cosmos*, April 2006, http://www.cosmosmagazine.com/node/348/full.

14. Mitch Jacoby, "Reintroducing Thorium," *Chemical & Engineering News*, November 16, 2009, http://pubs.acs.org/cen/email/html/8746sci2.html.

15. Seth Grae, "Thorium," *Chemical & Engineering News*, n.d. (copyright is 2003), http://pubs.acs.org/cen/80th/thorium.html.

16. Foreign Relations of the United States, Volume E-7, Document 298, "Indian Nuclear Developments and Their Likely Implications," n.d. (period covered is 1969–1976), http://www.state.gov/r/pa/ho/frus/nixon/e7txt/50225.htm.

17. Scott DiSavino, "Interview: Lightbridge Sees Nuclear Power Future in Thorium," December 11, 2009, http://in.reuters.com/article/worldNews/idINIndia-4465 9520091211?sp=true.

18. Lightbridge, "Corporate Overview," n.d., http://www.Ltbridge.com/images/ Thorium-Power-Ltd.-Information-Kit.pdf.

19. For more on other reactor designs, including the very high temperature reactor and the sodium-cooled fast reactor, see *The Economist*, "Nuclear's Next Generation," December 10, 2009, http://www.economist.com/sciencetechnology/tq/displayStory .cfm?story_id=15048703.

20. International Energy Agency, *World Energy Outlook 2009*, 160.

21. Richard Rhodes and Denis Beller, "The Need for Nuclear Power," *Foreign Affairs*, February 2000, http://www.iaea.org/Publications/Magazines/Bulletin/Bull422/ 42204784350.pdf, 50.

Chapter 29

1. International Energy Agency, *World Energy Outlook 2008*, 178.

2. Ibid., 159.

3. First Solar, "First Solar Passes $1 Per Watt Industry Milestone," February 24, 2009, http://investor.firstsolar.com/phoenix.zhtml?c=201491&p=irol-newsArticle& ID=1259614&highlight=.

4. ESolar, "Our Solution," http://www.esolar.com/solution.html.

5. Katie Howell, "Exxon Sinks $600M into Algae-Based Biofuels in Major Strategy Shift," *New York Times*, July 14, 2009, http://www.nytimes.com/gwire/2009/07/14/ 14greenwire-exxon-sinks-600m-into-algae-based-biofuels-in-33562.html.

6. Todd Woody, "Solar Power When the Sun Goes Down?" *New York Times*, November 3, 2009, http://greeninc.blogs.nytimes.com/2009/11/03/solar-power-when -the-sun-goes-down/.

7. Todd Woody, "Dow Unveils Solar Shingles," *New York Times*, October 7, 2009, http://greeninc.blogs.nytimes.com/2009/10/07/dow-unveils-solar-shingles/.

8. Jim Ostroff, "Eight Innovations That Will Change Your Future," Kiplinger Letter, December 9, 2009, http://www.kiplinger.com/businessresource/forecast/archive/eight -innovations-that-will-change-your-future.html.

9. Andrew Taylor, "Obama's Budget Cuts Take Aim at 121 Federal Programs," Associated Press, May 8, 2009, http://www.boston.com/news/nation/washington/articles/ 2009/05/08/obamas_budget_cuts_take_aim_at_121_federal_programs/.

10. *Mother Jones*, "Lie by Lie: The Mother Jones Iraq War Timeline (8/1/90– 2/14/08)," n.d., http://www.motherjones.com/bush_war_timeline.

11. Ibid.

12. Dafna Linzer, "Before Iraq War, US Ignored Work of UN Arms Inspectors, Panel Says," *Washington Post*, April 4, 2005, http://seattletimes.nwsource.com/html/ nationworld/2002230005_intel04.html.

13. International Atomic Energy Agency, "Transcripts of Interviews," September 1, 2009, http://www.iaea.org/NewsCenter/Transcripts/2009/bas010909.html.

14. *Time*, "10 Questions for Mohamed ElBaradei," August 17, 2009, http://www .time.com/time/magazine/article/0,9171,1914983,00.html.

15. Roger Cohen, "Bunkers or Breakthrough?" *New York Times*, November 5, 2009, http://www.nytimes.com/2009/11/06/opinion/06iht-edcohen.html?_r=2&adxnnl =1&adxnnlx=1258954347-wP/SZxb94bd5d5GepPd7rA.

16. International Atomic Energy Agency, "The 'Atoms for Peace' Agency," http:// www.iaea.org/About/index.html.

17. Ibid. In 2008, the agency's budget was 277 million euros.

18. White House, "Remarks by President Barack Obama," Prague, Czech Republic, April 5, 2009, http://www.whitehouse.gov/the_press_office/Remarks-By-President -Barack-Obama-In-Prague-As-Delivered/.

19. Wikipedia, "David E. Lilienthal," http://en.wikipedia.org/wiki/David_E. _Lilienthal.

20. The Acheson-Lilienthal Report, March 16, 1946, http://honors.umd.edu/ HONR269J//archive/AchesonLilienthal.html.

21. For further discussions of the caucus-versus-primary semantics, see Iowacaucus .org, http://www.iowacaucus.org/iacaucus.html, or American Political Science Association, "Why Should Iowa Remain the First Presidential Primary?" March 12, 2009, http://www.apsanet.org/content_63423.cfm.

22. Energy Policy Act, http://frwebgate.access.gpo.gov/cgi-bin/getdoc.cgi?dbname= 109_cong_bills&docid=f:h6enr.txt.pdf.

23. Energy Independence and Security Act, http://frwebgate.access.gpo.gov/cgi-bin/ getdoc.cgi?dbname=110_cong_bills&docid=f:h6enr.txt.pdf.

24. American Clean Energy and Security Act, http://frwebgate.access.gpo.gov/cgi -bin/getdoc.cgi?dbname=111_cong_bills&docid=f:h2454pcs.txt.pdf.

25. Peter Lattman, "Lawyers in the US Senate Quiz: The Answers," September 5, 2007, http://blogs.wsj.com/law/2007/09/05/lawyers-in-the-us-senate-quiz-the-answers/.

26. National Society of Professional Engineers, "Professional Engineers in Congress," n.d., http://www.nspe.org/GovernmentRelations/TakeAction/IssueBriefs/ib _pro_eng_congress.html.

27. Ibid. For more on Barton, see http://joebarton.house.gov/Back.aspx?Page= Biography.

28. Sara Ditta, "A Remedy for What Ails Us: More Doctors in 111th Congress," *Roll Call*, January 7, 2009, http://kagen.house.gov/index.php?option=com_content& view=article&id=299:a-remedy-for-what-ails-us-more-doctors-in-111th-congress& catid=66:in-the-news&Itemid=195.

29. Council of Bars and Law Societies of Europe, "Number of Lawyers in CCBE Member Bars," http://www.ccbe.org/fileadmin/user_upload/NTCdocument/table _number_lawyers1_1179905628.pdf.

30. Wikianswers, http://wiki.answers.com/Q/What_is_the_total_number_of_lawyers _in_the_US.

31. BP Statistical Review of World Energy 2009, http://www.bp.com/liveassets/ bp_internet/globalbp/globalbp_uk_english/reports_and_publications/statistical _energy_review_2008/STAGING/local_assets/2009_downloads/renewables_section _2009.pdf. Ahead of Iran, in order, are China, Indonesia, United Arab Emirates, Malaysia, Qatar, South Korea, and Iceland.

32. Central Intelligence Agency, World Factbook, "Middle East: Iran," https://www.cia.gov/library/publications/the-world-factbook/geos/ir.html.

33. Brett Jarman, "Iranian CNG Revolution Continues—1.5 Million NGVs and Growing," NGVGlobal.com, August 5, 2009, http://www.ngvglobal.com/iranian-cng-revolution-continues-1-5-million-ngvs-and-growing-0805#more-3640.

34. International Association for Natural Gas Vehicles, "Natural Gas Vehicle Statistics," http://www.iangv.org/tools-resources/statistics.html.

35. University of Missouri, "Corncobs Key to Licensing Agreement for Natural Gas Storage Tanks," May 13, 2009, http://munews.missouri.edu/news-releases/2009/0513-suppes-pfeifer-natural-gas-license.php. See also Green Car Congress, "COFs Among the Best Adsorbents for Storage of Hydrogen, Natural Gas and CO_2," June 7, 2009, http://www.greencarcongress.com/2009/06/cofs-20090607.html.

36. *National Post*, "Graphic: The State of Nuclear Power," July 31, 2009, http://network.nationalpost.com/np/blogs/posted/archive/2009/07/31/graphic-the-state-of-nuclear-power.aspx.

37. Areva, *All About Nuclear Energy: From Atom to Zirconium.* (Paris: Areva, 2008), 105.

38. For more on Taylor's views, see: Jerry Taylor, "Nuclear Energy: Risky Business," Reason.com, October 21, 2008, http://reason.org/news/show/1003239.html.

39. e360, "Leveling Appalachia," n.d., http://www.e360.yale.edu/content/feature.msp?id=2198.

40. The legal fights over this type of mining have been ongoing for years. For instance, see Ken Ward Jr., "4 Mining Permits Blocked," *Charleston Gazette*, March 24, 2007, http://wvgazette.com/News/MiningtheMountains/200703240003.

41. Eric Bontrager, "EPA Puts Brakes on 3 More Mountaintop Mining Permits," Greenwire, April 9, 2009, http://www.nytimes.com/gwire/2009/04/09/09greenwire-epa-puts-brakes-on-3-more-mountaintop-permits-10493.html?scp=26&sq=energy%20regulation%20and%20sierra%20club&st=cse.

42. Ken Ward Jr., "Exclusive: Patriot Coal Says—We Can Mine It Underground," *Coal Tattoo*, August 14, 2009, http://blogs.wvgazette.com/coaltattoo/2009/08/14/exclusive-patriot-coal-says-we-can-mine-it-underground/#more-1081.

43. World Bank, "World Bank Calls for More Results in Gas Flaring Reduction," December 5, 2008, http://web.worldbank.org/WBSITE/EXTERNAL/NEWS/0,,content-MDK:22003319~pagePK:34370~piPK:34424~theSitePK:4607,00.html. The World Bank puts total flared gas at 150 billion cubic meters. To convert meters to cubic feet, multiply by 35.3. For conversions, see BP Statistical Review of World Energy 2009.

44. Recall that 1 barrel of crude is equal to 5,487 cubic feet of gas.

45. World Bank, "World Bank Calls for More Results."

46. Committee on Foreign Relations, United States Senate, "Iraq: Assessment of Progress in Economic Reconstruction and Governmental Capacity," December 2005, http://frwebgate.access.gpo.gov/cgi-bin/getdoc.cgi?dbname=109_cong_senate_committee_prints&docid=f:24804.pdf, 5.

47. BP Statistical Review of World Energy 2009.

48. R3 Sciences, http://www.r3sciences.com/.

49. Exxon Mobil has begun marketing a technology that turns dimethyl ether into gasoline. See Robert Rapier, "Energy Potpourri," R-Squared Energy Blog, October 5, 2008, http://i-r-squared.blogspot.com/2009/10/energy-potpourri.html#links.

50. Monty Hoyt, "US Burning Its Way Toward Fuel Crisis," *Christian Science Monitor*, June 2, 1971, 1.

51. Matt LeTourneau, U.S. Chamber of Commerce, personal communication with author, August 12, 2009.

Chapter 30

1. Peter W. Huber and Mark P. Mills, *The Bottomless Well: The Twilight of Fuel, the Virtue of Waste, and Why We Will Never Run Out of Energy* (New York: Basic Books, 2005), xxvi.

2. Dennis P. Lockhart, "The US Economy and the Employment Challenge," fbr atlanta.org, August 26, 2009, http://www.frbatlanta.org/invoke.cfm?objectid=572D828F-5056-9F12-12AD9734C17D1C6A&method=display.

3. Associated Press, "Population Living in Poverty Increasing," August 19, 2009, http://www.cbsnews.com/stories/2009/08/19/national/main5253630.shtml.

4. Bob Herbert, "A Word, Mr. President," *New York Times*, November 9, 2009, http://www.nytimes.com/2009/11/10/opinion/10herbert.html?_r=1&ref=opinion.

5. David Streitfeld, "US Mortgage Delinquencies Reach a Record High," *New York Times*, November 19, 2009, http://www.nytimes.com/2009/11/20/business/20mortgage.html.

6. Rasmussen Reports, "56% Don't Want to Pay More to Fight Global Warming," July 1, 2009, http://www.rasmussenreports.com/public_content/politics/general_politics/56_don_t_want_to_pay_more_to_fight_global_warming.

7. *Business Week*, "The Rise and Fall of Dennis Kozlowski," December 23, 2002, http://www.businessweek.com/magazine/content/02_51/b3813001.htm.

8. Among the best short explanations of what happened is Michael Lewis's "The End," Portfolio.com, November 11, 2008, http://www.portfolio.com/news-markets/national-news/portfolio/2008/11/11/The-End-of-Wall-Streets-Boom/index.html.

9. Richard Harris, "Putting a Financial Spin on Global Warming," June 24, 2009, http://www.npr.org/templates/story/story.php?storyId=105834436&sc=fb&cc=fp.

10. William Stanley Jevons, *The Coal Question: An Inquiry Concerning the Progress of the Nation, and the Probable Exhaustion of Our Coal Mines* (London: Macmillan, 1866), available at http://books.google.com/books?id=cUgPAAAAIAAJ&dq=jevons+and+the+coal+question&printsec=frontcover&source=bl&ots=FR4jmtfbVI&sig=6rp-Zo-8w2TI1jCzUaHlDbHi4po&hl=en&sa=X&oi=book_result&resnum=1&ct=result#PPA2,M1.

11. The corollary to that statement is that it is the nature of liberalism to feel bad about it.

Epilogue

1. Harlan Kirgan, "Mississippi Sierra Club Calls Deepwater Horizon Oil Spill 'America's Chernobyl,'" *Mississippi Press*, May 2, 2010, http://blog.gulflive.com/mississippi-press-news/2010/05/spill_is_americas_chernobyl.html.

2. Elizabeth Weise, "New Microbe Chows Down on Spilled Oil," *USA Today*, September 1, 2010, http://www.usatoday.com/tech/science/2010–08–24-oil-microbe_N.htm.

3. Neela Banerjee, "U. S. Lifts Moratorium on Deep-Water Drilling in Gulf of Mexico," *Los Angeles Times*, October 13, 2010, http://articles.latimes.com/2010/oct/13/nation/la-na-oil-moratorium-20101013.

4. Energy Information Administration data, http://www.eia.doe.gov/dnav/pet/pet_crd_crpdn_adc_mbblpd_a.htm.

5. Brian Baskin, "China's Oil Demand Is Poised to Push Up Prices," *Wall Street Journal*, November 8, 2010, http://online.wsj.com/article/SB10001424052748704405704575596743559257682.html.

6. Grant Smith, "China to Drive Energy Surge Through 2035, Warming Planet, IEA Outlook Says," *Bloomberg*, November 9, 2010, http://www.bloomberg.com/news/2010–11–09/china-to-drive-energy-surge-through-2035-warming-planet-iea-outlook-says.html.

7. International Energy Agency, "World Energy Outlook 2010," 80. In the "Current Policies Scenario," coal becomes the world's dominant fuel in 2035, slightly eclipsing oil.

8. State Department, "Roundtable with Business Leaders Opening and Closing Remarks, Hillary Rodham Clinton, Governor's House, Lahore, Pakistan, October 29, 2009," http://www.state.gov/secretary/rm/2009a/10/131073.htm.

9. Energy Information Administration data, http://tonto.eia.doe.gov/country/country_time_series.cfm?fips=PK#elec.

10. Energy Information Adminstration data, http://tonto.eia.doe.gov/country/country_energy_data.cfm?fips=FR.

11. *People's Daily Online*, "2 bln Tons of Coal Reserves Discovered in Southern Pakistan," October 2, 2010, http://english.people.com.cn/90001/90777/90851/7156516.html.

12. Energy Information Administration, "Natural Gas Weekly Update," November 10, 2010, http://www.eia.doe.gov/oog/info/ngw/ngupdate.asp.

13. Energy Information Administration data, http://www.eia.gov/dnav/ng/hist/n9070us2m.htm.

14. For latest prices, see *Bloomberg*, http://www.bloomberg.com/energy.

15. Energy Information Administration data, http://www.eia.gov/dnav/ng/hist/n9070us2A.htm.

16. Tennessee Valley Authority, "TVA to Idle Nine Coal-Fired Units," August 24, 2010, http://www.tva.gov/news/releases/julsep10/coal_plants.html.

17. Whitehouse.gov, full transcript here: http://www.whitehouse.gov/the-press-office/2010/11/03/press-conference-president.

18. Daniel Fineren and Muriel Boselli, "IEA Sees Gas Glut Lasting Until 2020," Reuters, November 9, 2010, http://in.reuters.com/article/idINLDE6A815R20101109?loomia_ow=t0:s0:a54:g12:r1:c0.605462:b39103646:z3.

19. International Energy Agency, "World Energy Outlook 2010," 77. Note that these projections are based on what the IEA calls its "New Policies Scenario."

20. Ibid, 237.

21. Ibid, 239.

22. E-mail exchanges with Janet Warren by author, February 17 and 19, 2010.

23. Author interview with Tony Moyer via phone, February 17, 2010.

24. Gerry Vassilatos, "The Sonic Weapon of Vladimir Gavreau," *Journal of Borderland Research*, October 30, 1996, http://journal.borderlands.com/1996/the-sonic-weapon -of-vladimir-gavreau.

25. Anne Mostue, "Discontent of Mars Hill Residents Leads to Lawsuit Against First Wind," Maine Public Broadcasting Network, August 7, 2009, http://www.mpbn .net/Home/tabid/36/ctl/ViewItem/mid/3478/ItemId/8549/Default.aspx. Pennsylvania litigation information obtained from Brad Tupi, a Pittsburgh-based lawyer representing clients who are suing. Tupi was interviewed by phone on Feb 23, 2010. NZ litigation in ABC news item, February 19, 2010, http://www.abc.net.au/news/video/2010 /02/19/2825235.htm.

26. *Copenhagen Post*, "Dong Gives Up on Land-based Turbines," September 1, 2010, http://www.cphpost.dk/news/scitech/92-technology/49869-dong-gives-up-on -land-based-turbines.html.

27. See http://www.wind-watch.org/affiliates.php.

28. Wind Concerns Ontario, http://windconcernsontario.wordpress.com/member -pages.

29. Industrial Wind Action Group, http://www.windaction.org/orglist.

30. For more, see http://windfallthemovie.com/index_2.html.

31. Patricia Breakey, "Meredith Aims for Wind Ban," *The Daily Star,* November 8, 2007, http://thedailystar.com/local/x112888993/Meredith-aims-for-wind-ban.

32. A 2008 report by the Department of Energy estimated that an aggressive effort to have wind energy provide one-fifth of U. S. electricity needs by 2030 would "displace 50% of electricity generated from natural gas and 18% of that generated from coal." DOE Study on 20% Wind, 2008, 16, http://www1.eere.energy.gov /windandhydro/pdfs/41869.pdf.

33. Steve Hargreaves, "Pickens' wind plan hits a snag," CNNMoney.com, November 12, 2008, http://money.cnn.com/2008/11/12/news/economy/pickens/index.htm.

34. http://www.boonepickens.com/media_summary/030510.pdf.

35. http://www.bloomberg.com/markets/commodities/energy-prices.

36. http://blogs.barrons.com/stockstowatchtoday/2010/09/13/exxon-deutsche-cuts-to -hold-natural-gas-glut-through-2015.

37. Rebecca Smith, "The New Nukes," *Wall Street Journal*, September 8, 2009, 3, http://online.wsj.com/article/SB10001424052970204409904574350342705855178 .html. Wind figure is for Sheringham Shoal offshore wind farm in the UK. Estimated cost is 1 billion British Pounds. In mid-September 2009, that was equal to about $1.7 billion. See "Onshore Construction Begins for Sheringham Shoal Wind Farm," NewEnergyFocus.com, September 7, 2009, http://www.newenergyfocus.com/do /ecco.py/view_item?listid=1&listcatid=32&listitemid=2978&section=Wind.

38. http://www.boston.com/business/articles/2010/05/07/cape_wind_project_could _boost_prices.

39. Energy Iinformation Administration data, http://www.eia.doe.gov/cneaf /electricity/epm/table5_3.html.

40. American Wind Energy Association, "US Wind Industry Reports Slowest Quarter Since 2007, As China Installs Three Times As Much Wind-Powered Electricity," October 29, 2010, http://www.awea.org/rn_release_10–29–10.cfm.

41. Matthew L. Wald and Tom Zeller Jr., "Cost of Green Power Makes Projects Tougher Sell," *New York Times*, November 7, 2010, http://www.nytimes.com/2010 /11/08/science/earth/08fossil.html.

42. Associated Press, "New Deposits Mean Natural Gas Should Be Cheap for Decades," October 25, 2010.

43. Debra Kahn, "Energy Secretary Laments 'Law of Bureaucracy,'" E&E News, October 22, 2010, http://www.eenews.net/public/eenewspm/2010/10/22/4.

44. Tux Turkel, "Protesters Arrested at Lincoln Windfarm," *Portland Press Herald*, November 8, 2010, http://www.pressherald.com/news/Protesters-arrested-at-Lincoln -maine-windfarm.html.

45. Environmental Proptection Agency press release, October 13, 2010, http://yosemite.epa.gov/opa/admpress.nsf/d0cf6618525a9efb85257359003fb69d /bf822ddbec29c0dc852577bb005bac0f!OpenDocument.

46. FollowTheScience.org, "39 Groups Request Congressional Hearing on EPA Ethanol Decision," August 31, 2010, http://www.followthescience.org/2010/08/39-groups -request-congressional-hearing-on-epa-ethanol-decision

47. Robert Bryce, "Despite Billions in Subsidies, Corn Ethanol Has Not Cut U. S. Oil Imports," Manhattan Institute, October 2010, http://www.manhattan-institute.org /html/ib_07.htm.

48. For more on this, see my book *Gusher of Lies*, 184–185.

49. See the EPA's filings here: http://www.epa.gov/otaq/regs/fuels/additive/e15/, and here: http://www.regulations.gov/search/Regs/home.html#documentDetail?R =0900006480b80cca. Also see *Federal Register*, Volume 75, No. 213, November 4, 2010, 68096.

50. Nick Snow, "Proposed Ozone Standard Would Devastate U. S. Economy, API Warns," *Oil & Gas Journal*, September 27, 2010, http://www.ogj.com/index /article-display.articles.oil-gas-journal.volume-108.issue-36.general-interest.proposed -ozone-standard-would-devastate-us-economy-api-warns.html.html.

51. Environmental Protection Agency data on nitrogen oxide available at http://www.epa.gov/air/nitrogenoxides.

52. Bryant Furlow, "Sen. Bingaman Supports Push to Cut Ethanol Subsidies," *New Mexico Independent*, July 16, 2010, http://newmexicoindependent.com/59525 /sen-bingaman-supports-push-to-cut-ethanol-subsidies.

53. Greg Meyer, "U.S. Ethanol Exports Fuel European Unease," *Financial Times*, November 15, 2010, http://www.cnbc.com/id/40190117.

GLOSSARY

Barrel of oil: A common measure of energy that is equal to 42 gallons of crude, 5.8 million Btu, or 5.8 gigajoules. In power terms, production of 1 barrel of oil per day is equal to about 30 horsepower, or 22,150 watts.

Btu: British Thermal Unit, a unit of energy christened in 1849 by James Prescott Joule, who, like James Watt, later had a unit of energy named for him. One Btu is the quantity of heat required to raise the temperature of 1 pound of water by 1 degree Fahrenheit at the temperature at which water has its greatest density (39 degrees Fahrenheit). Put another way, 1 Btu is approximately equal to the energy released in the burning of a wooden match. Also, 1 Btu = 1.055 kilojoules.

Energy: The ability to do work; also, a quantity or volume of fuel. A kilowatt-hour is a measure of energy.

Energy density: The amount of energy that can be contained in a given unit of volume, area, or mass. Common energy density metrics include Btu per gallon and joules per kilogram.

Horsepower (hp): A unit of power coined by inventor James Watt, who deemed that 1 horsepower was equal to 33,000 foot-pounds per minute. A more modern definition: 1 horsepower = 746 watts. Car engine power ratings are often listed in both watts and horsepower.

Joule (J): A unit of energy, whereby 1 joule = 1 watt-second; the energy exerted by the force of 1 newton acting to move an object 1 meter. Depending on the amount of energy, one can use millijoules, kilojoules, megajoules, and so on, as with any SI unit.

Power: The rate at which work gets done; a measure of energy flow. Power = energy/time. A kilowatt is a measure of power.

Power density: The amount of power that can be harnessed in a given unit of volume, area, or mass. Examples of power density metrics include: horsepower per cubic inch, watts per square meter, and watts per kilogram. When comparing renewable energy sources to other sources, the most telling metric is perhaps watts per square meter.

Quad: A unit used in the United States to measure large quantities of energy. A quad is 1 quadrillion (10^{15}) Btu. Annual U.S. energy use totals about 100 quads, or 100 exajoules (EJ). In 2006, global energy use was about 472 quads (472 EJ).

For reference, 1 quad is approximately equal to 172 million barrels of oil equivalent, or 1 trillion cubic feet of natural gas, or 1 exajoule (EJ).

Ton of oil: A common measure of energy equal to 7.33 barrels of oil. A ton of oil is equal to approximately 40 million Btu, 42 gigajoules (GJ), or 12 megawatt-hours of electricity.

Watt (W): A unit of power. Car engine power ratings are often listed in both watts and horsepower. By definition, 1 watt = 1 joule per second. Depending on the amount of power, one can use milliwatts, kilowatts, megawatts, and so on, as with any SI unit.

BIBLIOGRAPHY

Areva. *All About Nuclear Energy: From Atom to Zirconium.* Paris: Areva, 2008.

Barre, Bertrand, and Pierre-Rene Bauquis. *Understanding the Future: Nuclear Power.* Strasbourg: Editions Hirlé, 2007.

Behravesh, Nariman. *Spin-Free Economics: A No-Nonsense Nonpartisan Guide to Today's Global Economic Debates.* New York: McGraw Hill, 2009.

Bradley, Robert L., Jr. *Capitalism at Work: Business, Government, and Energy.* Salem: M&M Scrivener Press, 2009.

———. *Oil, Gas & Government: The US Experience.* Lanham, MD: Rowman and Littlefield, 1996.

Bradley, Robert L., Jr., and Richard W. Fulmer. *Energy: The Master Resource.* Dubuque: Kendall/Hunt, 2004.

Brand, Stewart. *Whole Earth Discipline: An Ecopragmatist Manifesto.* New York: Viking, 2009.

Bryce, Robert. *Cronies: Oil, the Bushes, and the Rise of Texas, America's Superstate.* New York: PublicAffairs, 2004.

———. *Gusher of Lies: The Dangerous Delusions of "Energy Independence."* New York: PublicAffairs, 2008.

———. *Pipe Dreams: Greed, Ego, and the Death of Enron.* New York: PublicAffairs, 2002.

Burrell, Brian. *Merriam Webster's Guide to Everyday Math: A Home and Business Reference.* Springfield, MA: Merriam Webster, 1998.

Caro, Robert A. *The Path to Power: The Years of Lyndon Johnson.* New York: Vintage, 1983.

Collins, Jim. *Good to Great: Why Some Companies Make the Leap . . . and Others Don't.* New York: HarperCollins, 2001.

Crane, Hewitt, Edwin Kinderman, and Ripudaman Malhotra. *A Cubic Mile of Oil: Realities and Options for Averting the Looming Global Energy Crisis.* New York: Oxford University Press, 2010.

Cravens, Gwyneth. *Power to Save the World: The Truth About Nuclear Energy.* New York: Knopf, 2008.

Darley, Julian. *High Noon for Natural Gas: The New Energy Crisis.* White River Junction, VT: Chelsea Green, 2004.

Economides, Michael J., and Tony Martin, eds. *Modern Fracturing: Enhancing Natural Gas Production.* Houston: ET Publishing, 2008.

Freeman, S. David. *Winning Our Energy Independence: An Energy Insider Shows How.* Layton, UT: Gibbs Smith, 2007.

Freese, Barbara. *Coal: A Human History.* New York: Penguin, 2003.

Freund, Paul, and Olav Kaarstad. *Keeping the Lights On: Fossil Fuels in the Century of Climate Change.* Oslo: Universitetsflorlaget, 2007.

Goodell, Jeff. *Big Coal: The Dirty Secret Behind America's Energy Future.* New York: Houghton Mifflin, 2008.

Goodwyn, Lawrence. *Texas Oil, American Dreams: A Study of the Texas Independent Producers and Royalty Owners Association.* Austin: Texas State Historical Association, 1996.

Hayden, Howard C. *The Solar Fraud: Why Solar Energy Won't Run the World.* Pueblo, CO: Vales Lake Publishing, 2001.

Hefner, Robert A., III. *The GET: Grand Energy Transition.* Oklahoma City: Hefner Foundation, 2008.

Heinberg, Richard. *The Party's Over: Oil, War, and the Fate of Industrial Societies.* Gabriola Island, British Columbia: New Society Publishers, 2003.

Huber, Peter W., and Mark P. Mills. *The Bottomless Well: The Twilight of Fuel, the Virtue of Waste, and Why We Will Never Run Out of Energy.* New York: Basic Books, 2005.

Jevons, William Stanley. *The Coal Question: An Inquiry Concerning the Progress of the Nation, and the Probable Exhaustion of Our Coal Mines.* London: Macmillan, 1866.

Jones, Van. *The Green Collar Economy: How One Solution Can Fix Our Two Biggest Problems.* New York: HarperOne, 2008.

Klein, Maury. *The Power Makers: Steam, Electricity, and the Men Who Invented Modern America.* New York: Bloomsbury, 2008.

Kunstler, James Howard. *The Long Emergency: Surviving the Converging Catastrophes of the Twenty-First Century.* New York: Atlantic Monthly Press, 2005.

Linklater, Andro. *Measuring America: How the United States Was Shaped by the Greatest Land Sale in History.* New York: Plume, 2002.

Lovins, Amory B., E. Kyle Datta, Odd-Even Bustnes, Jonathan G. Koomey, and Nathan J. Glasgow. *Winning the Oil Endgame: Innovation for Profits, Jobs, and Security.* Snowmass, CO: Rocky Mountain Institute, 2004.

MacKay, David J.C. *Sustainable Energy—Without the Hot Air.* Cambridge: UIT Cambridge, 2009. Available at http://www.inference.phy.cam.ac.uk/sustainable/book/tex/cft.pdf.

Muller, Richard A. *Physics for Future Presidents: The Science Behind the Headlines.* New York: W. W. Norton, 2008.

Nye, David E. *Electrifying America: Social Meanings of a New Technology.* Cambridge: MIT Press, 1997.

Pimentel, David, ed. *Biofuels, Solar and Wind as Renewable Energy Systems.* Ithaca, NY: Springer, 2008.

Polimeni, John M., Kozo Mayumi, Mario Giampietro, and Blake Alcott. *The Jevons Paradox and the Myth of Resource Efficiency Improvements.* Sterling, VA: Earthscan, 2008.

Prindle, David E. *Petroleum Politics and the Texas Railroad Commission*. Austin: University of Texas Press, 1981.

Roberts, Paul. *The End of Oil: On the Edge of a Perilous New World*. New York: Houghton Mifflin, 2004.

Schumacher, E. F. *Small Is Beautiful: Economics as if People Mattered*. New York: Harper and Row, 1973.

Simmons, Matthew. *Twilight in the Desert: The Coming Saudi Oil Shock and the World Economy*. Hoboken, NJ: John Wiley and Sons, 2005.

Smil, Vaclav. *Creating the Twentieth Century: Technical Innovations of 1867–1914 and Their Lasting Impact*. New York: Oxford University Press, 2005.

———. *Energies: An Illustrated Guide to the Biosphere and Civilization*. Cambridge: MIT Press, 1999.

———. *Energy: A Beginner's Guide*. Oxford, UK: Oneworld Publications, 2006.

———. *Energy at the Crossroads: Global Perspectives and Uncertainties*. Cambridge: MIT Press, 2003.

———. *Global Catastrophes and Trends: The Next Fifty Years*. Cambridge: MIT Press, 2008.

———. *Oil*. Oxford, UK: Oneworld Publications, 2008.

Tertzakian, Peter, and Keith Hollihan. *The End of Energy Obesity: Breaking Today's Energy Addiction for a Prosperous and Secure Tomorrow*. New York: Wiley, 2009.

Tucker, William. *Terrestrial Energy: How Nuclear Power Will Lead the Green Revolution and End America's Energy Odyssey*. Savage, MD: Bartleby Press, 2008.

Yergin, Daniel. *The Prize: The Epic Quest for Oil, Money and Power*. New York: Simon and Schuster, 1991.

INDEX

ROBERT BRYCE has been producing industrial-strength journalism for two decades. His articles on energy and other subjects have appeared in dozens of publications, ranging from the *Wall Street Journal* to *Counterpunch* and *Atlantic Monthly* to *Oklahoma Stripper*. He is the author, most recently, of *Gusher of Lies: The Dangerous Delusions of "Energy Independence."* Bryce, the managing editor of *Energy Tribune*, lives in Austin with his wife, Lorin, their three children—Mary, Michael, and Jacob—and a bird dog named Biscuit.